Stress Corrosion
Cracking of Pipelines

WILEY SERIES IN CORROSION

R. Winston Revie, Series Editor

Corrosion Inspection and Monitoring · Pierre R. Roberge

Corrosion Resistance of Aluminum and Magnesium Alloys: Understanding, Performance, and Testing · Edward Ghali

Microbiologically Influenced Corrosion · Brenda J. Little and Jason S. Lee

Corrosion Resistance of Aluminum and Magnesium Alloys · Edward Ghali

Metallurgy and Corrosion Control in Oil and Gas Production · Robert Heidersbach

Green Corrosion Inhibitors: Theory and Practice · V. S. Sastri

Heterogenous Electode Processes and Localized Corrosion · Yongjun Tan

Stress Corrosion Cracking of Pipelines · Y. Frank Cheng

Stress Corrosion Cracking of Pipelines

Y. Frank Cheng
Professor and Canada Research Chair
in Pipeline Engineering
University of Calgary

WILEY
A John Wiley & Sons, Inc., Publication

Published by John Wiley & Sons, Inc., Hoboken, New Jersey.
Published simultaneously in Canada.

Library of Congress Cataloging-in-Publication Data

Cheng, Y. Frank, 1969–
Stress corrosion cracking of pipelines / Y. Frank Cheng.
pages cm
Includes bibliographical references and index.
ISBN 978-1-118-02267-2 (cloth)
1. Pipelines–Corrosion. 2. Pipelines–Cracking. 3. Steel–Corrosion. I. Title.
TJ930.C545 2013
621.8′672–dc23

2012035723

Printed in the United States of America

10 9 8 7 6 5 4 3 2 1

To my wife, Jianshu (Jenny), and my son, Winston

Contents

Foreword

In this book, Dr. Cheng presents a lively discussion of stress corrosion cracking, sharing with readers his insights into this complex failure mode. He has written this book for all those who would like an understanding of stress corrosion cracking as it pertains to the reliability of the pipeline infrastructure on which society relies to meet energy needs.

Dr. Cheng skillfully explains and juxtaposes the most fundamental aspects of this type of corrosion with the realities of engineering experiences in the pipeline industry, including case histories. He outlines the situations that can develop over many years during the pipeline operational lifetime, leading to initiation and growth of stress corrosion cracks. In nine chapters, he leads the reader through the science of stress corrosion cracking, as it is currently understood at an atomistic level, into discussions of soil environments and engineering aspects of constructing and maintaining welded steel pipelines, with the complexities that weld zones entail.

After reviewing the fundamental and engineering aspects, Dr. Cheng offers strategies for managing stress corrosion cracking by preventing, detecting, and monitoring to achieve the goal of zero failures—no failures—so that, indeed, *zero* means *zero*, and *no* means *no*, ensuring reliability of the pipeline, deliverability of the energy, protection of the environment, and safety of the public. Throughout the book, technologies for managing stress corrosion cracking are discussed as essential elements of maintaining pipeline integrity.

In summary, Dr. Cheng has provided an important service by writing this book, delivering a valuable source of knowledge and information on technologies for managing stress corrosion cracking and enhancing pipeline integrity. I commend this book to all those who have an interest in this exciting subject, this rapidly developing area of technology that is vital to all those who rely on energy for their standard of living and to fuel their ambitions and dreams for the future.

R. Winston Revie
Series Editor
Wiley Series in Corrosion

Ottawa, Ontario, Canada
July 2012

Preface

Pipelines sit at the nexus of national economies, of growing concerns for the natural environment, and of the global energy infrastructure. Environmental disasters such as the tragic explosion of the Deepwater Horizon drilling rig have diminished public tolerance for human activity that results in the release of hydrocarbons into the natural environment. After spending decades outside public focus, the pipeline industry now emerges at the center of a complex global debate that involves multiple interests.

First observed in pipelines in the United States during the 1960s and later reported in Canadian pipelines during the 1980s, stress corrosion cracking (SCC) has represented both a challenge to an industry that has grown increasingly concerned with the safe operation of pipelines as well as a source of scientific motivation for researchers trying to understand the detailed mechanisms behind this complex process. The main objective in writing this book is to provide a summative and, more important, up-to-date narrative of the current state of scientific understanding of and relevant engineering practice involving pipeline SCC. Moreover, preparation of this book is intended to pay tribute to the numerous researchers and engineers who have contributed to the body of knowledge in the field of pipeline SCC.

The nine chapters are designed to meet the needs of scientists, engineers, managers, technologists, students, and all of those requiring knowledge in this area. The book introduces pipelines, the development of the global pipeline industry, and the hazardous effects of SCC on the integrity of these systems. The second chapter explores the fundamentals of SCC in metals. Specifically, attention is given to (1) metal–environment combinations that give rise to SCC, (2) its metallurgical, mechanical, and environmental aspects, and (3) the various mechanisms that illustrate the initiation and propagation of stress corrosion cracks in metals. Moreover, the occurrence and characteristics of damage from hydrogen and corrosion fatigue are analyzed and compared to SCC. The role of microbiological activity in SCC processes is also discussed.

Chapters 3 through 6 describe SCC as a unique phenomenon and mechanism that can result in pipeline failure. Topics also cover a wide spectrum of environmental conditions that are relevant to pipeline operation, including nearly neutral pH, high-pH trapped electrolyte, and acidic soil environments. In addition, the primary characteristics of and contributing factors to pipeline SCC are summarized. This portrait of SCC includes both current theoretical and practical bodies of knowledge surrounding propagation kinetics, predictive methodologies, and of the crack initiation mechanism.

Chapter 7 focuses on corrosion and SCC that occurs at pipeline welds, with close connections to local steel metallurgy and electrochemical features. High-strength steel pipeline technology and the metallurgy of high-strength line pipe steels are discussed in Chapter 8. As advanced pipeline materials, high-strength steels distinguish themselves from conventional pipeline steels with unique metallurgical, mechanical, and microelectrochemical characteristics. All of these contribute to the occurrence of hydrogen damage, corrosion, and SCC. In particular, complex strains exerted on high-strength steel pipelines and the implications on corrosion of steels and the use of the mechanoelectrochemical effect theory for the prediction of defect propagation and evaluation on the remaining strength of steels are introduced. These discourses will serve as a reliable foundation for corrosion- and SCC-preventive strain-based design of pipelines. Chapter 9 reviews current industrial practices in the management of pipeline SCC, including prevention, monitoring, and mitigation techniques. Moreover, it is shown how SCC management has been integrated with broader integrity management programs in use for modern pipeline systems.

The uniqueness of this book does not lie in the fact that it is the first book especially contributing to pipeline SCC, but that it contains the latest research results and data relating to pipeline SCC. To assist the reader in understanding the scientific aspects associated with the phenomena of pipeline SCC, a number of theoretical models and concepts have been developed. Moreover, these conceptual and modeling results are supported by and based on advanced microscopic electrochemical measurements. In turn, effective integration of electrochemical, microelectrochemical, materials science, and surface analysis techniques helps to advance a fundamental understanding of the engineering phenomenon. Furthermore, the scientific concepts and models explored in the ensuing discussions provide reliable and accurate methodology for industry to predict, monitor, and manage pipeline SCC.

I am very pleased to express my deepest gratitude to Dr. Winston Revie, who wrote the Foreword for the book. In the past seven years, I have had the good fortune of frequent interactions with Winston on a wide variety of issues in the area of pipeline engineering. His guidance, encouragement, and mentorship have been and remain invaluable to me in the formation and evolution of my professional career.

I wish to thank Dr. Ron Hugo for his devoted support of my academic career at the University of Calgary. The role that he plays in the development of the Pipeline Engineering Centre at the university and in the local community cannot be overemphasized. These efforts have contributed to creating an ideal working environment for my research in pipelines.

I acknowledge the numerous fruitful discussions I have had with Drs. Bill Shaw, Jingli Luo, Fraser King, and many other colleagues and friends. I am also indebted to the dedicated and unfailing assistance provided by the numerous students and postdoctoral fellows that I have had the pleasure to supervise in my research group.

I thank Mr. Michael Leventhal of John Wiley & Sons for his patience and understanding of the lengthy time period required to prepare this book. Indeed, there have been frequent unexpected events interrupting the overall process. Michael's support was crucial to completing the project.

Research grants from the Canada Research Chairs Program, Natural Science and Engineering Research Council of Canada (NSERC), Canada Foundation for Innovation (CFI), Pipeline Engineering Center of the Schulich School of Engineering, University of Calgary, and a number of industrial organizations have created the favorable conditions that helped support an active research environment that has enabled the writing of this book. I am grateful and indebted to the assistance provided by these programs, agencies, and organizations.

Finally, I thank my wife and my son, who, in many ways, have provided encouragement and have supported the creation of this book.

Y. FRANK CHENG

Calgary, Alberta, Canada

List of Abbreviations and Symbols

AD	anodic dissolution
AF	acicular ferrite
API	American Petroleum Institute
ASME	American Society of Mechanical Engineers
BCC	body-centered cubic
bcf	billion cubic feet
BF	bainitic ferrite
CCT	continuous cooling transformation
CE	counter electrode
CEPA	Canadian Energy Pipeline Association
CF	corrosion fatigue
CGR	crack growth rate
CP	cathodic protection
CO_2	carbon dioxide
CSA	Canadian Standards Association
CSL	coincidence site lattice
DNV	Det Norske Veritas
DOS	degree of sensitization
DOT	Department of Transportation
DSAW	double submerged arc welding
EAC	environmentally assisted cracking
EAT	equivalent aging time
EBW	electron beam welding
ECDA	external corrosion direct assessment
EDX	energy-dispersive x-ray

EMAT	electromagnetic acoustic transducer
ERW	electric resistance welding
ESCM	electrochemical state conversion model
FBE	fusion-bonded epoxy
FCC	face-centered cubic
FEA	finite element analysis
FIB	focused ion beam
GBF	grain boundary ferrite
GPS	global positioning system
H_2S	hydrogen sulfide
HAGB	high-angle grain boundary
HAZ	heat-affected zones
HE	hydrogen embrittlement
HIB	hydrogen-induced blistering
HIC	hydrogen-induced cracking
HPCC	high-performance composite coating
HSSCC	hydrogen sulfide stress corrosion cracking
ICDA	internal corrosion direct assessment
IDQ	inter-ruptured direct quenching
IEA	International Energy Agency
IEAW	indirect electric arc welding
IGSCC	intergranular SCC
ILI	in-line inspection
IMP	integrity management program
IOB	iron-oxidizing bacteria
IRB	iron-reducing bacteria
ISO	International Standards Organization
LAGB	low-angle grain boundary
LAP	local additional potential
LEIS	localized electrochemical impedance spectroscopy
LPB	low-plasticity burnishing
MAG	metal active gas
MAOP	maximum allowable operating pressure
MFL	magnetic flux leakage
MIC	microbially influenced corrosion
MIG	metal inert gas
MnS	magnesium sulfide
M/A	martensite/austenite
MOC	management of change
MOP	maximum operating pressure
MPI	magnetic particle inspection
mpy	mils per year
NACE	national Association of Corrosion Engineers
NDT	nondestructive testing
NEB	National Energy Board

NRTC	NOVA Research and Technology Center
OPS	Office of Pipeline Safety
PE	polyethylene
PECPMS	pipeline environment and CP monitoring system
PIM	pipeline integrity management
PRCI	Pipeline Research Council International
PSB	persistent slip band
PWHT	postweld heat treatment
RE	reference electrode
ROW	right-of-way
RP	rolling plane
RRA	reduction-in-area
SAW	submerged arc welding
SBD	strain-based design
SCC	stress corrosion cracking
SCCDA	stress corrosion cracking direct assessment
SCE	saturated calomel electrode
SEM	scanning electron microscopy
SF	safety factor
SHE	standard hydrogen electrode
SKP	scanning Kelvin probe
SMYS	specified minimum yield strength
SOB	sulfide-oxidizing bacteria
SPH	smooth particle hydrodynamics
SRB	sulfate-reduced bacteria
SSC	sulfide stress cracking
SSCC	sulfide stress corrosion cracking
SSRT	slow strain rate tensile
SVET	scanning vibrating electrode technique
TFI	transverse field inspection
TGSCC	transgranular SCC
TM	thermomechanical
TMCP	Thermomechanical controlled processing
UT	ultrasonic tool
UTCD	ultrasonic crack detection
VSR	vibratory stress relief
YPE	yielding point elongation

a	crack length
A_1	areas of a crack tip
A_2	area of the region ahead of the crack
b_a	anodic Tafel slope
b_c	cathodic Tafel slope
C	capacitance of space-charge layer of passive film

C_{app}	hydrogen apparent solubility
d	initial depth of grain
da/dN	fatigue crack propagation rate per cycle
D_{eff}	hydrogen diffusivity
D_l	lattice diffusion coefficient of hydrogen
e	elongation
E	Young's modulus
E°	standard electrode potential
E_a	anodic potential
E_{app}	applied potential
E_{corr}	corrosion potential
E_{corr1}	corrosion potential at a crack tip
E_{corr2}	corrosion potential at a region adjacent to a crack tip
E_g	galvanic potential
E_0	electrode potential at an intact site
E_{pi}	potential at a local active site
E_{pit}	pitting potential
f	frequency
F	Faraday constant
G	formation free energy of individual species
H_{ave}	average depth of source dislocation
I_g	galvanic current
i_1	anodic current density at a crack tip
i_2	cathodic current density at the adjacent region from a crack tip
i^0	exchange current density
i_a	anodic current density
i_a^*	anodic current density immediately after the film rupture
i_{corr}	corrosion current density
i_{corr1}	corrosion current density at a crack tip
i_{corr2}	corrosion current density at the adjacent region from a crack tip
i_D	current density at defect
i_p	passive current density
i_{pit}	pitting current density
i_N	current density at a nondefect area (i.e., intact area)
i_{total}	total current density measured
J_H	hydrogen flux
$J_H L_H$	hydrogen permeation rate
k	constant
k_B	Boltzmann constant
k_H	effect of hydrogen on the anodic dissolution rate of steel
k_σ	effect of stress on anodic dissolution in the absence of hydrogen
$k_{H\sigma}$	synergistic effect of hydrogen and stress on the anodic dissolution at a crack tip
K_{ISCC}	minimum threshold stress intensity for SCC
K_{max}	maximum of stress intensity factor

L	initial size and depth of grain
L_H	thickness of the specimen for hydrogen permeation test
M	atomic weight
n	number of electrons exchanged in the electrode reaction
n_0	number of dislocations on the steel surface
n_d	number of dislocations in a dislocation pile-up
N	number of stress cycles to failure
N_0	initial density of dislocations prior to plastic deformation
N_{max}	maximum dislocation density
N_T	hydrogen trapping density
r	distance from the local charged point on the steel surface to the solution layer
r_0	atomic radius
R	ideal gas constant
$R^0_{ct,\sigma}$	charge-transfer resistances of steel without hydrogen charging
$R^H_{ct,\sigma}$	charge-transfer resistances of steel with hydrogen charging
q_i	charge of electrons
Q_F	electric charge passed between two successive film-rupture events
S	stress
t	time
t_L	time lag
t_0	start time that passive film ruptures
T	temperature
U	ultimate tensile stress
V_H	average volume of hydrogen in steel
V_m	molar volume of steel substrate
W	thickness of grain
W_m	molar weight
x	amount of hydrogen atoms permeating into the steel
y	amount of hydrogen atoms permeating stressed steel
Y/T	yielding strength/tensile strength ratio
z	repassivation exponent

$[H^+]$	concentration of hydrogen ions in solution
$[H^+_{CP}]$	concentrations of hydrogen ions in solution in the presence of CP
$[H^0_{ads}]$	subsurface concentration of hydrogen adsorbed in uncharged steel
$[H_{ads}]$	subsurface concentration of adsorbed hydrogen in charged steel
ΔE	LAP at defect
ΔG	change in free energy
ΔK	stress intensity factor
ΔK_{th}	threshold of stress intensity factor
ΔN	density of new dislocations during plastic deformation
ΔP	excess pressure
ΔS	entropy change

ΔU	change of internal energy
$\Delta\mu$	difference of chemical potentials of the steel in the presence and absence of hydrogen charging
$\Delta\varphi_e^0$	change of electrochemical corrosion potential of steel during elastic deformation
$\Delta\varphi_p^0$	change of electrochemical corrosion potential of steel during plastic deformation
$\Delta\varphi_T^0$	shift of total corrosion potential during tensile testing
$\Delta\tau$	hardening intensity
α	charge-transfer coefficient (cathodic)
β	charge-transfer coefficient (anodic)
β_c	cathodic Tafel slope
σ	stress
$\sigma_{0.2}$	applied stress level resulting in a 0.2% of total deformation and often used as an approximation of the proof stress of steels
σ_h	hoop stress
σ_V	volume stress
σ_{ys}	yield strength
ε	strain
ε_F	rupture ductility of passive film
ε_r	dielectric constant of water
ε_p	plastic strain
$\dot{\varepsilon}$	strain rate
$\dot{\varepsilon}_{ct}$	strain rate at a crack tip
ρ	density
v_{th}	minimum mobility velocity of dislocations
φ	electrode potential
φ^θ	standard equilibrium electrode potential
μ	chemical potential
i_p^∞	steady-state hydrogen permeation current density
I_ψ	ratio of reduction-in-area obtained in solution to that in air
ν	orientation-dependent factor

1

Introduction

Statistically, pipelines provide the safest and most economical form of transportation of crude oil, natural gas, and other petrochemical commodities compared to truck, rail cars, and tankers [Cheng, 2010]. There are about 2 million kilometers of transmission pipelines worldwide. These include natural gas, oil, condensates, petroleum gas, and other refined petroleum products, as well as carbon dioxide (CO_2) and hydrogen. The pipelines could be very large in diameter (e.g., a Russian pipeline system has a diameter of up to 1422 mm) and can be over several thousand kilometers in length [Hopkins, 2007]. Most pipelines are buried or under the sea, but some operate aboveground.

Liquids and gases have been transported by pipelines for thousands of years. Ancient Chinese and Egyptians used pipes to transport water, hydrocarbons, and even natural gases [Hopkins, 2007]. Most of the current pipeline industry was developed to transport oil, bringing considerable profits to energy producers and pipeline operators, and development is driven by expanding energy demands. Tens of thousands of kilometers of new pipelines are constructed every year. Pipelines have become one of the most environmentally friendly and safest means of oil and natural gas transportation and contribute to strong national economies. As a consequence, they have been integrated into the components of national security in most countries.

More than 90% of pipelines are made of steel, primarily carbon steel, with aluminum, fiberglass, composite, polyethylene, and other types making up the remaining 10% [Alberta Energy and Utilities Board, 2007]. Requirements for higher capacities and operating pressure and additional economic benefits have led to a demand for

Stress Corrosion Cracking of Pipelines, First Edition. Y. Frank Cheng.

higher-strength pipeline materials, especially high-strength steels, as well as new techniques for welding, construction, inspection, and pipeline integrity and maintenance programs.

1.1 PIPELINES AS "ENERGY HIGHWAYS"

Human beings need energy to survive. For today and tomorrow, fossil fuels, including oil and gas, are the predominant forms of energy consumed worldwide. In fact, "even if the use of renewable energies doubles or triples over the next 25 years, the world is likely to still depend on fossil fuels for at least 50 percent of its energy needs" [Chevron, 2012]. The International Energy Agency estimated in 2010 that the world oil supply rises by 85 million barrels per day and forecast that the global demand would average nearly 88 million barrels per day in 2011 [Whipple, 2010], which demonstrates a clear relationship between oil consumption and a country's economic situation.

Oil and gas are usually found in very remote regions that are different from the locations where they are processed and consumed. Pipelines provide the necessary transportation function for this form of energy. Pipelines are regarded as "energy highways" of the global oil and gas industry, and their impact on the energy industry and the general economy therefore cannot be underestimated. In North America, a total length of over 800,000 kms of transmission pipeline network transports 97% of Canadian crude oil and natural gas from the producing regions to markets throughout Canada and the United States. Statistics show [Canadian Energy Pipeline Association, 2007] that Canadian pipelines transport approximately 2.65 million barrels of crude oil and equivalent and 17.1 billion cubic feet (bcf) of natural gas daily. Moreover, virtually all oil and gas exports—worth $60 billion in 2009—are carried by pipelines [Canadian Energy Pipeline Association, 2012]. With an asset value of approximately $20 billion, the Canadian pipelines are anticipated to double in size by 2015 to meet the oil and gas production increases that are forecast. Among the world's nations, the United States and Canada have the largest networks of energy pipelines for both oil and natural gas.

Oil pipeline networks are classified into crude oil lines and refined product lines, and the crude oil lines are subdivided into gathering lines and trunk lines. Gathering lines are small pipelines, from 2 to 8 in. in diameter, and are used where crude oil is found deep within the Earth where it is impractical to use larger diameters [Alberta Energy and Utilities Board, 2007]. It is estimated that there are between 48,000 and 64,000 kms of small gathering lines in the United States. These small lines gather oil from many wells, both onshore and offshore, and connect to larger trunk lines ranging from 8 to 24 in. in diameter. Trunk lines include a few very large lines, such as the TransAlaska Pipeline System, which is 48 in. in diameter [Alberta Energy and Utilities Board, 2007]. There are approximately 89,000 km of crude oil trunk lines in the United States.

Gas gathering lines connect individual gas wells to field gas-treating and processing facilities or to branches of larger gathering systems. Most gas wells flow naturally with sufficient pressure to supply the energy needed to force the gas through the gathering line to the processing plant. Like crude oil trunk lines, gas transmission systems can cover large geographical areas and be several hundreds or thousands of miles long. One of the largest natural gas supplies is in western Siberia. A large-diameter pipeline system moves gas from that area, including a pipeline almost 4600 km long, to export gas to Western Europe [Hopkins, 2007]. These trunk lines, which have diameters ranging from 40 to 55 in., constitute an impressive pipeline network. Compared to crude oil pipelines, gas transmission lines operate at relative high pressures.

Oil and gas pipeline systems are remarkable for their efficiency and low cost. Compared to other conventional means of transportation, such as rail and trucks, pipelines provide a very cheap way to transport oil. For example, for every 1000 barrel-miles of transportation of petroleum, the cost by pipeline is between 4 and 12 cents, whereas those by rail and truck are 12 to 60 cents and 52 to 75 cents, respectively [Kennedy, 1993]. Oil and gas pipelines are also energy-efficient, consuming about 0.4% of the energy content of the crude oil or gas transported per 1000 km [Marcus, 2009].

1.2 PIPELINE SAFETY AND INTEGRITY MANAGEMENT

Pipeline integrity is maintained by coating and cathodic protection (CP) as well as by comprehensive pipeline safety maintenance programs generally called *pipeline integrity management* (PIM) programs. A PIM is a process to develop, implement, measure, and manage the integrity of a pipeline through assessment, mitigation, and prevention of risks to ensure safe, environmentally responsible, and reliable service [Nelson, 2002]. Integrity management of pipeline systems is essential to the safe and efficient transport of oil and natural gas on the basis of safety assessment and lifetime prediction. Attempts to define pipeline performance, structural strength, and lifetime spawn a number of specialized fields, including corrosion, materials science, fracture mechanics, nondestructive evaluation, electrochemistry, environmental science, and mathematical modeling on both microscopic and macroscopic scales.

The goal of a PIM program is to ensure that the risk is "as low as is reasonably practicable" [Nelson, 2002]. An integrity management program (IMP) is usually valid for two or three years and is then updated to include new or modified processes, developed during implementation of the PIM, through multiple time-driven integrity plans. A PIM program supports monitoring, inspection, and maintenance programs to reduce greatly the risk of failures that could cause disastrous consequences to human life, the environment, and business operations.

1.3 PIPELINE STRESS CORROSION CRACKING

A number of factors contribute to pipeline failures. Although corrosion is identified as the most common cause of oil and gas transmission pipeline failure [U.S. Department of Transportation, 2005], stress corrosion cracking has been identified as leading to a number of pipeline leaking and/or rupture events, with catastrophic consequences [National Energy Board, 1996].

Stress corrosion cracking (SCC) is a term used to describe service failure in engineering materials that occurs by slow, environmentally induced crack propagation. The crack propagation observed is the result of the combined and synergistic interactions of mechanical stress and corrosion reactions [Jones, 1992]. For a certain material, SCC occurrence depends on both an aggressive environment and a stress, especially a tensile stress. During the operation of pipelines in the field, line pipe steels are exposed to electrolytes trapped under disbonded coating, where solution chemistry or electrochemistry is developed to support pipeline SCC [Fu and Cheng, 2010]. The stress is due primarily to the internal operating pressure or pressure fluctuation of natural gas or liquid petroleum [Zheng et al., 1997]. Moreover, soil movement–induced longitudinal stress and strain contribute to the initiation and propagation of stress corrosion cracks in pipelines [Canadian Energy Pipeline Association, 1998]. A wide variety of factors experienced by pipelines during their operation have been demonstrated to affect and contribute to SCC at somewhat different levels, such as the steel metallurgy (chemical composition, grade, microstructure, heat treatment, alloying elements, impurities, and welding), environmental parameters (soil chemistry, conductivity, seasonal dry–wet cycle, temperature, humidity, CO_2 and gas conditions, and microorganisms), coatings, and CP (type, properties, failure mode, coating compatibility with CP, and CP potential/current), stressing condition (pressure, pressure fluctuation, residual stress, longitudinal stress, local stress–strain concentration), and corrosion reaction (corrosion pits, geometry of pits, hydrogen evolution, passivity and passive film formation, active dissolution, and mass transport) [Parkins, 2000].

Pipeline SCC incidents throughout North America and the world, including in Australia, Russia, Iran, Saudi Arabic, Brazil, and Argentina, have highlighted threats to pipelines from this problem. In Canada, two major ruptures and fires on the TransCanada Pipeline System in 1995, together with further evidence of the more widespread nature of SCC, led to the initiation of a national inquiry. This was the first comprehensive inquiry in the world on pipeline SCC and has been far-reaching across Canadian pipelines and extended to other countries [National Energy Board, 1996].

In the United States, the Williams 26-in. pipeline ruptured near Toledo, Washington in 2003, resulting in the shutdown of its trunk line from Canada to Oregon [Williams Pipeline, 2003]. This pipeline had also failed in 1992, 1994, and 1999, failures all attributed to SCC. With the occurrence of SCC-caused failures of gas and liquid pipelines, an advisory bulletin was issued in 2003 to remind owners and operators of gas transmission and hazardous liquid pipelines to consider SCC as a risk factor when developing and implementing integrity management plans [Baker, 2005]. It was commented that "SCC is a serious pipeline integrity issue of concern to operators of pipelines within the United States." When comparing the pipeline SCC statistics in the United States and Canada, it was pointed out that "the fact that SCC represents only 1.5 percent of reportable incidents in the United States versus 17 percent in Canada is due to the far greater occurrence of third party damage in the United States."

Research on pipeline SCC could be tracked back to the 1980s, and still has global interest. Management of SCC in the modern pipeline industry has been integrated with companies' integrity management programs. Our understanding of this important problem has evolved to the stage that comprehensive reviews describing the scientific, technical, and practical aspects of SCC in pipelines are common, all of which have facilitated the development of this book.

In addition to SCC fundamentals, such as the metallurgical, environmental, and mechanical aspects of SCC and the correlation with various hydrogen damage and corrosion fatigue, this book covers a wide spectrum of topics. Specifically, it includes the primary characteristics of and factors contributing to pipeline SCC and reports on progress to date on the investigation and understanding of SCC in pipelines occurring in nearly neutral–pH, high-pH, and acidic soil environments. The pipeline weld poses a sensitive region to SCC. Consequently, welding metallurgy is included, and the implications of corrosion and SCC are discussed. As advanced pipeline materials, high-strength steels distinguish themselves from conventional pipeline steels with unique metallurgical, mechanical, and metallurgical microelectrochemical characteristics. All of them contribute to the occurrence of hydrogen damage, corrosion, and SCC. The corrosion- and SCC-preventive strain-based design of high-strength steel pipelines is discussed based on the latest research in this area. Finally, industrial experience in the management of pipeline SCC, including prevention, monitoring, and mitigation as well as its integration with pipeline IMP, is incorporated.

REFERENCES

Alberta Energy and Utilities Board (2007) *Pipeline Performance in Alberta, 1990–2005*, Report 2007-A, AEUB, Calgary, Alberta, Canada.

Baker, M, Jr. (2005) *Final Report on Stress Corrosion Cracking Study*, Integrity Management Program Delivery Order DTRS56-02-D-70036, Office of Pipeline Safety, U.S. Department of Transportation, Washington, DC.

Canadian Energy Pipeline Association (1998) *Stress Corrosion Cracking Recommended Practices: Addendum on Circumferential SCC*, CEPA, Calgary, Alberta, Canada.

Canadian Energy Pipeline Association (2007) *CEPA Statistics*, CEPA, Calgary, Alberta, Canada.

Canadian Energy Pipeline Association (2012) http://www.cepa.com/about-cepa/industry-information/factoids.

Cheng, Frank Y (2010) Pipeline engineering, in *Pipeline Engineering*, Y.F. Cheng, Editor, Encyclopedia of Life Support Systems (EOLSS), developed under the auspices of UNESCO, EOLSS Publishers, Oxford, UK.

Chevron (2012) Energy supply and demands, http://www.chevron.com/globalissues/energy supplydemand/.

Fu, AQ, Cheng, YF (2010) Electrochemical polarization behavior of X70 steel in thin carbonate/bicarbonate solution layers trapped under a disbonded coating and its implication on pipeline SCC, *Corros. Sci.* 52, 2511–2518.

Hopkins, P (2007) *Oil and Gas Pipelines: Yesterday and Today,* Pipeline Systems Division, American Society of Mechanical Engineers, New York.

Jones, RH (1992) *Stress Corrosion Cracking: Materials Performance and Evaluation*, ASM, Materials Park, OH.

Kennedy, JL (1993) *Oil and Gas Pipeline Fundamentals*, 2nd ed., PennWell, Tulsa, OK.

Marcus, S (2009) Oil and gas pipeline in Canada, *J. Oil Gas* 2, 15.

National Energy Board (1996) *Stress Corrosion Cracking on Canadian Oil and Gas Pipelines*, MH-2-95, NEB, Calgary, Alberta, Canada.

Nelson, BR (2002) Pipeline integrity: program development, risk assessment and data management, *11th Annual GIS for Oil and Gas Conference*, Houston, TX.

Parkins, RN (2000) A review of stress corrosion cracking of high pressure gas pipelines, *Corrosion 2000*, Paper 363, NACE, Houston, TX.

U.S. Department of Transportation (2005) *Pipeline Accident Brief*, Research and Special Programs Administration, Office of Pipeline Safety, U.S. DOT, Washington, DC.

Whipple, T (2010) Peak oil review, *Energy Bull.*, Aug. 16.

Williams Pipeline, Gas pipeline SCC: catastrophic ruptures, 1 May and 13 December 2003, http://www.corrosion-doctors.org/Pipeline/Williams-explosion.htm.

Zheng, W, MacLeod, FA, Revie, RW, Tyson, WR, Shen, G, Shehata, M, Ray, G, Kiff, D, McKinnon, J (1997) *Growth of Stress Corrosion Cracks in Pipelines in Near-Neutral pH Environment: The CANMET Full-Scale Tests Final Report to the CANMET/Industry Consortium*, CANMET/MTL, Ottawa, Ontario, Canada.

2

Fundamentals of Stress Corrosion Cracking

2.1 DEFINITION OF STRESS CORROSION CRACKING

Cracking is possibly the most common mode of material failure. It may be the most dangerous failure mechanism since fracture can occur instantaneously and without advance warning. Harsh environments can compound the cracking problem and accelerate the rate of failure. *Environmentally assisted cracking* (EAC) is a generic term used to describe various mechanisms for this phenomenon. It can generally be classified into three different forms: stress corrosion cracking (SCC), corrosion fatigue (CF), and hydrogen-induced cracking (HIC). The three forms can appear to be very similar in nature despite some fundamental differences.

SCC is defined slightly differently in various sources. For example, Jones [1992] defines SCC as a term used to describe service failures in engineering materials that occur by slow, environmentally induced crack propagation. The crack propagation observed is the result of the combined and synergistic interaction of mechanical stress and corrosion reactions. According to Corrosion Doctors [online source 1], SCC is a cracking process of metals that requires the simultaneous action of a corrodent and sustainable tensile stress. In *Wikipedia*, online source SCC is defined as the unexpected sudden failure of normally ductile metals subjected to constant tensile stress in a corrosive environment. Finally, SCC could be defined as cracking due to a process involving conjoint corrosion and straining of a metal due to residual or applied stresses [UK National Physical Laboratory, 1982].

Stress Corrosion Cracking of Pipelines, First Edition. Y. Frank Cheng.

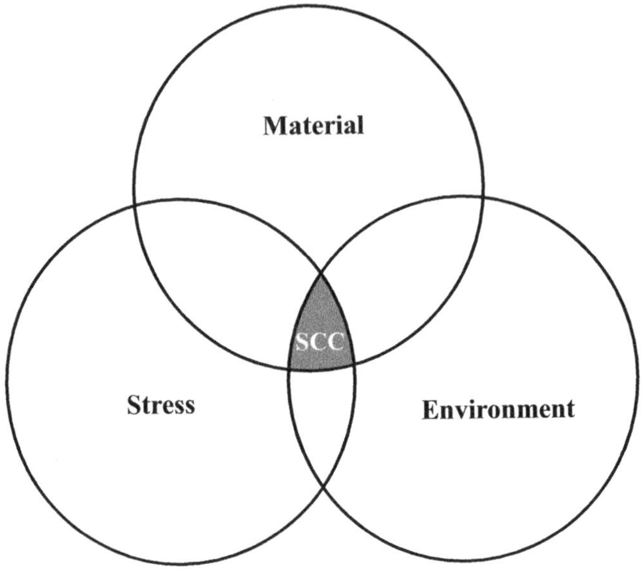

Figure 2-1 *Three essential factors required for the occurrence of SCC.*

Despite the difference in definitions, there is a common requirement for the occurrence of SCC. Three essential factors must happen simultaneously: a susceptible material, a corrosive environment, and sufficient tensile stress, as described in Fig. 2-1. Consequently, SCC is relatively rare, although failures can be very costly and destructive when they do occur.

The stresses that cause SCC are either generated as a result of the use of metallic components in service or as residual stress introduced during manufacturing, such as welding and bending. The stress required to cause SCC is small, usually below the macroscopic yield strength of a metal. However, stress concentration may develop locally since stress corrosion cracks frequently initiate at surface flaws that either preexist or are formed during service by corrosion, wear, or other processes. Moreover, the stress must be tensile in nature, and comprehensive stresses can be used to prevent SCC.

The corrosive environment may be a permanent service environment, such as seawater for an offshore platform structure, or temporary environments caused by a particular operation, such as electrolyte residue after cleaning rust from a metallic structure. Moreover, environments that cause SCC are usually aqueous and can be either condensed layers of moisture or bulk solutions. Cracking will not normally occur when there is a significant corrosion rate, and stress corrosion cracks can initiate and propagate with little outside evidence of corrosion. Generally, SCC is observed in metal–environment combinations that result in the formation of a film on a metal surface [Jones, 1992]. Thus, SCC is of great concern in corrosion-resistant metals exposed to aggressive environments.

There are two types of SCC modes in metals: intergranular or transgranular. In the former mode, cracks grow along the grain boundaries, whereas cracks following the latter mode grow across the grains.

2.2 SPECIFIC METAL–ENVIRONMENT COMBINATIONS

SCC is not an inevitable process. For most metals in most environments it will not occur. It can therefore identify specific combinations of metal and environment that are subject to the problem. Table 2-1 lists some combinations of metal and

TABLE 2-1 Some Combinations of Metals and Environments That Result in SCC

Metal	Environment
Aluminum alloys	$NaCl–H_2O_2$ solutions
	NaCl solutions
	Seawater
	Air or water vapor
Copper alloys	Ammonia vapor and solutions
	Amines
	Water or water vapor
Gold alloys	$FeCl_3$ solutions
	Acetic acid–salt solutions
Inconel	Caustic soda solutions
Lead	Lead acetate solutions
Magnesium alloys	$NaCl–Na_2CrO_4$ solutions
	Rural and coastal atmospheres
	Seawater
	Distilled water
Nickel	Fused caustic soda
Steels	NaOH solutions
	$NaOH–Na_2SiO_4$ solutions
	Calcium, ammonium, and sodium nitrite solutions
	Mixed acids ($H_2SO_4–HNO_3$)
	Acidic H_2S solutions
	Seawater
	Carbonate–bicarbonate solutions
Stainless steels	Acidic chloride solutions
	$NaCl–H_2O_2$ solutions
	Seawater
	H_2S
	$NaOH–H_2S$ solutions
	Condensing steam from chloride waters
Titanium alloys	Red fuming nitric acid
	Seawater
	Methanol–HCl

Source: Craig and Lane [2005].

environment that are most commonly associated with SCC. Some typical combinations that result in SCC are examined in detail as below.

1. *Brass in ammonia-containing environments*. This was first identified as an SCC problem when the brass cartridge cases used by the British Army in India were found to suffer from cracking, where the ammonia comes from the decay of organic material [*Wikipedia*, online source]. Since it usually occurred during the rainy season, or the stress corrosion cracks resembled those in seasoned wood, it is also called *seasonal cracking*. The cracking is intergranular.
2. *Chloride SCC of stainless steel*. Austenitic stainless steels suffer from SCC in hot solutions containing chloride, where a high chloride concentration is required. Even when the average amount of chloride in the environment is low, chloride can concentrate at heated surfaces, or by pitting or crevice corrosion, to cause cracking. The temperature usually needs to be above 70°C, although SCC can occur at lower temperatures in some situations, notably in connection with more acid solutions. Cracking continues at low stresses and commonly occurs as a result of residual stresses from welding or fabrication. Cracking is normally transgranular, although it may switch to an intergranular path as a result of sensitization of the steel [Cassagne, 2007].
3. *Carbon steels in passivating environments*. Carbon and low-alloy steels can suffer from SCC in a wide range of environments that tend to form a protective passive or oxide film. The environments that would passivate carbon steels have been found to cause SCC, including strong caustic solutions, phosphates, nitrates, carbonates, and high-temperature water. The problems are important for both economic and safety reasons, due to the extensive use of carbon steels. For example, caustic cracking of steam-generating boilers was a serious problem in the late-nineteenth century, where the necessary strong caustic solution was produced by evaporation of the very dilute solution inside the boiler as it escaped through leaks in the riveted seams [Brown, 1972]. More recently, gas transmission pipelines have cracked in carbonate–bicarbonate solutions trapped under disbonded coatings in the presence of nonappropriate CP, resulting potentially in pipeline leaking, rupture, and even more devastating consequences [National Energy Board, 1996].
4. *Hydrogen embrittlement and HIC of high-strength steels*. All steels are affected by hydrogen, as evidenced by the influence of hydrogen on SCC growth and the occurrence of HIC under the influence of very high hydrogen concentrations [Bernstein and Thompson, 1980]. However, hydrogen embrittlement (HE) under static load is experienced only in steels of relatively high strength. There is no clearly defined limit for the strength level above which HE will be experienced, since this problem will be a function of the amount of hydrogen in the steel, the stress applied, the severity of the stress concentration, and the composition and microstructure of the steel. Hydrogen may be introduced into the steel by a number of routes, including welding, pickling, electroplating, exposure to hydrogen-containing gases, and corrosion in service. Hydrogen penetrating steel may be released or may get trapped inside the metal.

Since the first SCC documented in a boiler explosion, which was attributed to carbon steel caustic cracking, in the UK in 1865 [Galvele, 1999], the scope of environmentally induced cracking has increased considerably. Moreover, not only metals show SCC; it has been found [Wiederhorn and Bolz, 1970] that glasses experience cracking in the presence of water and that SCC can occur on ceramics and polymeric materials in corrosive environments [Jones, 1992].

2.3 METALLURGICAL ASPECTS OF SCC

The exact alloy composition, microstructure, and heat treatment can have marked effects on the SCC performance of a metal. Actually, stress corrosion cracks are usually initiated at the metallurgical defects contained in steels, which also affect the crack propagation mechanism and kinetics.

2.3.1 Effect of Strength of Materials on SCC

There is a common misunderstanding regarding how the strength of metals correlates with SCC (i.e., how an increase in strength of materials increases susceptibility to SCC). Actually, there have been few general rules governing the influence of material strength on SCC susceptibility [UK National Physical Laboratory, 1982]. For example, for the hydrogen embrittlement process, high strength normally increases the material susceptibility. However, the SCC process that replies on plastic strain at the crack tip will be easier for lower-strength materials.

2.3.2 Effect of Alloying Composition on SCC

Quite small changes in the composition of an alloy can have a marked influence on SCC behavior [UK National Physical Laboratory, 1982]. The effects of alloying additions are not necessarily consistent from one environment to another. For example, a high molybdenum content improves the resistance of a low-alloy steel to SCC in carbonate–bicarbonate solutions, but makes it more susceptible to caustic cracking [UK National Physical Laboratory, 1982].

2.3.3 Effect of Heat Treatment on SCC

Changes in the heat-treatment processes of an alloy can change its sensitivity to SCC, the mode of fracture, and even the fracture mechanism. Austenitic stainless steels suffer from transgranular SCC in chloride solutions. However, in steels heat-treated appropriately, SCC cracks could become intergranular [Alyousif and Nishimura, 2008]. Furthermore, if the same alloy is rolled, a certain amount of strain-induced martensite will be formed. This, combined with the high strength of work-hardened material, leads to a susceptibility to hydrogen embrittlement.

2.3.4 Grain Boundary Precipitation

A typical example of grain boundary precipitation is sensitization of austenitic stainless steels, where carbide precipitates at the grain boundaries, causing a depletion of chromium adjacent to the boundaries [Callister and Rethwisch, 2009]. Sensitization is an important metallurgical phenomenon occurring in austenitic stainless steels. It usually occurs in the temperature range 500 to 850°C, with the rate of precipitation controlled by the chromium diffusion. The chromium depletion zone can have a minimum chromium concentration of 8 to 10%, with a width of 10 nm to hundreds of nanometers. Due to the different chromium contents between grain boundaries and the grains, there exists a galvanic effect, where the grain boundaries serve as an anode and the grains as a cathode. A preferential dissolution of grain boundaries would result in the propagation of cracks along the grain boundaries, causing intergranular SCC. SCC susceptibility and crack growth rate can be described as the degree of sensitization, which is expressed as the width of the chromium depletion zone and the chromium concentration. The most common method of reducing the possibility of developing a sensitized microstructure is to reduce the carbon concentration in the steel or to control the thermal history of the material.

In addition to stainless steels, chromium carbide precipitation and chromium depletion can also occur in nickel-based alloys, such as alloy 600, but it does not connect to intergranular SCC of the alloy. Intermetallic precipitate can also form in aluminum alloys, resulting in intergranular SCC [Speidel, 1975], where the galvanic effect between the precipitates and the aluminum alloy matrix is important for cracking initiation and propagation. Sometimes, the precipitate is anodic to the matrix; in others, it is cathodic. In general, hydrogen evolution always accompanies the anodic reactions that occur on aluminum alloys. Since the passive layer on aluminum alloys is solvable in high-pH solutions, cathodic polarization that causes reduction of oxygen or water and the generation of hydroxyl ions stimulates dissolution of aluminum-passive film as well as hydrogen evolution. Thus, it is difficult to distinguish between anodic dissolution and hydrogen embrittlement for aluminum alloys in aqueous solutions.

2.3.5 Grain Boundary Segregation

Carbon and/or nitrogen segregation in grain boundaries plays an important role in the SCC of carbon steels. It was found that approximately 0.01% carbon is required to cause SCC of steels in nitrate or caustic environments [Parkins, 1990]. Actually, grain boundary enrichment of impurities can contribute to the intergranular SCC of iron-based alloys, austenitic stainless steels, and nickel-based alloys. The extent of their effect depends on the electrochemical potential, corrosive environment, and the type, shape, and concentration of impurities on the grain boundary.

Although segregation of substitutional elements to grain boundaries can strongly affect SCC of steels, it does not occur at all potentials or in all solutions. For example, phosphorus segregation promotes intergranular SCC of low-alloy steels in caustic or water environments at relatively oxidizing potentials [Burstein and Woodward, 1981;

Bandyopadhyay and Briant, 1983]. However, phosphorus does not affect SCC at low potentials in caustic solutions [Bandyopadhyay and Briant, 1983] and has only modest effects in carbonate–bicarbonate solutions [Stenzel et al., 1986].

2.4 ELECTROCHEMISTRY OF SCC

A complete description of SCC must treat both the thermodynamic requirements and kinetic aspects of cracking. While knowledge of the thermodynamic conditions helps to determine whether cracking is feasible, kinetic information describes the rate at which cracks propagate.

2.4.1 SCC Thermodynamics

The thermodynamic conditions for anodic dissolution–based SCC include two aspects: Dissolution or oxidation of the metal in the electrolyte must be thermodynamically possible, and a protective film formed on the crack wall must be thermodynamically stable.

The first condition requires that without oxidation, crack advance by dissolution would not occur. This implies that the crack propagation of dissolution-based SCC is through the anodic dissolution process at the crack tip. Thus, for dissolution-based SCC, the crack propagation rate can be calculated by the sum of the total number of coulombs of charges exchanged at the crack tip. It is important to note that if the brittle crack-advance process is initiated and then controlled by anodic dissolution, the crack growth rate will be zero if the anodic current density is zero and will increase with the increasing dissolution current density.

The second condition requires that the ratio of corrosion currents from the crack wall relative to the crack tip be a critical parameter. This ratio must be substantially less than 1 for a crack to propagate. Otherwise, the crack will blunt or the crack-tip solution will saturate. The reduced dissolution rate of the crack wall is dependent on the formation of passive film, while the crack tip is in the bare steel state. Moreover, crack initiation is also controlled by this ratio—a corrosion pit with a high rate of wall corrosion will result in uniform corrosion rather than cracking.

The electrode potential of a metal has a notable effect on the tendency for SCC to occur. A typical example is that for hydrogen embrittlement of high-strength steels, a more negative potential tends to increase the rate of hydrogen evolution, and thereby the susceptibility to HE. SCC processes that do not involve hydrogen typically occur over a certain potential range (i.e., the critical potential for SCC). Therefore, identification of the SCC critical potential has led to the use of various electrochemical methods, such as potentiodynamic polarization curve measurement, for assessing SCC susceptibility.

A potential–pH diagram provides a direct, powerful tool to determine the thermodynamic stability of specific phases of metals under certain potential–pH combinations. The effects of a variety of environmental conditions, such as solution pH, oxygen concentration, and temperature, on the thermodynamic conditions for SCC can be related to their effects on the potential–pH diagram or on the electrode potential relative to the various stability regions. It is generally believed that both high pH, used to favor film formation, and low pH, which increases uniform corrosion, tend to decrease the SCC susceptibility of metals. For metals in which SCC occurs by a hydrogen-induced crack growth mechanism, the thermodynamic requirement for cracking is governed by the hydrogen reduction line in potential–pH diagram.

There are several factors that limit the application of potential–pH diagrams in SCC research. Potential–pH diagrams for a complex solution mixture and for the temperature of interest are frequently unavailable. Moreover, there is a substantial deviation of the chemical and electrochemical conditions of a crack or a pit from the bulk solution chemistry and electrochemistry. Finally, the electrochemical potential of metals at the crack tip can differ from that at the free surface of the material since efforts to measure the crack-tip chemistry and potential are restricted by the crack size, which is usually smaller than 1 μm.

2.4.2 SCC Kinetics

The life of a structural component depends on the kinetics of crack growth. The crack-tip reactions and rate-determining steps controlling the crack growth rate are specific to alloy–environment combinations. For a dissolution-based crack growth mechanism, the crack propagation rate is a function of the total charge transfer at the crack tip and can be given by the crack-tip current density [Ford, 1996]:

$$\frac{da}{dt} = \frac{i_a M}{nF\rho} \qquad (2\text{-}1)$$

where a is the crack length, t the time, da/dt the crack propagation rate, i_a the anodic dissolution current density at the crack tip, M the atomic weight, n the number of electrons exchanged, F is Faraday's constant, and ρ is density.

There usually exists a deviation between the calculated crack propagation rate by Eq. (2-1) and the actual cracking speed determined in the field. A *negative deviation* occurs when the actual crack propagation rate is smaller than the value calculated. Processes such as crack deflection, branching, or bridging would reduce the stress intensity at the crack tip and slow down crack propagation. Moreover, the crack-wall corrosion effect may alter the corrosion rate at the crack tip: for example, the crack tip is covered by a film for some fraction of time, or the diffusion rate of species into and out of the crack is limited. The crack propagation rate may decrease. When the actual crack propagation rate is higher than the rate calculated, it is called *positive deviation*. Positive deviation in the crack velocity results from crack jumps produced by mechanical factors. Hydrogen-enhanced crack growth resulting from hydrogen evolution and permeation during corrosion is the most common form of this type of mechanical fracture.

For a mechanical fracture-based SCC process, the total crack advance exceeds the total charge transfer at the crack tip, but the crack velocity may still be controlled by the crack-tip current density. A number of technical challenges must still be addressed to determine the crack growth rate quantitatively. For example, the identification of a corrosion product or layer that initiates the cleavage cracking process is still speculative. The electrochemical conditions controlling the brittle SCC process have not been fully understood. The greatest uncertainty involves the possibility of a transition between dissolution and mechanical SCC processes that is affected by electrochemistry, materials chemistry and microstructure, and mechanical stress.

2.5 SCC MECHANISMS

An SCC mechanism illustrates the combined effects of mechanical, physical, and chemical/electrochemical factors causing the separation of metal bonds at the crack tip, thereby advancing the crack [Newman and Procter, 1990]. There is no universal mechanism that can describe all SCC cases. The SCC process is usually divided into three stages [Jones, 1992]: crack initiation, steady-state crack propagation, and final failure, with identical or different mechanisms controlling each stage.

2.5.1 SCC Initiation Mechanisms

Generally, initiation of a crack is associated with microscopic crack formation at localized corrosion or mechanical defect sites, and is related to pitting, intergranular attack, scratches, weld defects, or design notches [Ford, 1996]. Actually, stress corrosion cracks can be initiated by a number of mechanisms.

1. *Crack initiation at surface discontinuities.* Stress corrosion cracks are often initiated at various discontinuities on a metal surface, such as scratches, grooves, or mechanical dents. These discontinuities could be preexisting or introduced by corrosion during service. The local environment and stress conditions favor an enhanced dissolution, or formation of a poor film, resulting in electrochemical activity that differs from that in adjacent regions.
2. *Crack initiation at metallurgical defects.* A variety of metallurgical defects, such as inclusions, grain boundaries, and voids, are formed or introduced during metal manufacturing. There are different electrochemical activities between the defects and the adjacent metal matrix, resulting in a galvanic effect and preferential dissolution of one phase, which could be the defect or the matrix. Moreover, depending on the stability of grains and grain boundaries, cracking could be intergranular or transgranular. A sensitized stainless steel usually has an intergranular SCC, where the grain boundaries become active due to local chromium depletion [Cassagne, 2007; Callister and Rethwisch, 2009]. In a nearly neutral–pH bicarbonate solution, the grain of X70 pipeline steel is more anodic relative to the grain boundary. As a consequence, preferential dissolution and corrosion pits are formed on grains rather than at grain boundaries [Liu et al., 2010]. This serves as evidence illustrating the transgranular mode of nearly neutral–pH SCC of pipelines.

3. *Crack initiation at corrosion pits*. Stress corrosion cracks can be initiated at corrosion pits formed during exposure to the service environment or during cleaning operations, such as pickling. Actually, pitting corrosion is often the first step in SCC in gas pipelines [National Energy Board, 1996]. Transition of a corrosion pit toward a crack depends on several conditions. First, electrochemistry at the base of the pit must be more active than that of the pit wall, which usually shows somewhat passive behavior. Second, a 10 : 1 ratio of pit depth to width would be the geometric condition for the transition. In particular, a penetration/lateral corrosion ratio of 1 indicates uniform corrosion, and a ratio of 1000 is observed for a growing stress corrosion crack. Third, the pit–crack transition depends on the material chemistry and microstructure. For example, an active grain boundary favors the development of a cracking path, and thus the transition. Finally, the local stress or strain rate at the base of the pit is critical in enhancing the local dissolution, and thus the transition of corrosion pits to cracks. However, it is worth pointing out that electrochemistry of the pit base is more important than local stress–strain concentration with respect to pit–crack transition, as evidenced by many preexisting pits that do not initiate cracks. Moreover, a preexisting pit may not develop the same local electrochemistry as that of a pit grown during service, because the development of a concentration cell depends on the presence of an actively corroding pit that meets the geometrical requirement and establishes anion and cation flow.

To date, there have been few well-developed quantitative models for crack initiation, in part because a crack initiation event is difficult to measure experimentally. Moreover, crack initiation has not been defined precisely. It is difficult to determine at what point a pit develops into a crack and when intergranular corrosion becomes intergranular SCC.

Fracture mechanics gives a critical "pit" dimension that must be exceeded to transition into a crack [Corrosion Doctors, online source 2]:

$$a_0 = \frac{1}{\pi}\left(\frac{\Delta K_{\text{th}}}{c\,\Delta\sigma_0}\right)^2 \qquad (2\text{-}2)$$

where a_0 is the critical crack depth, ΔK_{th} the stress intensity threshold, C is a constant, and $\Delta\sigma_0$ is the alternating surface stress. It should be noted that this model lacks sufficient reliability to predict the pit–crack transition because it is based on the assumption that the pitting corrosion condition is identical to the metal surface condition. In almost all cases, this is not true. Actually, this model does not describe the transition from a corrosion pit to a crack because it treats the pit as a small crack in which the crack depth (pit depth) is affected by corrosion.

2.5.2 Dissolution-Based SCC Propagation

The basic assumption in all the SCC propagation mechanisms proposed for ductile alloys in aqueous solutions is that the crack tip must propagate faster than the crack sides so that the crack does not degrade into a blunt notch [Ford, 1982]. There are basically two categories of SCC propagation mechanisms: dissolution-based and mechanical fracture–based.

The essence of the dissolution-based SCC propagation mechanism is that a crack advances by preferential dissolution at its tip. For intergranular cracking, the presence of segregated solutes or precipitated phases at grain boundaries results in electrochemical heterogeneity locally. In the presence of appropriate environments, preferential dissolution occurs in the region. In the absence of stress, the initial attack (corrosion) may not extend far, due to film formation. However, in the presence of stress, repetitive rupture of film can sustain the dissolution reaction, causing cracks to continue to grow. Two models have been proposed to explain the crack growth modes.

1. *Slip-dissolution (or film rupture) model* [Galvele, 1995]. Tensile stress ruptures film at a crack tip, and the crack grows rapidly from the exposed bare metal until the crack tip repassivates, in some cases, or in other cases, grows continuously to failure. Consequently, two different cracking modes are generated from this model: a periodic growth mode where the crack tip is repassivated completely and is ruptured periodically by the emergence of new slip steps, and a continuous growth mode, where once crack propagation begins, the crack tip remains bare due to the higher rate of film rupture at the crack tip than the rate of repassivation.

   While the film rupture model has been accepted as a reliable mechanism for intergranular SCC, it was later modified as a mechanism of transgranular SCC. The mechanism assumes a protective film being ruptured by a slip step, with the exposed reactive metal attacked preferentially by the environment. The re-formation of film at the slip step causes crack growth to cease until further slip breaks that film, and dissolution is continued. The objection to this transgranular dissolution mechanism that is raised is that it is difficult to account for the matching opposite fracture surfaces [Beavers and Pugh, 1980].

   It has been stated that one of the characteristics of SCC is that the cracks show a high aspect ratio: the ratio of the crack length or depth to the crack opening. However, low aspect ratios are considered in the slip-dissolution model. Moreover, it is suggested [Galvele, 1999] that while slip steps are probably important for crack initiation, it is questionable if they are important for crack propagation. Furthermore, the slip-dissolution model assumes that there is an intermediate repassivation rate typical of SCC. However, when repassivation rates for a large number of systems were measured, no intermediate repassivation rate for SCC has been recorded [Carranza and Galvele, 1988a,b].

2. *Active-path model.* Intergranular SCC can occur by an active-path process that results from a difference in the microchemistry of the material at the grain or interface boundary [Jones et al., 1989]. The characteristics of active-path intergranular SCC can be attributed to grain boundary segregation and precipitation, resulting in different microchemistry and electrochemical activity at the boundaries. A typical example to follow this model is the SCC of sensitized stainless steels.

2.5.3 Mechanical Fracture–Based SCC Propagation

Mechanical fracture mechanisms originally assumed that stress concentration developed at the base of corrosion pits or mechanical defects increases to the point of ductile deformation and fracture of metals or alloys. Moreover, the crack propagates by dissolution and then fails by mechanical fracture [Harwood, 1956].

1. *Corrosion tunnel model* [Swann and Pickering, 1963]. Corrosion tunnels are formed by preferential corrosion along emerging slip steps on the metal surface. These tunnels grow in diameter and length until the stress in the remaining ligaments causes ductile deformation and fracture. The crack propagates by alternating tunnel growth and ductile fracture. Cracks propagating by this model result in a grooved fracture surface with evidence of microvoid coalescence on the peaks of the grooves. The width of the corrosion slots approaches atomic dimensions, and close correspondence of matching fracture surfaces would be possible. Transgranular SCC can be explained in terms of the formation and mechanical separation of corrosion slots.
2. *Adsorption-enhanced plasticity model* [Lynch, 1981, 1985]. The adsorption of specific ions reduces the critical resolved shear stress for dislocation mobility. Dislocations are then easier to move locally under the influence of the tensile stress compared to conditions in the absence of ion adsorption. Chemisorption of environmental species would facilitate the nucleation of dislocations at the crack tip, promoting the shear processes responsible for brittle, cleavagelike fracture. Cleavage fracture is not an atomically brittle process, but occurs by alternate slip at the crack tip in conjunction with the formation of very small voids ahead of the crack. This mechanism could explain many similarities among SCC, liquid–metal embrittlement, and hydrogen embrittlement.
3. *Adsorption-induced brittle fracture model* [Hart, 1968]. This model is based on the assumption that adsorption of environmental species lowers the strength of interatomic bonds and thus the stress required for cleavage fracture. During the electrochemical process atoms are absorbed on the metal surface to weaken the metallic bonds. The stress required for initiation of a crack then decreases. The crack grows until it is blunted by plastic deformation or grows out of the region adsorbed. This model predicts that cracks propagate in a continuous manner and does not explain how the crack maintains an atomically sharp tip in a normally ductile material. It does not apply to the discontinuous nature of crack propagation. This model is different from the adsorption-enhanced plasticity model in that the cohesive strength, rather than the resolved shear stress, decreases due to ion adsorption.
4. *Tarnish-rupture model* [Forty and Humble, 1963; McEvily and Bond, 1965]. This model was originally proposed to explain transgranular SCC. A brittle surface film forms on the metal, fracturing under the applied stress. Fracture of the film exposes the underlying bare metal, which reacts rapidly with the environment to re-form the surface film. The crack propagates by alternating

film growth and fracture. The model was later modified to explain intergranular SCC based on the assumption that the oxide film penetrates along the grain boundary ahead of the crack tip. Again, crack propagation consists of alternating periods of film growth and brittle-film fracture. However, the crack arrest markings are not invariably observed on the intergranular fracture surface. It assumed penetration of the film into the grain boundary ahead of the crack tip, which may not be the case for all systems. To date there have been insufficient experimental results to confirm or refute this model.

5. *Film-induced cleavage model* [Harwood, 1956]. In this model, the surface film grows and may elevate internal stress with its thickness. This, combined with the applied tensile stress, induces brittle failure in the film, which propagates across into the metal and provides a period of crack growth. Loss of the film stress and the plasticity in the metal then blunts the crack and stops its growth, resulting in periods of crack growth followed by rest while the film grows back to the conditions for cleavage. This model has the unique ability to explain the crack arrest markings, the cleavagelike facets on the fracture surface, and the discontinuous nature of crack propagation. The nature of the embrittling film that initiates cleavage is likely to be a dealloyed layer or an oxide film. The critical point in this model is the hypothesis that a brittle crack will continue to propagate after it has entered the normally ductile metal matrix. This results in a thin surface layer to induce brittle crack propagation over distances much greater than the film thickness. There is evidence that a brittle crack can propagate into a ductile matrix if the crack is sharp and propagates at high velocity before entering the ductile metal matrix.
6. *Localized surface plasticity* (LSP) [Jones, 1996]. This mechanism involves the contribution of LSP to SCC. The film rupture by plastic slip, anodic dissolution occurring at the film rupture sites by galvanic coupling to the surrounding passive surface, and LSP at the film-rupture slip-band surface sites account for the initiation of cracks. Cracks grow under triaxial stress conditions that suppress further slip at the crack tip initiated. However, the presence of combined shear-tensile loading contributes to multiple initiation sites and subsequent characteristic branching of secondary cracks off the main crack. Vacancies generated during anodic dissolution may result in LSP by locally relieving strain hardening at the film-rupture surface sites. Vacancies may interact with dislocations to increase the dislocation mobility or may simply weaken interatomic bonds. Vacancies may also alter the surface plasticity and near-surface ductility by weakening the lattice.
7. *Surface mobility model* [Galvele, 1993]. According to this model, environmentally induced crack propagation is due to the capture of vacancies by the stressed lattice at the crack tip. The rate-controlling step is the rate of movement of the vacancies along the surface of the crack. The environment plays a dual role. It increases the surface self-diffusivity of the metal or alloy and simultaneously assures a free supply of vacancies to the metal surfaces by selective dissolution of the alloy. The crack propagation rate is dependent on the surface

self-diffusive coefficient and diffusion distance of vacancies, surface stress at the crack tip, atomic size, and temperature. When a process slower than surface diffusion becomes the rate-controlling step, the crack velocity becomes smaller. For example, diffusion of halide ions in the solution inside a crack could be the rate-controlling step. Moreover, a slow supply of vacancies could be a rate-controlling step. This model has been considered premature because the ohmic and diffusive difficulties that may control the propagation of cracks depend crucially on the crack geometry and dissolving surface area, which are not considered in the model [Newman and Procter, 1990].

Although a number of mechanisms and models have been developed to explain SCC phenomena, many of them cannot be verified easily by experimental tests. It is suspected that there is no universal model for SCC initiation and propagation. Each specific combination of metal, environment, and stress can produce different fracture evidence. Moreover, even the SCC fracture surfaces generally look brittle in nature, with evidence of noncontinual crack growth, and cross sections of specimens show some evidence of local ductility or plastic flow. This adds to the confusion surrounding mechanistic understanding of the SCC process.

2.6 EFFECTS OF HYDROGEN ON SCC AND HYDROGEN DAMAGE

Hydrogen plays a role in a number of cracking mechanisms, including SCC. Actually, hydrogen-induced crack growth can be a form of SCC in many metal–environment combinations. However, there are hydrogen-induced cracking phenomena that occur under constant stress, but the corrosion process is absent. These are considered separate cracking mechanisms from SCC. Furthermore, hydrogen can induce material cracking even in the absence of external stress [Jin and Cheng, 2010].

2.6.1 Sources of Hydrogen

Some of the wide variety of sources capable of producing hydrogen that potentially penetrates metals are noted below.

- *Electroplating*. Hydrogen atoms can be "plated out" on a metal surface, along with a reduction of metallic cations during electroplating, and enter the metal. This phenomenon is particularly prevalent in chromium plating [Demakis, 2002].
- *Acid pickling*. When steel is acid-pickled, hydrogen is generated by reduction of hydrogen ions in the corrosion reaction and may enter the steel. Electrolytic pickling can also introduce hydrogen into the steel by cathodic charging.
- *Heat treatment*. The furnace atmosphere of hydrogen can result in hydrogen entry during the heat treatment of steel.

- *Welding*. Hydrogen pickup in welding can lead to cracks in either the weld metal or in heat-affected zones. Moisture in welding electrodes is a common source of such hydrogen.
- *Services*. Certain service environments can introduce hydrogen into steel; these include a high-pressure hydrogen environment, the reaction of steam with iron at high temperatures, and the corrosion of steel, where hydrogen evolution is the reductive reaction.

2.6.2 Characteristics of Hydrogen in Metals

Before hydrogen can enter a metal, it must first be adsorbed on the metal surface as atomic hydrogen. Thus, newly created atomic hydrogen, such as that produced during a corrosion reaction, can quite possibly enter a metal. Atomic hydrogen can diffuse through a metal lattice because of its small size, but hydrogen molecules cannot, due to their larger size. Furthermore, atomic hydrogen diffusing through a metal lattice can combine to form molecular hydrogen within structural defects such as voids, microcracks, or discontinuities around inclusions.

The solubility of hydrogen in metals is a function of their physical form (solid or liquid), crystalline lattice, alloy content, temperature, and the hydrogen partial pressure in the environment. Generally, an elevated temperature causes an increase in hydrogen solubility. Particularly, at the melting point when liquid iron forms, the hydrogen solubility increases significantly. Moreover, there is higher hydrogen solubility for α-Fe with a body-centered-cubic (BCC) structure than for γ-Fe, which has a face-centered-cubic (FCC) lattice structure [Gibala and Kumnick, 1984]. According to an empirical relation, hydrogen solubility is approximately dependent on the square root of the hydrogen partial pressure [Suh and Eagar, 1998]. Moreover, steels with a low solubility for hydrogen may have a strong tendency toward embrittlement [Oriani, 1984].

The permeability of hydrogen through iron or steel depends on both its solubility and its diffusivity in materials. Generally, the permeability of hydrogen increases with rising temperature; however, the increasing permeation rate of hydrogen with temperatures from 100 to 400°C can also lead to more effective escape of hydrogen from the metal to the environment, where the hydrogen pressure is low or negligible [Hirth, 1980]. For example, corrosion reactions at atmospheric pressure and at temperatures above 150°C are unlikely to charge significant quantities of hydrogen into steel. At the same temperature, the permeation rate of hydrogen in γ-Fe or FCC alloys is markedly lower than in α-Fe or BCC alloys [Hirth, 1980]. Furthermore, a small amount of cold work in steel will increase the permeability of hydrogen and possibly lead to detrimental effects.

2.6.3 The Hydrogen Effect

The effects of hydrogen on the mechanical properties of steels are generally controlled by two factors: the form of hydrogen in the steel (atomic or molecular), and the strength level and stress state of the steel. Generally, molecular hydrogen forming in steels creates blisters and does not cause embrittlement of the steel. In annealed steels stressed below their yield strengths, the most likely damage to occur from hydrogen charged into the steel is blistering, whereas hardened, cold-rolled, or highly stressed steels are more likely to be mechanically embrittled or cracked as a result of hydrogen entry.

Once hydrogen has been absorbed by a material, its effect is essentially identical, whether it is from a gaseous or an aqueous source. However, there are three primary differences between the gaseous and cathodic hydrogen-absorption processes [Hirth, 1980]. First, cathodic hydrogen is present on the metal surface as atomic hydrogen, which is reduced electrochemically from the oxidative phase, while gaseous hydrogen absorbs in the molecular form and must dissociate to form atomic hydrogen before entry. Desorption of loosely bound molecular hydrogen is relatively easy, while the dissociation of adsorbed molecular hydrogen can be the rate-limiting process in hydrogen absorption and embrittlement. Therefore, desorption and absorption rates of gaseous and cathodic hydrogen would be substantially different at equal hydrogen fugacity. Second, hydrogen concentration produced by cathodic hydrogen can be quite large and is dependent on the electrode potential of steels and the corrosion reaction kinetics, while the gaseous hydrogen pressures are generally much lower. Finally, the surface of the material may be substantially different under corrosion conditions where cathodic hydrogen is generated in the presence of gaseous hydrogen containing substantial quantities of other gases. As a consequence, hydrogen adsorption and entry become significantly different.

Metallic materials can suffer from a number of different forms of hydrogen damage, resulting from the combined factors of stress, hydrogen form, and material properties. Table 2-2 lists the general types of metals that are typically susceptible to various forms of hydrogen damage.

1. *Hydrogen embrittlement.* HE is a process by which various metals, particularly high-strength steels, lose ductility and become brittle and fracture following exposure to hydrogen. The HE susceptibility of metals and alloys increases with decreasing strain rate and increasing hydrogen pressure and purity.

TABLE 2-2 Types of Metals That Are Susceptible to Hydrogen Damage

HIC	Hydrogen Effect	Loss in Tensile Ductility	Blistering	Flakes or Fish-Eyes	Loss in Flow Properties	Metal Hydride Formation
Steels	Carbon steels	Steels	Steels	Steels (forging and casting)	Iron	V
Nickel alloys	Low-alloy steels	Nickel alloys	Coppers		Steels	Nb
Stainless steels		Aluminum alloys	Aluminums		Nickel alloys	Ti
Titanium alloys		Bronze				Zr

Source: Craig and Lane [2005].

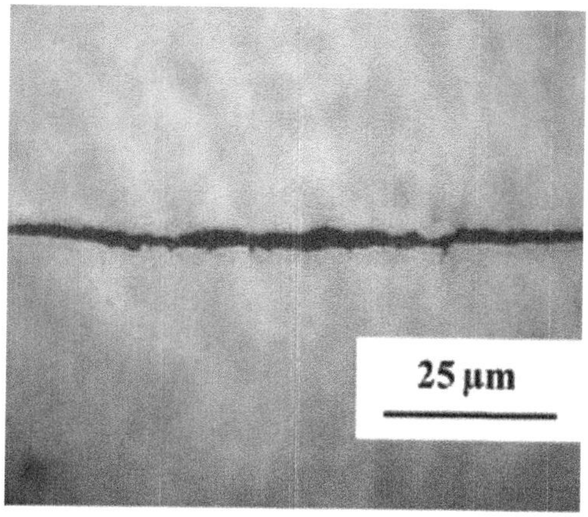

Figure 2-2 *Optical view of a hydrogen-induced crack formed in X100 pipeline steel upon hydrogen charging in a soil solution. (From Jin et al. [2010].)*

2. *Hydrogen-induced cracking*. HIC refers to the brittle fracture of a ductile alloy under loading in the presence of hydrogen. Particularly, fracture occurs at stresses below the yield strength of the material. Figure 2-2 shows a hydrogen-induced crack in X100 high-strength pipeline steel upon hydrogen charging in a soil solution. For many steels there exists a threshold stress below which HIC does not occur. This threshold stress or stress intensity is not a material property because it depends on the strength level of the steel and the environmental effect. Generally, the threshold stress decreases as the strength of an alloy increases. Furthermore, HIC is associated with the absorption of hydrogen and a delayed time to failure when hydrogen atoms diffuse to regions with high triaxial stresses. HIC is always featured with sharp singular cracks rather than extensive branching or secondary cracks.

   Furthermore, cracking of steels can occur from the surface in the presence of hydrogen sulfide (H_2S) under applied stress, called *sulfide stress cracking* (SSC) [Szklarz, 1999; Leyer et al., 2005]. For SSC, the environmental severity depends primarily on pH and H_2S partial pressure [National Association of Corrosion Engineers, 2003].
3. *Loss in tensile ductility*. This is hydrogen damage that causes significant decreases in elongation and reduction in the area of alloys. It is often observed in low-strength alloy, and the hydrogen content affects the extent of the ductility loss. Generally, this behavior is more apparent when the strain rate decreases.
4. *Hydrogen attack*. This is a high-temperature form of hydrogen damage that occurs in carbon and low-alloy steels exposed to high-pressure hydrogen at a high temperature for an extended period. Hydrogen enters the steel and reacts with carbon to form methane gas, resulting in the formation of cracks and fissures, or decarburizing the steel and lowering its strength. The threshold temperature for hydrogen attack is approximately 200°C.

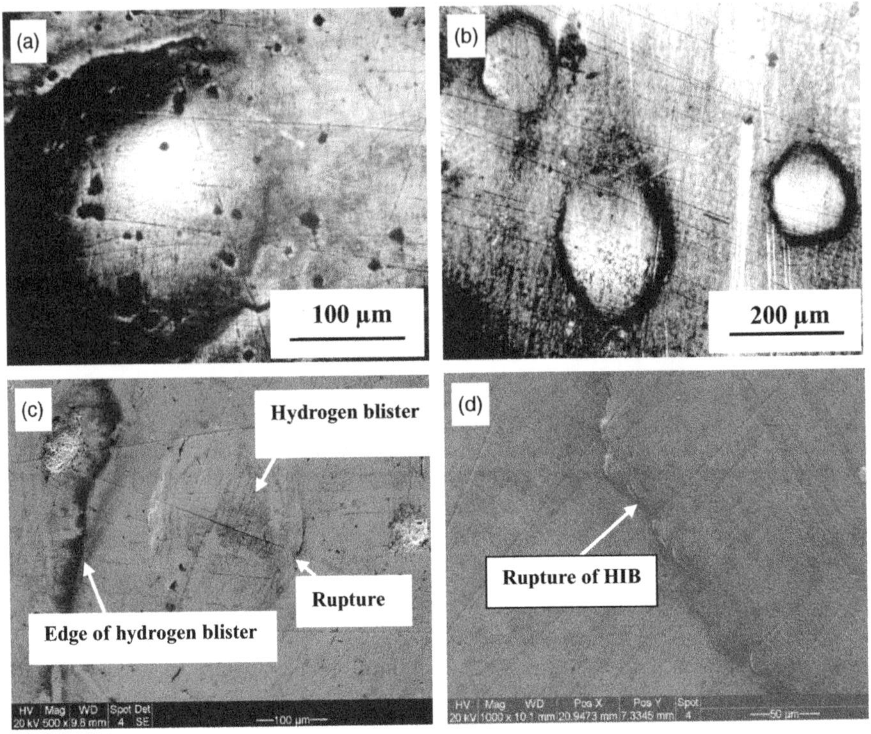

Figure 2-3 Surface morphology of hydrogen blisters on X100 steel after 20 h of charging at (a) 50 mA/cm^2 and (b) 100 mA/cm^2 as well as ruptured blisters (c) and (d), where HIB refers to hydrogen-induced blistering. (From Jin et al. [2010].)

5. *Blistering*. Hydrogen blistering occurs when atomic hydrogen diffusing through metals collects at internal defects such as inclusions and microcracks, and form molecular hydrogen. High pressures can be built up at such locations due to the continued absorption of hydrogen, leading to blister formation, growth, and eventual bursting of the steel. Blisters are often found in low-strength steels that are exposed to corrosive environments. Hydrogen-induced blistering and blister cracking have also been observed in high-strength pipeline steels, as shown as in Fig. 2-3.
6. *Flakes and fish-eyes*. Flakes and fish-eyes are common to hydrogen damage in forging, weldments, and casting, due to hydrogen pickup during melting operations, because the metal melt has a high solubility for hydrogen. During cooling, hydrogen diffuses to and precipitates in voids and discontinuities, forming flakes or fish-eyes due to decreased solubility in the solid metal.
7. *Loss in flow properties*. This occurs in iron and steel at ambient temperature and in several alloy systems at elevated temperatures. Steady-state creep under constant load is observed in the presence of hydrogen.

8. *Hydride formation.* The hydride formation embrittles metals such as magnesium, niobium, vanadium, uranium, thorium, zirconium, titanium, and their alloys. Hydride formation is enhanced for some metal–hydrogen systems by the application of stress.

2.6.4 Mechanisms of Hydrogen Damage

With the numerous forms of hydrogen damage, many theories and models have been developed to explain the various forms of hydrogen-induced degradation of materials. Some primary concepts are discussed below.

1. *Decohesion model.* The decohesion model proposed in 1960 is one of the oldest models used to explain the change in properties of materials as a result of atomic hydrogen [Oriani, 1990]. It is based on the increased solubility of hydrogen in a triaxial stress field at the tip of a crack. The accumulation of hydrogen results in a decrease in the atom-bonding forces of the metal lattice. The influence of stress results in a brittle fracture either along the grain boundaries (intergranular cleavage) or across grains (transgranular cleavage), owing to the decrease in bonding forces.

   This mechanism is supported primarily by the observation that in some non-hydride-forming systems, HE appears to occur in the absence of significant local deformation. However, direct evidence for this mechanism has not been obtained. Furthermore, considerating the decreasing bonding force and energy change during segregation of hydrogen to interfaces and surfaces, it would be expected that only a small amount of hydrogen-induced decrease in the separation energy occurs during transgranular fracture, but that there would be a substantially large decrease during intergranular fracture in those cases where hydrogen segregation to the grain boundaries has taken place. Similarly, direct experimental characterization of this effect is currently lacking.
2. *Hydrogen-enhanced local plasticity model* [Beachem, 1972]. This model was suggested based on the observation of very small dimples that are normally associated with microvoid coalescence on the fracture surface of metals such as aluminum alloys [Lynch, 1986] in the presence of hydrogen. The prerequisite for this mechanism is the accumulation of hydrogen in the field of stress, for example, in the vicinity of the tip of cracks or in the stressing areas of dislocations. During the initiation of a dislocation movement by external stresses, hydrogen decreases considerably the resistance to dislocation motion through shielding the stress field of the dislocations against each other. Consequently, a local dislocation movement occurs at low levels of shearing stress, accompanying material softening behavior. This works as an enhanced dislocation injection at surfaces and the promotion of shear instabilities.

   Since hardening effects are also observed with the hydrogen entry, the idea of softening is not a general one. Furthermore, hydrogen and dislocations are required to nucleate cracks at inclusions, and the enhanced dislocation motion is not a model in itself but a contributing factor to an overall degradation mechanism.

3. *Hydride formation mechanism.* The hydride formation and cleavage mechanism is a well-established HE mechanism with extensive experimental and theoretical support, where a hydride forms and cracks near the crack tip. The nucleation and growth of hydrides ahead of a crack has been observed in α-Ti charged with hydrogen from the gas phase [Teter et al., 2001]. It was found that the hydrides nucleated in the stress field of the crack and then grew to large sizes not by the growth of individual hydrides but by the nucleation and growth of new hydrides in the stress field of the others. The small hydride grew together to form large hydrides. This autocatalytic process of hydride nucleation and growth together with the brittle nature seems to be the main cause of embrittlement of typical hydride-forming elements (e.g., V, Nb, Ti, and Zr) [Parkins, 1992].
4. *Internal pressure theory.* This theory involves the collection of hydrogen at internal defects such as voids, and then internal crack growth from hydrogen gas at large internal pressures in the voids. The pressure developed by this process decreases the fracture stress required by adding to the applied stress. Dislocation transport could have generated large internal supersaturations and hence large internal pressures even when the external pressure is low. However, such a mechanism cannot be general since supersaturation is usually greatly overestimated, and a low-pressure crack propagation phenomenon is also frequently observed. Pressure-enhanced void growth can occur, but only in high-fugacity environments, as demonstrated by blister formation in the absence of external stress. At low temperatures, large dislocation-enhanced supersaturations in voids are possible.
5. *Surface adsorption theory.* Hydrogen adsorbs on the surface adjacent to the crack tip, decreasing the surface free energy and thus the work of fracture [Hirth, 1984]. The crack propagation is enhanced at stress levels below those typically experienced for this alloy in a hydrogen-free environment. The primary concern for this theory is that it underestimates the work of fracture and does not apply for the discontinuous crack growth that has been observed for HIC.
6. *Hydrogen trapping.* One of principal factors that determine hydrogen damage susceptibility is the presence of various traps in metals, especially steels. *Hydrogen trapping* refers to the binding of hydrogen atoms to impurities, structural defects, or microstructural constituents in the alloy. Binding is attributed to the local stress field, temperature gradient, chemical potential gradient, or physical trapping. The hydrogen traps may be mobile, such as dislocations, or stationary, such as grain boundaries and solute atoms. They may also be reversible or irreversible. The former term refers to short-duration trapping of hydrogen and features low binding energy; the latter is characterized by a long residency time for hydrogen characterized by high binding energy.

2.7 ROLE OF MICROORGANISMS IN SCC

2.7.1 Microbially Influenced Corrosion

Microbially influenced corrosion (MIC) is a phenomenon in which corrosion is initiated or accelerated by the activities of microorganisms [Revie, 2000]. It is an electrochemical process where the participation of microorganisms is able to initiate, facilitate, and/or accelerate corrosion reactions without changing the electrochemical nature. Generally, microorganisms located on or adjacent to the metal do not attack the metal directly or cause a unique form of corrosion. Their by-products from metabolism induce or promote corrosion, such as pitting corrosion, crevice corrosion, and underdeposit corrosion [Little and Lee, 1997]. Typically, the by-products of a growing microbiological colony accelerate the corrosion process either by interacting with the corrosion product to prevent formation of protective film that would inhibit further corrosion, or by providing an additional reductive reaction that accelerates the corrosion process [Little and Lee, 1997]. Microorganisms are often associated with localized forms of corrosion, including pitting corrosion, crevice corrosion, galvanic corrosion, intergranular corrosion, dealloying, and SCC.

Microorganisms can influence a corrosion process by one or a combination of the following mechanisms.

1. Patchy microbial deposits, colonies, tubercles, or biocorrosion products form discontinuous biofilms on a metal surface, creating conditions for the formation of new galvanic cells and/or altering the conditions in existing galvanic cells, and thus affecting profoundly both general and localized corrosion of metals [Zhang et al., 2011]. Biofilms can develop in both anaerobic and aerobic regions.
2. The metabolic processes of microorganisms destroy the existing protective film on a metal surface. For example, sulfidization of passive film can depolarize the cathodic reaction and significantly enhance corrosion even at low oxygen levels [Antony et al., 2007]. Moreover, pitting corrosion is initiated following the breakdown of passive film. It was found [Franklin et al., 1991] that in complex sterile aerated media, pit initiation is followed by repassivation. However, when an aerobic heterotrophic bacterium is added to a medium, pits do not repassivate but continue to grow.
3. Most end products of MIC are concentrated short-chain fatty acids such as acetic acid, which are corrosive to steels [Evans et al., 1973]. In the absence of oxygen, the acids are reduced and contribute to corrosion. In the presence of oxygen, the area adjacent to sediment is regarded as the small anode relative to the surrounding cathode. A trace amount of oxygen is reduced cathodically, contributing to the metal dissolution.
4. A high concentration of ferrous ions in the metabolism of sulfate-reducing bacteria (SRB) is favorable for anaerobic corrosion, and the corrosion rate of steels depends on the concentration of H_2S produced by SRB [King et al., 1973]. The anaerobic corrosion of SRB is also the result of a phosphide with high activity

and volatility generated by SRB metabolism. The phosphide could interact with steel to generate ferrous phosphide, which could enhance corrosion. The corrosion product formed by SRB metabolism would also accelerate localized corrosion.

Generally, microorganisms play an important role in the process leading to biofilm formation. Biofilms, as well as biofouling deposits, drastically modify the characteristics of the metal–solution interface where corrosion reaction takes place. Microorganisms within a biofilm can accelerate a corrosion reaction by several mechanisms, such as establishment of concentration cells, production of aggressive metabolites (i.e., acids, sulfide, and/or phosphate), and direct metal oxidation and reduction. A biofilm can alter the solution chemistry at the metal–solution interface by, for example, lowering the solution pH or catalyzing oxidation–reduction reactions.

Bacteria often exist together with other microorganisms present in the natural environments where MIC occurs. They form complex biofilms, with the local chemistry changing with time. The chemistry developed under a biofilm features characteristics that usually differ from those of the bulk solution. The development of biofilm is a dynamic and continuous process, increasing in thickness and heterogeneity with time, and usually resulting in the occurrence of extensive micropitting corrosion [Sheng et al., 2007; Yuan and Pehkonen, 2007]. For example, the biofilm formed by the SRB undergoes attachment, growth, subsequent detachment, and reattachment and shows different morphology and polarization resistance.

The corrosion of steels progresses as a result of various species induced by bacteria, low pH, and other specific environmental conditions. Moreover, the cohesion of corrosion products to a metal surface is also important to support or inhibit further corrosion and hydrogen permeation [Yo1e, 1998]. Generally, MIC and microorganisms that affect SCC depend on the type and number of bacteria present and their elective metabolic activity as a hydrogen source. The morphologies that characterize MIC are usually dendritic with branched tunnels, which have subsurface cavities with a series of pits branching off each other, with the deposits distributed heterogeneously. This has a considerable effect on the nutritional conditions for bacterial growth on metal corrosion. It was reported [Padilla-Viveros et al., 2006] that under nutritionally rich conditions, the average corrosion rate of a carbon steel is low; it never exceeds 1.2 mils per year (mpy). In contrast, a significant corrosion rate of 8 mpy is obtained under oligotrophic conditions.

The SRB or "desulfovibrio" bacteria, a heterogeneous group of anaerobic heterotrophic bacteria that use inorganic sulfate as a final electron acceptor in respiration, are the main bacteria responsible for the numerous MIC cases. The overall corrosion process on steels effected by the SRB includes a series of reactions: anodic dissolution of steels; dissociation of water under anaerobic conditions when oxygen present in the biofilm is consumed; cathodic hydrogen evolution; cathodic depolarization with the generation of sulfur compounds, due to the ability of SRB to remove adsorbed hydrogen from a metal surface through the enzyme hydrogenase [Thierry and Sand, 1995]; and production of iron sulfides and iron hydroxides, which are corrosion products in MIC. Indeed, sulfide is often detected in corrosion products formed during

MIC. An enzyme (hydrogenase) is generated by SRB to facilitate the reduction of sulfate and sulfide formation [Stott et al., 1988; Javaherdashti, 1999]. Enhancement of the corrosion rate of steels and the resulting increase in hydrogen absorption may be due largely to the presence of these sulfides [Little and Lee, 1997]. Furthermore, sulfide is known to promote the entry of hydrogen atoms into steels by preventing the atomic hydrogen recombination reaction generated at the metal surface. However, there is no direct relationship between the sulfide produced by some SRB strains and the corrosion rate [Taheri et al., 2005].

While the individual bacterium type accelerates the corrosion of metals, the combination of SRB and iron-oxidizing bacteria (IOB) demonstrates higher corrosion rates than those observed in the presence of SRB or IOB alone [Xu et al., 2008]. This synergistic effect is further enhanced by chloride ions that would be involved in the corrosion, especially the pitting corrosion, process.

Microbial activity has also been found to inhibit corrosion, as shown by the lower corrosion rate of mild steel in the presence of the bacterial culture *Pseudomonas flava* [Gunasekaran et al., 2004]. A mild steel surface contains a phosphate layer covered with bacteria, and the microbes contribute to corrosion-inhibitive film formation. It is expected that any inhibitive action developed by bacteria may be accomplished within the complex biofilm–corrosion product interactions occurring on a biofouled metal surface. Generally, biocorrosion and its counter process, microbial-induced corrosion inhibition, are rarely linked to a single mechanism or to a single species of microorganism. Microorganisms induce corrosion inhibition via two general mechanisms or their combination: neutralizing the action of corrosive species present in the environment, and forming a protective film or stabilizing a preexisting protective film on a metal [Videla and Herrera, 2009]. However, the inhibitive action of bacteria can be reversed to a corrosive action in bacterial consortia located within a biofilm thickness.

2.7.2 Microorganisms Involved in MIC

A wide variety of microorganisms are involved in the MIC of metals. Many microorganisms have the ability to substitute alternative oxidizing compounds for oxygen as terminal election acceptors when oxygen becomes depleted in the environment, permitting them to be active over a wide range of conditions for metal corrosion. The principal types of bacteria associated with corrosion are discussed below.

1. *Sulfate-reducing bacteria.* SRB are anaerobic microorganisms that have been found to be involved in numerous MIC problems, affecting a variety of metals and alloys [Zhang et al., 2011]. The SRB chemically reduce sulfates to sulfides, producing compounds such as H_2S, which can influence the anodic and cathodic processes significantly and ultimately enhance the corrosion of materials or, in the case of ferrous metals, produce corrosion-product iron sulfide. The most common strains of SRB exist in the temperature range 25 to 35°C, although some can function at temperatures up to 60°C. They can be detected through

the presence of black precipitates in liquid media or surface deposits as well as by a characteristic H_2S smell.

SRB have been considered prominent aggressive microorganisms causing unexpected failure of materials [Seed, 1990; Videla, 1990; Little et al., 1991]. The biofilm formed in the presence of SRB generally has a great influence on the enhancement of corrosion of steels. Moreover, SRB biofilm is electronically conductive, due to the highly conductive iron sulfide precipitates embedded in the biofilm, which shields the potential differences of the heterogeneous corrosion beneath the biofilm, leading to very flat potential distributions irrelevant to localized corrosion [Dong et al., 2011].

2. *Sulfur/sulfide-oxidizing bacteria.* Sulfide-oxidizing bacteria (SOB) are aerobic species that oxidize sulfide or elemental sulfur into sulfates. Some species oxidize sulfur into sulfuric acid (H_2SO_4), leading to a highly acidic environment ($pH < 1$). The high acidity has been associated with the degradation of coatings in a number of applications and with the corrosion of metals. They are found primarily in mineral deposits and are common in wastewater systems.
3. *Iron-oxidizing bacteria.* IOB have been found in conjunction with MIC and are typically located in corrosion pits on steels. Microscaled corrosion pits are initiated on stainless steel in IOB-containing solutions, and no pits are detected in a sterile medium [Xu et al., 2008]. Some species are known to accumulate iron compounds resulting from the process of oxidation. For example, the low corrosion resistance of 304L stainless steel to localized corrosion attack was detected in 3% NaCl solution containing an IOB culture. Further decrease in corrosion resistance was found when steel specimens were covered by a layer of biogenic hydroxide sediments. Pitting corrosion attack occurred on the surface of specimens under the inner layer of deposit. Moreover, it was proposed that MIC of stainless steels in the presence of IOB is associated with crevice corrosion caused by biogenic ferric oxide deposits precipitated on a steel surface [Starosvetsky et al., 2008]. Furthermore, the presence of IOB increases strongly in cathodic current after its exposure to a solution augmented with culture media since an alternative cathodic process is induced, which can provide enhanced steel corrosion. This alternative cathodic process takes place through the cathodic reduction of a corrosion product deposited on the specimen surface [Starosvetsky et al., 2001].
4. *Iron-reducing bacteria* (IRB). The involvement of IRB in the corrosion of steel is a controversial subject. Whereas it has been proposed that IRB are able to induce protection of carbon steel, other researchers suggest enhanced corrosion through the reduction and removal of passive films on the metal surface [Herrera and Videla, 2009]. The usual mechanism of IRB to induce corrosion is by reduction and dissolution of Fe^{3+} corrosion products, exposing the metal surface to the corrosive medium. Moreover, IRB are able to create anaerobic zones that favor the growth of SRB within biofilms, where both types of bacteria are present. Conversely, the mechanism of corrosion inhibition by IRB is that the biofilm impedes the dissolution of Fe^{2+} corrosion products,

creating a barrier between the metal and the environment. The environmental characteristics of the metal–biofilm–solution interface and its surroundings (e.g., pH, ionic composition, oxygen levels) control the nature and structure of protective layers and may change the microbial effects on metal from corrosive to protective.

5. *Organic acid–producing bacteria.* Some anaerobic microorganisms produce organic acids. These bacteria are more apt to be found in closed systems, including gas transmission pipelines and, sometimes, closed water systems.
6. *Acid-producing fungi.* Some fungi produce organic acids that attack metals. They can create environments suitable for anaerobic species. The widespread corrosion problems observed in aluminum fuel tanks in aircraft have been attributed to these organisms.

2.7.3 Role of MIC in SCC Processes

The environment involving MIC supports the corrosion and cracking processes in numerous circumstances. For example, tensile stresses in SCC and biological activity in MIC would be responsible for producing an embrittlement source, generally hydrogen in the material [Biezma, 2001]. The metallurgical feature and mechanical strength of materials as well as the hydrogen solubility and diffusivity in the materials would contribute to MIC and SCC. Since bacteria can change the local chemistry at the crack tip, they are thus expected to affect the development of conditions that result in SCC or corrosion.

Soils are always enriched by various elements, such as H, C, N, O, P, S, Cl, Na, K, Ca, and Mn. These elements are essential for the activity of microorganisms. Since the environment supports the corrosion reaction and biological activity that results in the MIC of materials, a combination with tensile stress and the metallurgical features of the material will be responsible for the susceptibility of the material to SCC.

Crack initiation is a crucial first step in SCC, and the formation of corrosion pits is usually the first step for crack initiation. Hence, factors that cause the surface electrochemical heterogeneity to lead to pit formation or to alter the existing pit chemistry will influence SCC. The role of MIC in increasing the pitting susceptibility of carbon and stainless steels has received much attention [Chamritski et al., 2004]. For example, slimes formed by bacteria introduce sites for initiation of pitting corrosion on stainless steel in the presence of seawater or fresh water. Moreover, there is evidence to indicate that aerobically produced biofilm catalyzes the cathodic reduction of dissolved oxygen, raising the corrosion potential of stainless steels above the pitting potential in the presence of chloride.

Furthermore, MIC is closely related to the HE or HIC of a variety of metals since microorganisms serve as a source of hydrogen by decatalyzing the hydrogen recombination reaction on the metal surface. The influence of microorganisms on the development of electrochemical conditions for hydrogen evolution alters the SCC process or even mechanism, changing completely from a dissolution-based mechanism to an HIC case. For example, bioelectrochemical and electrochemical approaches to the understanding of SRB-influenced corrosion and stress corrosion suggest [Javaherdashti et al., 2006] that in the presence of SRB, the cathodic reaction involving hydrogen proceeds at a faster rate, since SRB retards the recombination of hydrogen atoms. The "poisoning" influence of sulfide in the recombination of hydrogen atoms is significant, and sulfide ions produced as a result of the SRB-containing mixed bacterial culture and sulfate-containing nutrients would reduce the rate of recombination of electrochemically produced hydrogen atoms and thus facilitate their diffusion into steels to cause embrittlement. Hence, it is expected that hydrogen-assisted damage is enhanced in the presence of SRB-influenced corrosion. Actually, HIC has been proposed as a mechanism for environmentally assisted cracking of mild steels in biotic environments, where hydrogen permeation has been addressed as a possible mechanism for failure of material in environments with high SRB activity [Benson and Edyvean, 1998].

2.8 CORROSION FATIGUE

Corrosion fatigue (CF) is a conjoint action of a cyclic stress and a corrosive environment that results in material failure. The basic role of the corrosive environment is to decrease the life of the component. The fatigue and corrosion fatigue strength of a material is often determined by the *S–N* curve, that is, the stress (*S*)–number of stress cycles to failure (*N*) curve, which is a stress against the logarithm of the number of stress cycles to a failure curve, as shown in Fig. 2-4. When a corrosive environment is present, the ability of the material to resist applied stress a particular number of cycles of stress is decreased; that is, the stress required for both crack initiation and propagation is lower in corrosive environments. The crack growth rate can be much higher in a corrosive environment than it is in a noncorrosive environment. Therefore,

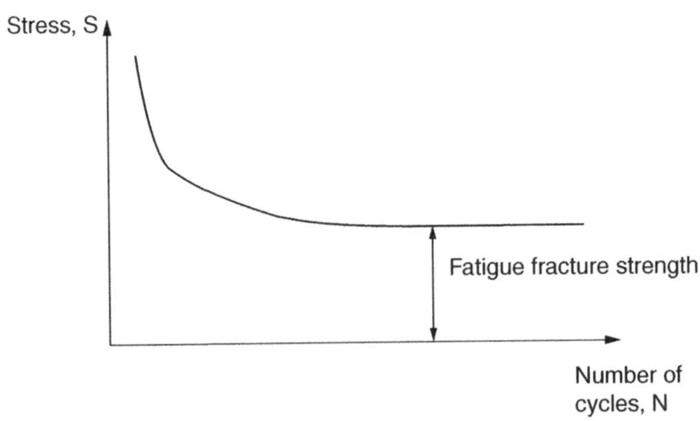

Figure 2-4 *Typical S–N curve for fatigue and corrosion fatigue measurements of a material.*

the fatigue life of a material is reduced if it is exposed simultaneously to a corrosive environment and fatigue stress conditions.

2.8.1 Features of Fatigue Failure

Fatigue cracks are usually initiated on the surface of a structural component. The crack initiation time varies and may be from 25 to 50% of the total number of cycles to failure. The initiation stage is called stage I in the fatigue process. It is usually accompanied by plastic deformation at 45° to the applied stress axis, as shown in Fig. 2-5. These are called *persistent slip bands* (PSBs). The crack length in stage I fatigue is usually very small, usually grain size.

At the end of stage I, the crack propagates macroscopically at 90° to the applied stress axis in stage II of the process. Multiple slip processes occur, leading to blunting of the crack tip and the formation of striations on the fracture surface. The striations are perpendicular to the crack growth direction and look like waves on the ocean surface. The striations represent one burst of crack growth but may take more than one cycle to form. Striations are characteristic of the fatigue process. The time for crack growth in stage II may take from 75 to 50% of the fatigue life.

Stage III occurs when the remaining cross section of material is too small to withstand the applied stress. Failure occurs, forming a final shear lip at 45° to the surface. The shear lip size depends on the strength of the material and the stress range applied. The features on the surface in stage III consist of tensile shear dimples.

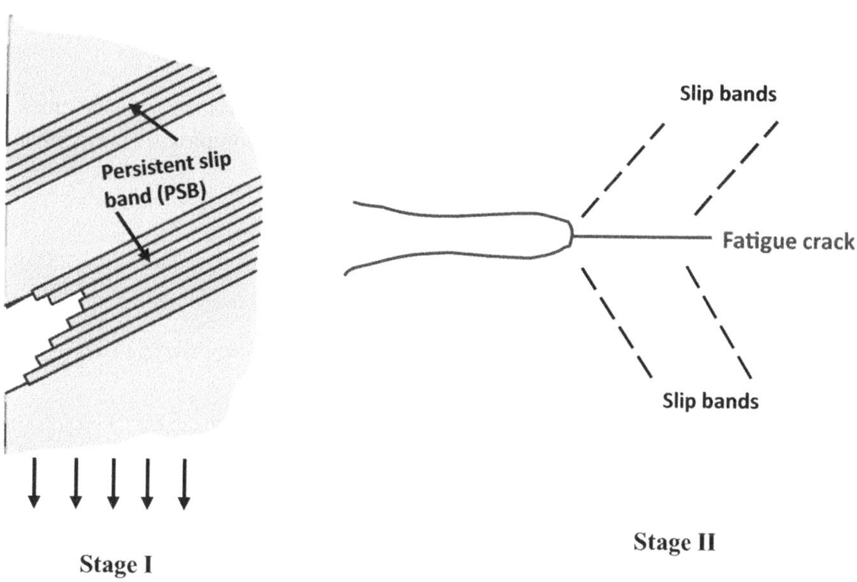

Figure 2-5 *Schematic diagram of different stages that occur during fatigue fracture of metals.*

2.8.2 Features of Corrosion Fatigue

The effect of a corrosive environment on the fatigue of metals will be felt in stages I and II of the process, and stage III is usually a mechanical fracture process.

- *Stage I.* In the absence of a corrosive environment, localized plastic deformation in the form of PSB will initiate cracks. Corrosion reaction would reduce the time of the initiation stage by introducing additional crack initiation mechanisms, including the local galvanic effect between phases, the dissolution of metal after mechanical disruption of passive film by PSBs, metal dissolution due to active sites at PSBs, and pitting corrosion [Zhao et al., 2012]. In addition to the stress concentration developing at the bottom of pits, the corrosion product formed may penetrate the steel surface, reducing the cohesive strength of the steel substrate on both sides of the slip planes and accelerating the initiation of corrosion fatigue cracks. Moreover, the fatigue cracks propagate easily along slip planes.

  Furthermore, corrosion fatigue is a frequency-dependent phenomenon. The lower the frequency of the cyclic stress, the more effective the corrosive environment is in decreasing the fatigue life of the component.
- *Stage II.* Effects of the corrosive environment on stage II of fatigue can be separated out using precracked specimens to conduct tests, and the data are plotted as the crack propagation rate per cycle (da/dN) versus the logarithm of the stress intensity factor ($\log \Delta K$), as shown in Fig. 2-6, where ΔK_{th} is the threshold of stress intensity factor. As the crack length must be measured during the test, the crack propagation rate per cycle can also be determined by measuring the slope of the crack length (a) against the number of cycles (N).

The usual effect of corrosion on material fatigue is to produce a faster crack propagation rate, and the associated mechanisms are quite complex. It may either

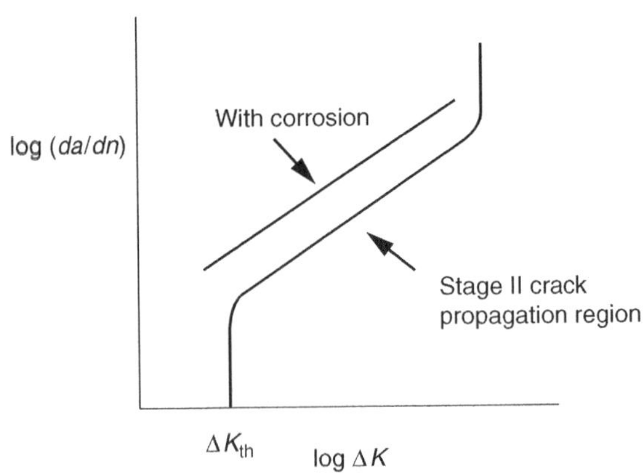

Figure 2-6 *Effect of corrosion on fatigue crack propagation in stage II.*

increase the crack growth rate by accelerating the normal crack growth or by invoking modified mechanisms of crack growth.

Corrosion fatigue cracks are usually of the transgranular type, but exceptions do exist. The general role of the corrosive environment at the crack tip is to accelerate the crack growth rate. One mechanism is the general corrosion of the crack tip in extending the crack length at all times in addition to the increase in crack length from mechanical stressing. The crack tip can then be considered to be extending, with an increase in the crack growth rate [Revie and Uhlig, 2008]. A second mechanism proposed is that the corrosive environment oxidizes the surface of the metal rapidly when the crack opens up. This would avoid the "self-healing" of the crack tip and maintain a longer crack than that under nonoxidizing conditions [Magnin, 2002]. If the corrosion product behind the crack tip builds up sufficiently, a wedging effect may occur in which the crack is held open longer at the tip as the corrosion product stops complete closure [Magnin, 2002]. This will also tend to increase the crack growth rate.

2.8.3 Factors Affecting CF and CF Management

A number of factors affect the CF processes. Corrosion in particular, especially pitting corrosion, causes stress raisers in the vicinity of the pit, leading to crack initiation at such a low stress that a material will not crack in the absence of pits. Moreover, the crack will propagate at a faster rate under the effect of corrosion reaction. Furthermore, metallurgical, mechanical, and environmental factors all affect the occurrence and propagation of corrosion fatigue cracks.

Selection of materials with low fatigue crack propagation rates is very important in CF prevention. Generally, materials withstand fatigue by two methods. The first is to resist initiation but have a rapid growth rate. The second is to allow easy crack initiation but have a slow crack growth rate. For corrosion fatigue, the second mechanism, that is, a high resistance to crack propagation, is more important because the crack initiation is easily accelerated for many reasons, such as pitting and scratching, in CF. Moreover, CF can also be prevented and controlled effectively by a number of methods, including:

- Reducing stress by using thicker sections
- Choosing heat treatments that reduce residual stresses
- Inducing compressive stresses
- Using inhibitors to stop corrosive action
- Applying cathodic protection or coatings for corrosion control

2.9 COMPARISON OF SCC, HIC, AND CF

SCC, HIC, and CF are all environmentally assisted cracking phenomena that can occur in a wide variety of industrial applications. However, there are essential differences among them. A complete understanding of these differences would enable

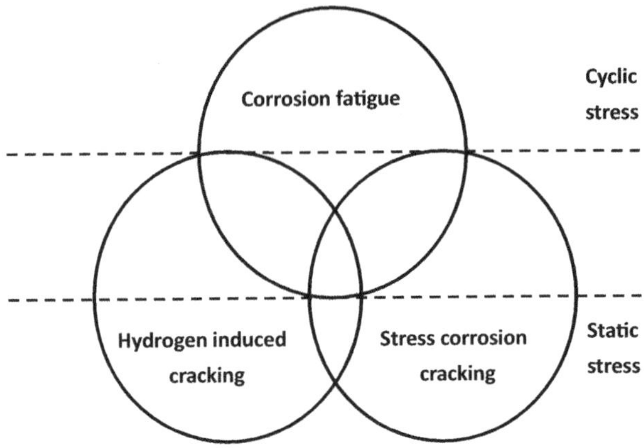

Figure 2-7 *Schematic diagram of interactions of SCC, HIC, and CF.*

selection of appropriate preventive and controlling techniques to mitigate and avoid these problems. The primary difference between SCC and CF is that SCC occurs in a corrosive environment under static tensile stress, whereas CF results from the combined effects of a corrosive environment and cyclic stress. If the cycle frequency is sufficiently low, CF can be treated approximately indistinguishably from SCC. For example, natural gas pipelines crack under pressure fluctuations. Although it is proposed that this problem belongs to the general type of CF, it has been attributed to SCC with slow fluctuations of internal pressure [National Energy Board, 1996]. Moreover, stress corrosion cracks often have many small branches, whereas corrosion fatigue cracks have little or no branching. Since CF can rupture oxide film more rapidly to expose bare steel to a corrosive environment, the critical stress intensity factor for CF is usually lower than that for SCC. HIC or hydrogen embrittlement

TABLE 2-3 Specific Features of SCC, CF, and HIC

	SCC	CF	HIC
Stress	Static, tensile	Cyclic tensile	Static tensile
Corrosive agent	Specific to the metal	Any	Any
Temperature rise	Accelerates	Accelerates	$<$ Ambient, increases
Pure metal	Resistant	Susceptible	Susceptible
Crack morphology	Transgranular, intergranular, branched	Transgranular, unbranched, blunt tips	Transgranular, intergranular, unbranched, sharp tip
Corrosion products in cracks	Absent, present	Present	Absent
Fracture surface	Cleavage-like	Beach marks, striations	Cleavage-like
Cathodic protection	Suppresses	Suppresses	Accelerates

Source: Craig and Lane [2005].

can overlap with SCC and CF in some cases, but it can also have its own intrinsic mechanism. Metals may crack in the presence of hydrogen without the existence of corrosive environments or even in the absence of external static stress.

The relationships among SCC, CF, and HIC are complex. A schematic diagram indicating the interrelationship among them is shown in Fig. 2-7. The figure represents areas involving two failure mechanisms and the interaction of all three phenomena. Furthermore, the region between the dashed lines indicates the transition between cyclic and static stresses. Table 2-3 details the specific features of SCC, CF, and HIC.

REFERENCES

Alyousif, OM, Nishimura, R (2008) On the stress corrosion cracking and hydrogen embrittlement of sensitized austenitic stainless steels in boiling saturated magnesium chloride solutions: effect of applied stress, *Corros. Sci.* 50, 2919–2926.

Antony, PJ, Chongdar, S, Kumarb, P, Raman, R (2007) Corrosion of 2205 duplex stainless steel in chloride medium containing sulfate-reducing bacteria, *Electrochim. Acta* 52, 3985–3994.

Bandyopadhyay, N, Briant, CL (1983) Caustic stress corrosion cracking of NiCrMoV rotor steels: the effects of impurity segregation and variation in alloy composition, *Metall. Trans. A* 14, 2005–2019.

Beachem, CD (1972) A new model for hydrogen-assisted cracking (hydrogen embrittlement), *Metall. Trans.* 3, 441–455.

Beavers, JA, Pugh, EN (1980) The propagation of transgranular stress corrosion cracks in admiralty metal, *Metall. Trans. A* 11, 809–820.

Benson, J, Edyvean, RGJ (1998) Hydrogen permeation through protected steel in open seawater and marine mud, *Corrosion* 54, 732–739.

Bernstein, IM, Thompson, AW (1980) *Hydrogen Effects in Metals*, AIME, New York.

Biezma, MV (2001) The role of hydrogen in microbiologically influenced corrosion and stress corrosion cracking, *Int. J. Hydrogen Energy* 26, 515–520.

Brown, BF (1972) A preface to the problem of stress corrosion cracking, in *Stress Corrosion Cracking of Metals: State of the Art*, STP 518, American Society for Testing and Materials, Philadelphia.

Burstein, GT, Woodward, J (1981) Effects of segregated phosphorus on stress corrosion cracking susceptibility of 3Cr–0.5Mo steel, *Met. Sci.* 15, 111–116.

Callister, WD, Rethwisch, DG (2009) *Materials Science and Engineering: An Introduction*, 8th ed, Wiley, Hoboken, NJ.

Carranza, RM, Galvele, JR (1988a) Repassivation kinetics in stress corrosion cracking, Part 1: Type AISI 304 stainless steel in chloride solutions, *Corros. Sci.* 28, 233–249.

Carranza, RM, Galvele, JR (1988b) Repassivation kinetics in stress corrosion cracking: Part 2. Brass in nonammoniacal solutions, *Corros. Sci.* 28, 851–865.

Cassagne, T (2007) Stress corrosion cracking of stainless steels in chloride environments, *Joint EFC-NACE Workshop*, Feiburg, Germany.

Chamritski, G, Burns, GR, Webster, BJ, Laycock, NJ (2004) Effect of iron-oxidizing bacteria on pitting of stainless steel, *Corrosion* 60, 658–669.

Corrosion Doctors, online source 1, http://www.corrosion-doctors.org/Forms-SCC/scc-definitions.htm.

Corrosion Doctors, online source 2, http://corrosion-doctors.org/Failure-Analysis/12-fracture-mechanics.htm.

Craig, BD, Lane, RA (2005) Environmentally assisted cracking: comparing the influence of hydrogen, stress and corrosion on cracking mechanisms, *AMPTIAC Q.* 9, 17–24.

Demakis, J (2002) Hydrogen embrittlement: how small details can have large effects, *Moldmak. Technol.* 4, 1–2.

Dong, ZH, Shi, W, Ruan, HM, Zhang, GA (2011) Heterogeneous corrosion of mild steel under SRB-biofilm characterized by electrochemical mapping technique, *Corros. Sci.* 53, 2978–2987.

Evans, TE, Chart, A, Skedgell, AN (1973) The colored film on stainless steel, *Trans. Inst. Metal Finish.* 51, 108–112.

Ford, FP (1982) Stress corrosion cracking, in *Corrosion Processes*, R.N. Parkins, Editor, Applied Science, London.

Ford, FP (1996) Quantitative prediction of environmentally assisted cracking, *Corrosion* 52, 375–395.

Forty, AJ, Humble, P (1963) The influence of surface tarnish on stress-corrosion of-brass, *Philos. Mag.* 8, 247–264.

Franklin, MJ, White, DC, Isaacs, HS (1991) Pitting corrosion by bacteria on carbon steel determined by scanning vibrating electrode technique, *Corros. Sci.* 32, 945–952.

Galvele, JR (1993) Surface mobility mechanism of stress corrosion cracking, *Corros. Sci.* 35, 419–434.

Galvele, JR (1995) Electrochemical aspects of stress corrosion cracking, in *Modern Aspects of Electrochemistry*, Vol. 27, R.W. White, J.O'M. Brockris, and B.E. Conway, Editors, Plenum Press, New York, p. 233.

Galvele, JR (1999) Past, present and future of stress corrosion cracking, *Corrosion* 55, 723–731.

Gibala, R, Kumnick, AJ (1984) Hydrogen trapping in iron and steels, in *Hydrogen Embrittlement and Stress Corrosion Cracking*, R. Gibala and R.F. Hehemann, Editors, ASM, Metals Park, OH.

Gunasekaran, G, Chongdar, S, Gaonkar, SN, Kumar, P (2004) Influence of bacteria on film formation inhibiting corrosion, *Corros. Sci.* 46, 1953–1967.

Hart, EW (1968) *Surface and Interface II*, Syracuse University Press, Syracuse, NY.

Harwood, JJ (1956) In *Stress Corrosion Cracking and Embrittlement*, W.D. Robertson, Editor, Wiley, New York.

Herrera, LK, Videla, HA (2009) Role of iron-reducing bacteria in corrosion and protection of carbon steel, *Int. Biodeterior. Biodegrad.* 63, 891–895.

Hirth, JP (1980) Effects of hydrogen on the properties of iron and steel, *Metall. Mater. Trans. A* 11, 861–890.

Hirth, JP (1984) Theories of hydrogen induced cracking of steels, in *Hydrogen Embrittlement and Stress Corrosion Cracking*, R. Gibala and R.F. Hehemann, Editors, ASM, Metals Park, OH.

Javaherdashti, R (1999) A review of some characteristics of MIC caused by sulfate-reducing bacteria: past, present and future, *Anti-Corros. Methods Mater.* 46, 173–180.

Javaherdashti, R, Raman, RKS, Panter, C, Pereloma, EV (2006) Microbiologically assisted stress corrosion cracking of carbon steel in mixed and pure cultures of sulfate reducing bacteria, *Int. Biodeterior. Biodegrad.* 58, 27–35.

Jin, TY, Liu, ZY, Cheng, YF (2010) Effects of non-metallic inclusions on hydrogen-induced cracking of API5L X100 steel, *Int. J. Hydrogen Energy* 35, 8014–8021.

Jones, DA (1996) Localized surface plasticity during stress corrosion cracking, *Corrosion* 52, 356–362.

Jones, RH (1992) *Stress Corrosion Cracking: Materials Performance and Evaluations*, ASM, Metals Park, OH.

Jones, RH, Arey, BW, Baer, DR, Friesel, MA (1989) Grain boundary chemistry and intergranular stress corrosion cracking of iron alloys in calcium nitrate, *Corrosion* 45, 494–502.

King, RA, Miller, JDA, Smith, JS (1973) Corrosion of mild steel by iron sulfides, *Br. Corros. J.* 8, 137–141.

Leyer, J, Sutter, P, Marchebois, H, Bosch, C, Kulgemeyer, A, Orlans-Joliet, BJ (2005) SSC resistance of a 125 ksi steel grade in slightly sour environments, *Corrosion 2005*, Paper 05088, NACE, Houston, TX.

Little, BJ, Lee, JS (1997) *Microbiologically Influenced Corrosion*, Wiley, New York.

Little, BJ, Wagner, P, Mansfeld, F (1991) Microbiologically influenced corrosion of metals and alloys, *Int. Mater. Rev.* 36, 253–272.

Liu, ZY, Li, XG, Cheng, YF (2010) In-situ characterization of the electrochemistry of grain and grain boundary of an X70 steel in a near-neutral pH solution, *Electrochem. Commun.* 12, 936–938.

Lynch, SP (1981) In *Hydrogen Effects in Metals*, A.W. Thompson and I.M. Bernstein, Editors, Metallurgical Society of AIME, Warrendale, PA, pp. 863–871.

Lynch, SP (1985) Mechanisms of stress corrosion cracking and liquid metal embrittlement in Al–An–Mg bicrystals, *J. Mater. Sci.* 20, 3329–3338.

Lynch, SP (1986) A fractographic study of hydrogen-assisted cracking and liquid-metal embrittlement in nickel, *J. Mater. Sci.* 21, 692–704.

Magnin, T (2002) Corrosion fatigue mechanism in metallic materials, in *Corrosion Mechanisms in Theory and Practice*, P. Marcus, Editor, CRC Press, Boca Raton, FL.

McEvily, AJ, Bond, PA (1965) On the initiation and growth of stress corrosion cracks in tarnished brass, *J. Electrochem. Soc.* 112, 141–148.

National Association of Corrosion Engineers (2003) *Petroleum and Natural Gas Industries—Materials for Use in H_2S Containing Environments in Oil and Gas Production*, Part 2: *Cracking-Resistant Carbon and Low Alloy Steel, and the Use of Cast Iron*, MR0175/ISO 15156-2, NACE Houston, TX.

National Energy Board (1996) *Stress Corrosion Cracking on Canadian Oil and Gas Pipelines*, MH-2-95, NEB, Calgary, Alberta, Canada.

Newman, RC, Procter, RP (1990) Stress corrosion cracking 1965–1990, *Br. Corros. J.* 25, 259–269.

Oriani, RA (1984) Hydrogen embrittlement of steels, in *Hydrogen Embrittlement and Stress Corrosion Cracking*, R. Gibala and R.F. Hehemann, Editors, ASM, Metals Park, OH.

Oriani, RA (1990) Hydrogen effects in high-strength steels, in *Environment-Induced Cracking of Metals*, R.P. Gangloff and M.B. Ives, Editors, NACE, Houston, TX, pp. 439–447.

Padilla-Viveros, A, Garcia-Ochoa, E, Alazard, D (2006) Comparative electrochemical noise study of the corrosion process of carbon steel by the sulfate-reducing bacterium *Desulfovibrio alaskensis* under nutritionally rich and oligotrophic culture conditions, *Electrochim. Acta* 51, 3841–3847.

Parkins, RN (1990) Stress corrosion cracking, in *Environmental-Induced Cracking of Metals*, R.P. Gangloff and M.B. Ives, Editors, NACE, Houston, TX.

Parkins, RN (1992) Mechanistic aspects of stress corrosion cracking, *Parkins Symposium on Stress Corrosion Cracking*, Metallurgical Society, Warrendale, PA.

Revie, RW (2000) *Uhlig's Corrosion Handbook*, 2nd ed., The Electrochemical Society, Pennington, NJ.

Revie, RW, Uhlig, HH (2008) *Corrosion and Corrosion Control*, 4th ed., Wiley, Hoboken, NJ.

Seed, LJ (1990) The significance of organisms in corrosion, *Corros. Rev.* 9, 3–101.

Sheng, X, Ting, YP, Pehkonen, SO (2007) The influence of sulfate-reducing bacteria biofilm on the corrosion of stainless steel AISI 316, *Corros. Sci.* 49, 2159–2176.

Speidel, MO (1975) Stress corrosion cracking of aluminum alloys, *Metall. Trans. A* 6, 631–651.

Starosvetsky, D, Armon, R, Yahalom, J, Starosvetsky, J (2001) Pitting corrosion of carbon steel caused by iron bacteria, *Int. Biodeterior. Biodegrad.* 47, 79–87.

Starosvetsky, J, Starosvetsky, D, Pokroy, B, Hilel, T, Armon, R (2008) Electrochemical behaviour of stainless steels in media containing iron-oxidizing bacteria (IOB) by corrosion process modeling, *Corros. Sci.* 50, 540–547.

Stenzel, H, Vehoff, H, Neumann, P (1986) In *Modeling Environmental Effects on Crack Growth Process*, R.H. Jones and W.W. Gerberich, Editors, AIME, Warrendale, PA, p. 225.

Stott, JFD, Skerry, BS, King, RA (1988) Laboratory evaluation of materials for resistance to anaerobic corrosion caused by sulfate reducing bacteria: philosophy and practical design, in *The Use of Synthetic Environments for Corrosion Testing*, P.E. Francis, and T.S. Lee, Editors, ASTM STP 970, American Society for Testing and Materials, Philadelphia, pp. 98–111.

Suh, D, Eagar, TW (1998) Mechanistic understanding of hydrogen in steel welds, *Proc. International Workshop Conference on Hydrogen Management for Welding Applications*, Ottawa, Ontario, Canada.

Swann, PR, Pickering, HW (1963) Implications of stress aging yield phenomena with regard to stress corrosion cracking, *Corrosion* 19, 369–372.

Szklarz, KE (1999) Sulfide stress cracking of a pipeline weld in sour gas service, Corrosion 1999, Paper 99428, NACE, Houston, TX.

Taheri, RA, Nouhi, A, Hamedi, J, Javaherdashti, R (2005) Comparison of corrosion rates of some steels in batch and semi-continuous cultures of sulfate-reducing bacteria, *Asian J. Microbiol. Biotechnol. Environ. Sci.* 7, 5–8.

Teter, DF, Robertson, IM, Birnbaum, HK (2001) The effects of hydrogen on the deformation and fracture of beta-titanium, *Acta Mater.* 49, 4313–4323.

Thierry, D, Sand, W (1995) In *Corrosion Mechanics in Theory and Practice*, M.J. Oudar, Editor, Marcel Dekker, New York.

UK National Physical Laboratory (1982) *Guide to Good Practice in Corrosion Control: Stress Corrosion Cracking*, NPL, London.

Videla, HA (1990) Sulphate reducing bacteria and anaerobic corrosion, *Corros. Rev.* 9, 103–141.

Videla, HA, Herrera, LK (2009) Understanding microbial inhibition of corrosion: a comprehensive overview, *Int. Biodeterior. Biodegrad.* 63, 896–900.

Wiederhorn, SM, Bolz, LH (1970) Stress corrosion and static fatigue of glass, *J. Am. Ceram. Soc.* 53, 543–548.

Wikipedia online source, http://en.wikipedia.org/wiki/Stress_corrosion_cracking.

Xu, C, Zhang, Y, Cheng, G, Zhu, W (2008) Pitting corrosion behavior of 316L stainless steel in the media of sulphate-reducing and iron-oxidizing bacteria, *Mater. Charact.* 59, 245–255.

Yo1e, P (1998) MIC investigation of damage of tubes fed by seawater in the coast power plants. *Proceedings of Corrosion and the Environment Conference*, Bath, UK, p. 163.

Yuan, SJ, Pehkonen, SO (2007) Microbiologically influenced corrosion of 304 stainless steel by aerobic *Pseudomonas* NCIMB 2021 bacteria: AFM and XPS study, *Colloids Surfact. B* 59, 87–99.

Zhang, C, Wen, F, Cao, Y (2011) Progress in research of corrosion and protection by sulfate-reducing bacteria, *Proc. Environ. Sci.* 10, 1177–1182.

Zhao, W, Wang, Y, Zhang, T, Wang, Y (2012) Study on the mechanism of high-cycle corrosion fatigue crack initiation in X80 steel, *Corros. Sci.* 57, 99–103.

3

Understanding Pipeline Stress Corrosion Cracking

3.1 INTRODUCTION

Over 98% of oil and gas transmission pipelines are buried. No matter how well these pipelines are designed, constructed, and protected, they are subjected to environmental attack, external damage, coating disbondment, inherent material defects, soil movement and instability, and third-party damage once in use in the field. Pipeline SCC occurs due to a combination of appropriate environmental (soil, coating, CP, temperature), stress (hoop stress, longitudinal and/or residual stress, fluctuating internal pressure), and material (grade and chemistry, microstructure, metallurgical defects, surface roughness, welds) factors.

According to statistics released by the National Energy Board of Canada (NEB) [National Energy Board, 2008], approximately 63% of the primary causes of pipeline failure occurring in NEB-regulated pipelines in Canada are due to corrosion and metal loss, including cracking. In particular, SCC-induced failures account for 10 to 13% of corrosion failures. Among other occurrences, five major SCC failures in TransCanada pipelines in 1985, 1991, and 1992, and a major explosion due to SCC in 1995 in Manitoba prompted Canadian pipeline regulatory organizations such as the NEB and the Canadian Energy Pipeline Association (CEPA) to develop processes to control SCC.

Of various reports, case summaries, and regulations regarding pipeline SCC, the most comprehensive and prominent is "Public Inquiry Concerning Stress Corrosion Cracking on Canadian Oil and Gas Pipelines," published by the NEB in 1996 [National Energy Board, 1996a]. Moreover, the Office of Pipeline Safety of the U.S.

Stress Corrosion Cracking of Pipelines, First Edition. Y. Frank Cheng.

Department of Transportation (DOT) and Michael Baker Jr., Inc. completed the report "Stress Corrosion Cracking Study" in 2005 [Baker, 2005]. CEPA released "Stress Corrosion Cracking: Recommended Practices," detailing the management of longitudinal, transgranular SCC in 1997 [Canadian Energy Pipeline Association, 1997] and 2007 [Canadian Energy Pipeline Association, 2007].

3.2 PRACTICAL CASE HISTORY OF SCC IN PIPELINES

The first recognized case of pipeline SCC was recorded in 1965 when a gas transmission pipeline ruptured due to SCC in Louisiana [Leis et al., 1996]. However, similar metallographic and fractographic information archived in pipeline failure analysis reports at Battelle indicated that the first SCC failure probably occurred in 1957 [Leis and Eiber, 1997].

Between March 1985 and March 1986, TransCanada Pipelines had three failures on its pipelines in northern Ontario [Fang et al., 2003]. These failures were attributed, for the first time, to pipeline SCC in Canada. Figure 3-1 shows the chronology of pipeline SCC events in Canada up to 1996, before the NEB released its inquiry report on pipeline SCC in 1996.

Comparing the pipeline SCC incidents occurring in Canada and the United States, the statistics showed that, whereas 17% of the reported pipeline failures were

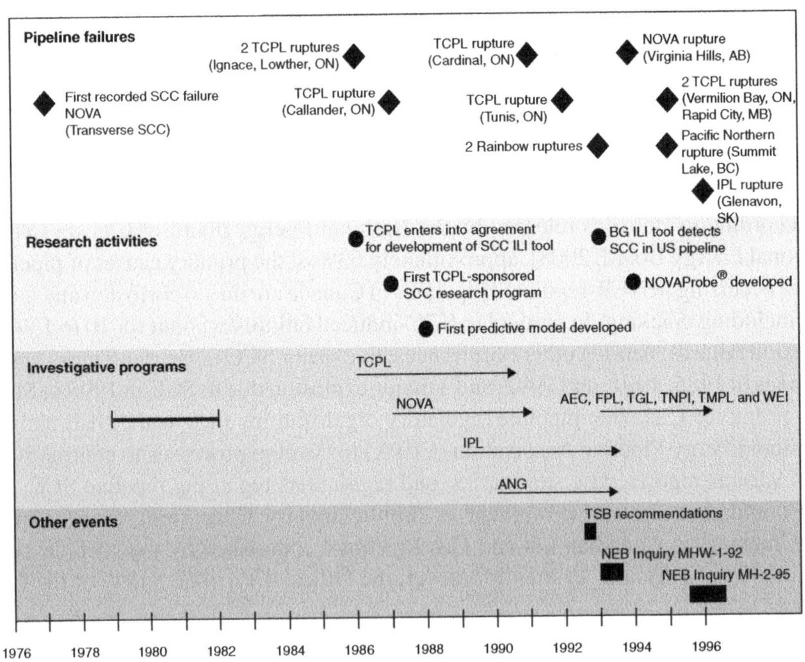

Figure 3-1 *Chronology of SCC events in Canadian pipelines up to 1996. (From National Energy Board [1996a].)*

attributed to SCC in Canada, SCC accounted for only 1.5% of reportable incidents for pipelines in the United States [Baker, 2005]. It seems that pipeline SCC is a more serious problem in Canada than in the United States. However, the report prepared by the Office of Pipeline Safety of U.S. DOT pointed out that "the fact that SCC represents only 1.5% of reportable incidents in the United States versus 17% in Canada is due to the far greater occurrence of third party damage in the United States." It thus concluded that "SCC is a serious pipeline integrity issue of concern to operators of pipelines within the United States" [Baker, 2005]. Furthermore, pipeline SCC is a widespread phenomenon. In addition to Canada and the United States, it has been detected in pipeline systems in Russia, Australia, Iran, Iraq, Italy, Pakistan, Saudi Arabia, and other countries.

3.2.1 Case 1: SCC of Enbridge Glenavon Pipelines (SCC in an Oil Pipeline)

On February 27, 1996, Enbridge's (formerly, IPL) main crude oil pipeline, designated as Line 3, suddenly experienced a simultaneous loss of operating pressure and an increase in the crude oil flow rate at both the Glenavon and Langbank pump stations, located near the towns of Glenavon and Langbank, Saskatchewan, Canada [Transportation Safety Board of Canada, 1999a]. Line 3 carries heavy sulfur-rich crude from northern Alberta to refineries in the midwestern United States. The line ruptured near Glenavon, Saskatchewan, a village 120 km east of Regina, resulting in a 800-m^3 crude oil release but no injuries. It was stated that "damage to Line 3 consisted of 1.870 meters (approximately 5.8 feet) of ruptured pipe, which had split open in the longitudinal direction, in proximity to the longitudinal seam weld" [Transportation Safety Board of Canada, 1999b].

The pipeline was installed in 1968, with polyethylene (PE) tape coated on the exterior surface. Soil analysis showed a high level of dissolved solids, gypsum based with slightly alkaline pH (7.35 to 7.95). Metallurgical examination of the fracture area indicated that the pipe fractured along the longitudinal seam and opened up over a length of 1.87 m. Moreover, the tape wrap showed poor bonding in some areas, with wrinkles and tape seam separations. Removal of the tape revealed general corrosion along the longitudinal seam weld, with the heaviest corrosion corresponding to the center of the failure. Nondestructive testing of the failure found 27 regions of indications of small colonies of cracking within 150 mm of the longitudinal seam weld, many within the corroded areas of the pipe. Eight similar regions of colonies of cracking were found adjacent to the downstream circumferential weld. All the cracks within the colonies were oriented in the axial direction. Several corroded areas were found to contain shallow colonies of cracks. Two large flat areas corresponding to the areas of deepest corrosion were observed on the fracture surface, which revealed some intergranular corrosion features as well as secondary cracking. The material analysis indicated that the primary cracking mode was transgranular. The field and lab analyses concluded that a nearly neutral–pH SCC developed under a disbonded PE exterior coating in the presence of a carbonate–bicarbonate solution, resulting in rupture of the pipeline.

3.2.2 Case 2: SCC of Williams Lake Pipelines (SCC in a Gas Pipeline)

On December 13, 2003, the Williams Lake 26-in. pipeline ruptured near Toledo, Washington. On December 19, 2003, the pipeline company that supplied most of Washington's natural gas was ordered to shut down its trunk line from Canada to Oregon after federal safety inspectors determined that frailties in the 268-mile (431-km) pipe would "likely result in serious harm to life, property and the environment" [Welch, 2003].

This rupture was actually the second failure of the Williams Lake 26-in. pipeline. On May 1, 2003, there was a rupture in the line, after which the associate administrator for pipeline safety issued a corrective action order, restricting the operating pressure by 20% and requiring the Williams Lake gas pipeline to reevaluate in-line inspection surveys, carry out a geotechnical evaluation of the area, and to take appropriate remedial action.

Factors leading to SCC of the Williams Lake pipeline include acidic soils, tar coating applied in 1957, operating pressure, imposed loads leading to stress, and susceptible steel. Investigation following the corrective action order revealed that the failed section had a dark stain on the edge of the rupture area, which indicates the occurrence of corrosion. Moreover, moisture between the tar coating and the pipe showed support corrosion, especially pitting corrosion. Visual examination also revealed significant longitudinal cracking of the pipe body. Metallurgical analysis confirmed that the cause of the failure was SCC [Corrosion Doctors, online source]. The pipeline was located in an area of geotechnical land movement, but this was not considered to be the cause of the failure.

Prior to this incident, the same pipeline also failed in 1992, 1994, and 1999. These early failures were all attributed to SCC. Moreover, it should be noted that this pipeline had been tested hydrostatically less than a year before the burst. However, hydrostatic testing does not guarantee that problems do not exist and cannot reveal stress corrosion cracks. It is suggested that the more complex integrity management techniques, such as electromagnetic acoustic transducer (EMAT) technology, and strict quality control are required to detect SCC.

After the failure of December 2003, Williams ramped down the gas pressure to 100 psi, 12% of its capacity, which essentially takes the line out of service. The company was ordered to improve all lines in populated areas within three years and to improve the entire system within a decade. The case of the Williams Lake pipeline provided a good example of the difficulty when trying to mitigate SCC, since standard methods of testing, such as hydrotesting, do not necessarily detect it. This case also reminded pipeline operators to design their integrity management systems to prevent harm to life, property, and the environment.

3.3 GENERAL FEATURES OF PIPELINE SCC

The SCC of pipelines is a type of environmentally assisted cracking (EAC), where the cracking results from various factors, including the environment surrounding the pipelines, mechanical stresses exerted on the line pipe steel, and the sensitive metallurgical features of the steel. Pipeline SCC has been categorized into two types: high-pH SCC and nearly neutral–pH SCC, based on the electrolyte contacting the line pipe steel. Thus, the pH refers to the environment at the crack location rather than the soil pH.

3.3.1 High-pH SCC of Pipelines

When the applied cathodic protection (CP) penetrates the coating to reach the pipeline and generate a cathodic potential, a concentrated carbonate–bicarbonate electrolyte is generated, which has been identified as the environment responsible for high-pH SCC. This environment may develop as a result of the interaction between hydroxyl ions produced by the cathodic reaction (i.e., reduction of water) and carbon dioxide in the soil generated by the decay of organic matter. CP current causes the pH of an electrolyte trapped under disbonded coating to increase, and carbon dioxide readily dissolves in the elevated pH electrolyte, resulting in the generation of a concentrated carbonate–bicarbonate solution with a pH between 8 and 10.

The fractured surface of cracks formed under high-pH conditions normally exhibits a dark, discolored layer of oxidized material, primarily magnetite. The last portion of the pipe wall to fracture (i.e., the rapid-fracture region) remains a shiny silver color. Metallographic examination of a section across the crack shows the fracture path to be intergranular, often with small branches, as shown in Fig. 3-2.

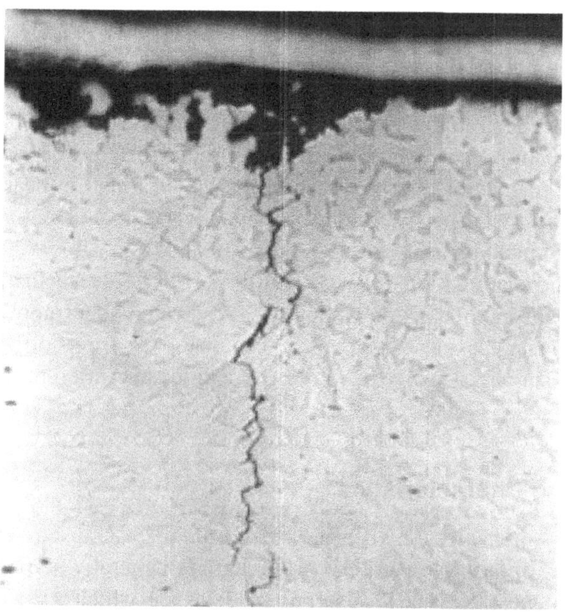

Figure 3-2 *Intergranular SCC in pipelines under high-pH conditions (magnification: 250 times). (From National Energy Board [1996a].)*

Laboratory simulation with small test specimens indicates that the high-pH SCC of pipelines is temperature sensitive and occurs more frequently at locations with temperatures above 38°C. In the field there is a greater likelihood of high-pH SCC immediately downstream from compressor stations, where the operating temperature might reach 65°C [National Energy Board, 1996a].

3.3.2 Nearly Neutral–pH SCC of Pipelines

Nearly neutral–pH SCC was not documented until the mid-1980s and was first identified on buried pipelines in Canada [National Energy Board, 1996a]. The form of SCC is associated with a diluted bicarbonate electrolyte with a pH value between 5.5 and 7.5 trapped under disbonded coating, tape primarily. Carbon dioxide is dissolved in the electrolyte to balance the pH, and the source of carbon dioxide is typically the decay of organic matter and geochemical reactions in the soil. Nearly neutral–pH cracking of pipelines occurs under conditions where little, if any, CP current reaches the pipe surface over a prolonged service period, either because of the presence of a shielding coating, a highly resistive soil, or ineffective CP [Parkins et al., 1994]. Typically, SCC colonies are initiated at locations on an outside pipe surface where there is already pitting corrosion.

Microbial sulfate reduction is not a common feature of nearly neutral–pH SCC sites across Canada [Delanty and O'Beime, 1992]. In many sites, iron(II) carbonate is the sole corrosion deposit. However, the presence of an active microbial process under a disbonded coating could influence a nearly neutral–pH SCC scenario by altering the anion ratio through sulfate reduction or by the generation of trace levels of sulfide [Jack et al., 2000]. Both of these consequences may favor cracking, as does an elevation in carbon dioxide levels [Wilmott et al., 1996; Van der Sluys et al., 1998].

Furthermore, differences between the trapped electrolyte under a disbonded coating and the adjacent groundwater are relatively minor. The microbial activity on components of coatings may contribute to this marginal difference [Jack et al., 2000]. However, little evidence of microbial activity has been found in nearly neutral–pH SCC sites. It is thus believed that the microbial activity is not essential to the occurrence of SCC, but may affect its severity.

Metallographic examination of nearly neutral–pH SCC reveals that the cracks are predominately transgranular, as shown in Fig. 3-3 and are wider (more open) than the high-pH form. The crack sides have experienced metal loss from corrosion. Thus, the fracture mechanism under nearly neutral–pH conditions is different from that under high-pH conditions.

3.3.3 Cracking Characteristics

There are many similarities between the two forms of SCC discussed above. Both occur as colonies of multiple parallel cracks that are generally perpendicular to the direction of the highest stress on the external pipe surface, which is usually hoop stress resulting from the internal pressure. These cracks can vary in depth and length and grow in two directions [i.e., axial (length) and transverse (depth) directions]. They

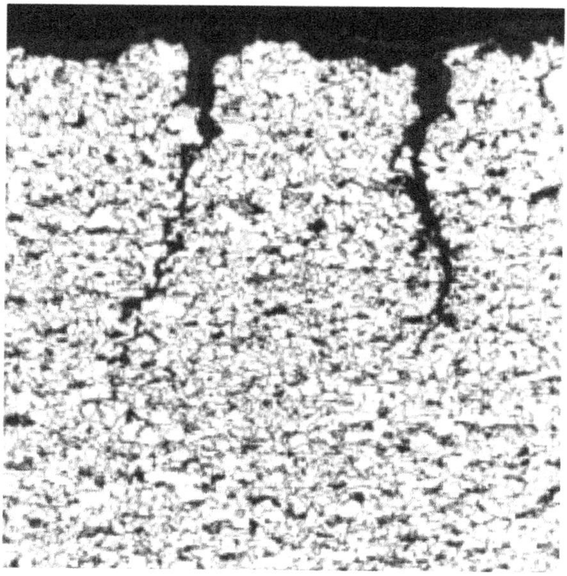

Figure 3-3 *Transgranular stress corrosion cracking in pipelines (magnification: 250 times). (From National Energy Board [1996a].)*

tend to coalesce or link together to form longer cracks. At some point these cracks may reach a critical depth and length combination that can result in a rupture. Stress corrosion cracks do not need to fully penetrate a pipe wall for a rupture to occur (i.e., a shallow crack may reach a length that becomes critical). The strength and ductility of the remaining wall determines the critical size at which crack behavior changes from a slowly growing stress corrosion mechanism to an extremely rapid brittleness. A leak will occur if a crack grows through the pipe wall before it reaches a critical length for rupture.

Crack coalescence is a common phenomenon in pipeline SCC. Crack coalescence results from joining individual cracks at the crack tips to form longer cracks. Coalescence can occur throughout the SCC life cycle. Depending on the size of a crack, either an environmental or a mechanical factor is capable of causing cracks to grow during stage III in Fig. 3-4. In stage IV, coalescence occurs by tearing [Canadian Energy Pipeline Association, 1996a], where mechanical loading has a stronger role in producing crack growth. The geometry of a crack colony is critical in determining whether the cracks will coalesce and grow to failure as well as the failure mode. Colonies of cracks that are long in the longitudinal direction but narrow in the circumferential direction are more dangerous to pipeline integrity than are colonies of cracks that are similar in length and width. For the former colonies, individual cracks can link up and lead to rupture. However, for colonies of cracks with the latter geometry, growth occurs primarily on the edge of the colony [Canadian Energy Pipeline Association, 1996b]. Moreover, cracks located deep within colonies are shielded from stress and become dormant [Canadian Energy Pipeline Association, 1996a].

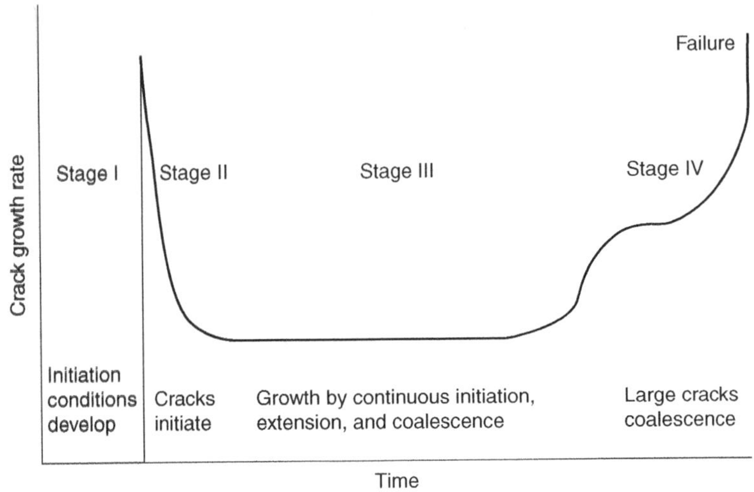

Figure 3-4 *Schematic diagram of the life cycle of pipeline SCC. (From National Energy Board [1996a].)*

Furthermore, the cracks in colonies can be closer together or farther apart, which makes the crack spacing either sparse or dense. Circumferential spacing equal to 20% of the wall thickness between cracks is used as the criterion that distinguishes between sparse and dense. Generally, cracks spaced circumferentially closer than 20% of the wall thickness tend to go dormant, whereas cracks spaced farther apart (at distances greater than 20% of the wall thickness) can continue to grow [Leis, 1995]. Thus, crack coalescence poses a potential mechanism affecting crack growth.

The most obvious differences between the two forms of SCC include the temperature sensitivity of high-pH SCC, the fracture morphology, and the pH of the environment supporting pipeline SCC. The primary characteristics of both types of pipeline SCC are summarized in Table 3-1.

3.4 CONDITIONS FOR PIPELINE SCC

Like all other SCC events, pipeline SCC occurs under three conditions that are met simultaneously: a corrosive environment, susceptible pipeline material, and tensile stress. If any one of these is eliminated, pipeline SCC can be prevented. Compared to Generally, the environment has been more commonly controlled to prevent SCC than have material and stress factors.

3.4.1 Corrosive Environments

An environment that results in pipeline SCC generally relates to electrolytes trapped between a disbonded coating and the line pipe steel. Thus, a corrosive environment may differ from the environment in the surrounding soil. For example, an environment

TABLE 3-1 Comparison of Characteristics of Two Types of Pipeline SCC

Factor	Nearly Neutral–pH SCC	High-pH SCC
Location	Associated with specific terrain conditions, often alternate wet–dry soils and soils that tend to disbond or damage coatings.	Typically within 20 km downstream of a pump or compressor station; the number of failures fall markedly with increased distance from a compressor or pump and lower pipe temperature.
Temperature	No apparent correlation with pipe temperature; may occur more frequently in colder climates where the CO_2 concentration in groundwater is higher.	Growth rate increases exponentially with temperature increase.
Associated electrolyte	Dilute bicarbonate solution with neutral pH in the range 5.5 to 7.5.	Concentrated carbonate–bicarbonate solution with an alkaline pH above 9.3.
Electrochemical potential	−760 to −790 mV($Cu/CuSO_4$); cathodic protection does not reach pipe surfaces at SCC sites.	−600 to −650 mV ($Cu/CuSO_4$); cathodic protection contributes to achieving these potentials.
Crack path and morphology	Primarily transgranular; wide cracks with evidence of substantial corrosion of crack sidewalls.	Primarily intergranular; narrow tight cracks with almost no evidence of secondary corrosion of crack walls.

Source: National Energy Board [1996a].

that supports nearly neutral–pH SCC may contain diluted bicarbonate ions and other species, such as chloride and sulfate ions, while high-pH SCC commonly occurs in environments that contain concentrated carbonate and bicarbonate ions [National Energy Board, 1996a]. These environments result from the combined effect of coating performance failure, CP penetration level, and soil conditions. However, when coating is damaged or missing, a pipeline is exposed to soil directly. The soil environment may result in pipeline SCC.

Carbonate and bicarbonate ions are generated from the dissolution of CO_2 in the soil in groundwater. The environment for nearly neutral–pH SCC can develop only in the absence of CP due to shielding disbondment of an impermeable coating such as PE tape. If CP penetrates the coating, which is usually permeable in nature (e.g., asphalt, coal tar) and reaches the pipe surface, a high-pH carbonate or bicarbonate environment will be generated. The role of soil in CP permeation is important. The high resistivity of dry soils may prevent CP from reaching the pipe and contribute to generation of a nearly neutral–pH environment. Furthermore, temperature affects the solubility of CO_2 in water. At a low temperature, the CO_2 solubility increases and the pH of the trapped water decreases. Actually, the environmental conditions for transgranular SCC (TGSCC) and intergranular SCC (IGSCC) on pipelines are related and may change from one to the other due to fluctuations or changes in the cathodic potential applied [Asher et al., 2007].

Four factors control a pipeline's electrochemical potential environment resulting in pipeline SCC: type and performance condition of the coating, CP level at the pipe steel surface, resistivity of soil, and temperature. The factors affecting coatings, CP, and temperature during SCC development are explored in depth in subsequent chapters. The change in soil resistivity occurring during a seasonal dry–wet cycle would affect the corrosion of steel and its SCC susceptibility, with full details given through the introduction of an advanced technique called a scanning Kelvin probe (SKP) used for that purpose.

The scanning Kelvin probe technique is a nonintrusive electrochemical method that permits the measurement of localized variation in potential in situ over the surface of an electrochemically active specimen [Grundmeier et al., 2000; Leng et al., 1999]. By measuring the difference in work functions of the working and reference electrodes (the latter a Kelvin probe made of elemental tungsten), the surface conditions and electrochemical activity of the target electrode can be determined accurately. Figure 3-5 shows the SKP measurements for a coated X65 pipeline steel electrode in a trapped concentrate or bicarbonate solution experiencing a wet–dry cycle, where the more negative potential region in the central part of the scanning area (the light grey area) is the disbonded area. With evaporation of the trapped solution, the solution layer changes from "thick" to relatively "thin." Correspondingly, the Kelvin potential is shifted negatively (dark grey area as marked "thin").

During a wet–dry cycle, the thickness of the solution layer trapped under a disbonded coating decreases, due to the evaporation of water. Reduction in the solution thickness facilitates oxygen diffusion and reduction on the steel surface. Simultaneously, the solution concentration increases with the thinning solution thickness, resulting in a reduction in oxygen solubility. The former enhances steel corrosion by increasing the cathodic partial reaction, resulting in a negative shift in the corrosion potential of the steel, while the latter probably causes a positive shift of corrosion potential due to the increasing oxidizing ability of a concentrated carbonate

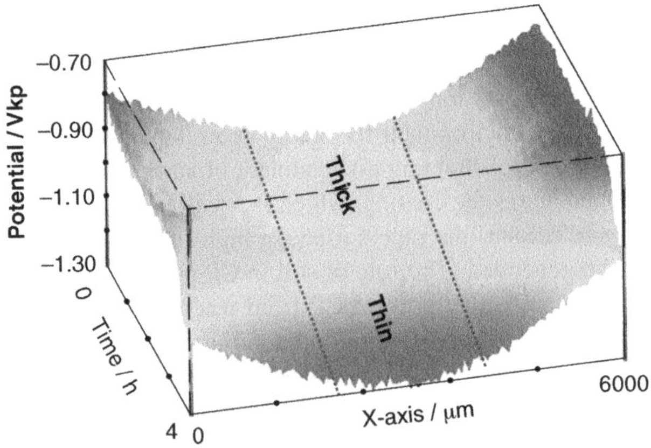

Figure 3-5 *Kelvin potential measured on an X65 pipeline steel electrode under a disbonded coating during a wet–dry cycle. (From Fu and Cheng [2009].)*

solution. There exists competition between the two effects as the solution layer thickness changes. The Kelvin potential measured shifts negatively during a wet–dry cycle (Fig. 3-5), indicating that the effect associated with accelerated oxygen diffusion and reduction is favored during the corrosion of steel in the trapped solution. Therefore, during the drying of soil, the decreasing solution layer thickness favors corrosion of the steel and shifts the steel potential negatively. Although a dry soil may "shield" pipeline from CP penetration, the negative potential of the steel that is achieved during seasonal drying may cause pipeline in the sensitive potential range to exhibit high-pH SCC.

Although microbial activity affects pipeline corrosion and SCC, it is worth pointing out that soil sample collection and soil solution preparation could be sufficient to permit the laboratory simulation of the soil SCC of pipelines, whereas attempting to reproduce the possible effect of microbial activity may be difficult or even impossible [Bueno et al., 2004].

3.4.2 Susceptible Line Pipe Steels

A susceptible pipeline material is another condition for developing SCC. Statistics show that SCC has occurred on a wide variety of grades of pipeline steels, from X25 to X65. Moreover, SCC has occurred on pipes with diameters from 114 to 1067 mm, and wall thicknesses ranging from 2.0 to 9.5 mm. Both electric resistance–welded (ERW) and double submerged arc-welded (DSAW) pipes were found to be involved in SCC-related failures [National Energy Board, 1996a].

Generally, SCC does not appear to be associated with a particular pipe manufacturing method or manufacturer. However, it was found [Canadian Energy Pipeline Association, 1996c] that the ERW longitudinal seam of pipes manufactured in the 1950s used in southern Ontario had a lower resistance to nearly neutral–pH SCC than that of the base steel. This is probably due to the low fracture toughness of the welding region or a higher-than-normal residual stress [TransCanada Pipelines, 1996]. Furthermore, the coarse-grained heat-afftected zone (HAZ) adjacent to the DSAW is significantly more susceptible to cracking than is the base material in a nearly neutral–pH environment [Canadian Energy Pipeline Association, 1996b]. The average crack velocities are about 30% higher in the HAZ than in the base steel. However, for high-strength X100 pipeline steel, there is a higher local corrosion rate at the base steel than at the weldment in a nearly neutral–pH solution, which is associated with the softening of the HAZ [Zhang and Cheng, 2010]. For low-grade steel such as X70 steel, there is a higher cracking rate at the HAZ than that for the base steel [Zhang and Cheng, 2009].

Pipe grades from X35 to X70 and X100 steels have been found to be susceptible to nearly neutral–pH SCC. In particular, high-strength steels such as X80 and X100 are more susceptible to hydrogen-induced SCC in nearly neutral–pH solutions. It was found [Jin et al., 2010; Xue and Cheng, 2011] that even in the absence of external stress, hydrogen, once exceeding a threshold amount, would result in cracking of both X80 and X100 steels. For high-pH SCC, no correlation has been seen between the range of strength of the steel pipes and SCC susceptibility [Parkins, 1994].

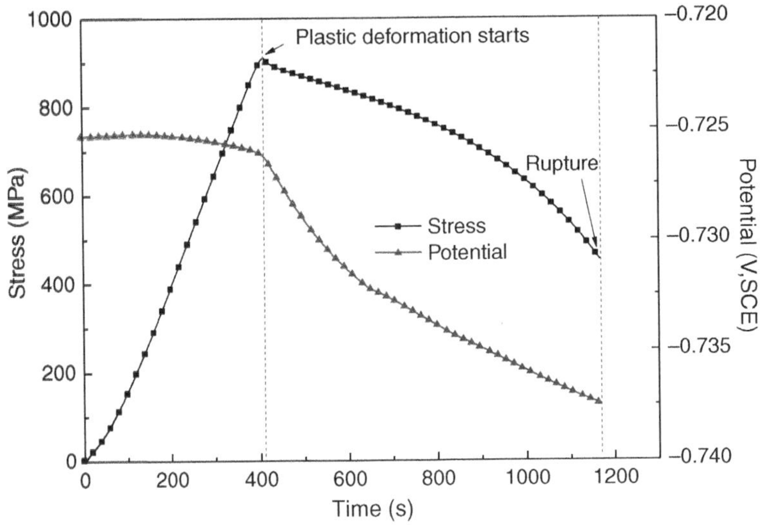

Figure 3-6 *Time dependence of stress and corrosion potential of 3.918% prestrained X100 steel during tensile testing at a strain rate of* $1 \times 10^{-4} s^{-1}$ *in a nearly neutral–pH solution. (From Xu and Cheng [2012].)*

Local plastic deformation on pipelines increases the susceptibility of steels to SCC. Xu and Cheng's latest research [2012] found that both yield strength and ultimate tensile strength increase, and the fracture strain decreases with the prestrain. Moreover, while an elastic strain causes a slight drop of corrosion potential by 1.1 mV(SCE) only for the 3.918% prestrained X100 steel specimen in a nearly neutral–pH solution, the corrosion potential shifts negatively to 11.2 mV(SCE) in the plastic deformation stage for the specimens, as shown in Fig. 3-6. It indicates that there is a more remarkable effect of plastic deformation than of elastic deformation on the corrosion potential of the steel; that is, while the mechanoelectrochemical effect of elastic deformation is limited and usually negligible, that of plastic deformation on steel corrosion is significant. Consequently, the conditions that produce and/or enhance local plasticity either on the pipe surface or at a crack tip would favor SCC.

Normally, the strains that would produce local plastic deformation do not occur at stress levels below the proportional limit of the steel, which is usually in the range 85 to 95% of the yield strength of the steel. There are two situations in which local plastic deformation would occur at stress levels below the proportional limit of the steel. One is deformation of the surface layer of the pipe that occurs at low stress levels before the bulk of the wall thickness [Leis, 1990], and the other is under cyclic loading [Canadian Energy Pipeline Association, 1996a].

Furthermore, there is a strong effect of nonmetallic inclusions on the susceptibility of pipeline steels to SCC. In X70 pipeline steel, the primary inclusions are enriched in aluminum oxides, which are brittle and incoherent in the steel matrix. Microcracks and interstices are easily formed at the boundary between inclusions and steel [Liu et al., 2009a]. In X100 pipeline steel, at least four types of inclusions are identified: elongated magnesium sulfide (MnS) inclusions and spherical Al-, Si-, and Ca–Al–O–S-enriched inclusions. In particular, the majority of inclusions in the steel are aluminum-enriched. Upon hydrogen charging, hydrogen blisters and HIC could be caused, with the cracks associated primarily with aluminum- and silicon-enriched inclusions, rather than the elongated MnS inclusions [Jin et al., 2010]. Furthermore, the size of inclusions is important in crack initiation. If the length of surface defects that are governed by the length of nonmetallic inclusions is smaller than 200 to 250 μm, longitudinal cracks would not form. Moreover, inspection of field pipes with a cracking history found [TransCanada Pipelines, 1996] that uncracked pipes contain longer inclusions than do cracked samples.

To date, no systematic relationship has been found between the chemical composition of steels and the nearly neutral–pH SCC susceptibility. Actually, it is generally believed that the microchemistry of pipeline steels has only a limited effect on SCC occurrence. For high-pH SCC, it is accepted that adding 2 to 6% chromium, nickel, or molybdenum to steel improves resistance to high-pH SCC. This is due primarily to the improved passivity of steels in high-pH environments. However, additions at such high levels make the steel too expensive to produce [Parkins, 1994].

The presence of mill scale on a pipe surface would contribute to SCC processes. It has been suggested [Canadian Energy Pipeline Association, 1996d] that at stress levels above the yield strength, the threshold stress on a smooth, clean steel surface is different from that on a surface that has seen years of service. Actually, for pipeline steels in nearly neutral–pH solutions, the mill scale affects both the cathodic hydrogen evolution and the anodic dissolution of steels [Meng et al., 2008]. Figures 3-7 and 3-8 show localized electrochemical impedance spectroscopy (LEIS) and scanning vibrating electrode technique (SVET) maps measured on an X70 steel electrode partially covered with a corrosion product layer formed in a nearly neutral–pH solution, where zones A and B refer to a corrosion product–covered surface and a bare steel surface, respectively, and line l indicates the boundary between the two zones. There is a higher impedance value in zone B, the bare steel zone, than in zone A, the corrosion product–covered zone. The average impedance value of zone B is 1330 Ω, while that of zone A is about 1190 Ω. In addition to the low-impedance area (zone A) and the high-impedance area (zone B), there is a transition area with intermediate impedance adjacent to the boundary line. Moreover, the SVET current density map contains three different zones (A, B, and C in Fig. 3-8). The current density measured in zone A, where the electrode surface is covered with a corrosion product layer, is higher (about 70 to 80 μA/cm^2) than in the other two zones. Zone B, which is adjacent to both zones A and C, is characterized by an intermediate current density of about 55 μA/cm^2. The smallest current density, about 35 to 40 μA/cm^2, is in zone C.

Current density mapping over the entire steel electrode shows that the corrosion activity of the corrosion product layer is not uniform, due to the porous, noncompact structure of the layer [Meng et al., 2008]. It is reasonable to assume that anodic reaction is concentrated at pores that are characterized by a high dissolution current density, and that the cathodic hydrogen evolution occurs primarily on the deposit

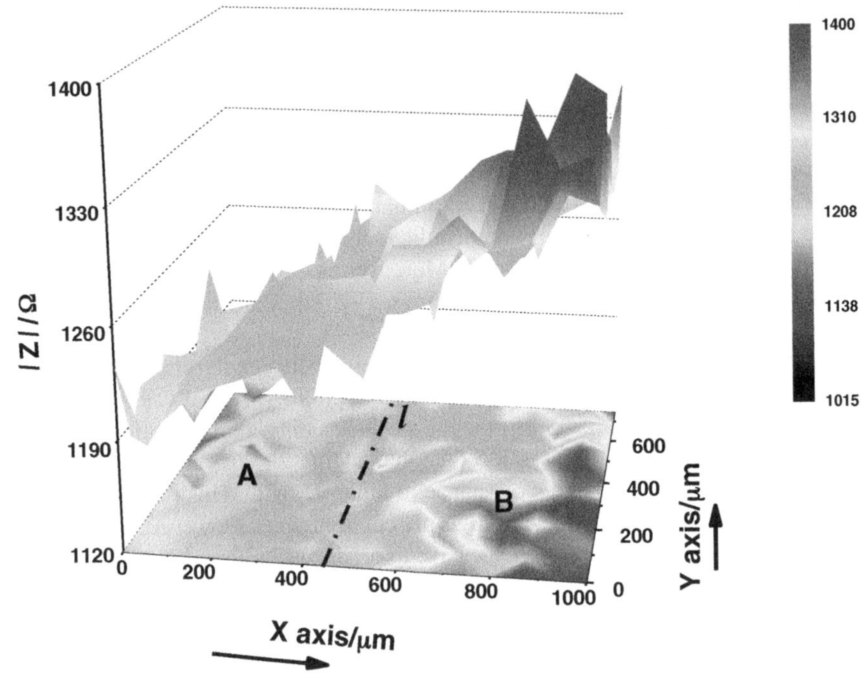

Figure 3-7 LEIS map measured on X70 steel electrode in a nearly neutral–pH solution. Zones A and B refer to a corrosion product–covered surface and a bare steel surface, respectively, and line l is the boundary of the two zones. (From Meng et al. [2008].)

surface. A small anode/big cathode geometry is thus generated to accelerate localized anodic dissolution of the steel, and the electrons generated are consumed by the cathodic water reduction. Consequently, the cathodic reaction is enhanced to maintain the charge neutrality. Therefore, the porous deposit layer enhances hydrogen evolution and thus hydrogen-induced SCC susceptibility.

Furthermore, both LEIS and SVET measurements show that the presence of a layer of corrosion product on an electrode surface enhances anodic dissolution of steel. It is attributed primarily to the increasing effective surface area for adsorption of intermediate species, such as $Fe(I)_{ad}$ or $Fe(II)_{ad}$, on the deposit [Keddam et al., 1981; Modiano et al., 2008]. A transition zone is observed in both measurements. It is believed that the steel adjacent to the corrosion product layer is somewhat oxidized to generate deposit scale, where the adsorption of intermediate species is enhanced, but is not as strong as the preformed deposit zone. As a consequence, the "self-catalytic" dissolution effect is less than it is in zone A but much higher than in zone C.

The occurrence of pipeline SCC is usually associated with the presence of corrosion product and mill scale on a pipe surface. The deposit and scale would accelerate corrosion of the steel. In particular, the small anode/big cathode geometry resulting from the porous structure of the deposit scale could result in pitting corrosion, as indicated by the quite high local dissolution current density at individual sites on the

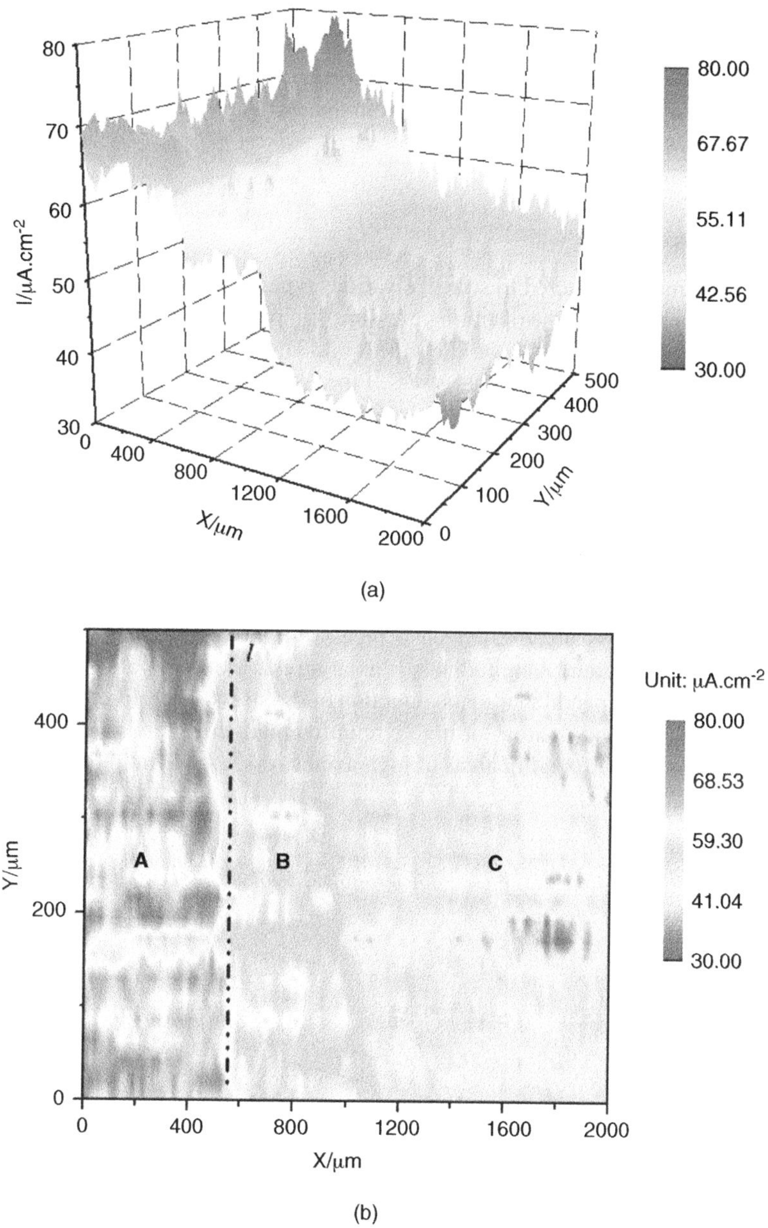

Figure 3-8 *SVET current density map measured on the same scanning area as in Fig. 3-7: (a) three-dimensional map, (b) two-dimensional contour map. (From Meng et al. [2008].)*

electrode surface (Fig. 3-8). Corrosion pits might be the most common sites for crack initiation. Therefore, in the presence of a deposit layer and mill scale on the pipe surface, pipeline corrosion, especially pitting corrosion, is expected to be enhanced. Stress corrosion cracks could be initiated from the corrosion pits that develop under the deposit and scale.

3.4.3 Stress

A buried pipeline is subjected to stress of various types from various sources, all of which contribute to SCC. The operating pressure of pipelines is normally the greatest source of stress—hoop stress—on the pipe wall. The soil that surrounds the pipe can move and generate longitudinal stress. Pipe manufacturing processes, such as welding, can create residual stresses.

An important effect of stress on crack initiation and propagation determines whether there is a threshold stress level below which cracks will not grow, whether increasing the stress in a pipeline causes cracks to grow faster, and how often hydrostatic retests should be scheduled. Generally, cracks grow in two ways: in length and depth due to dissolution and hydrogen embrittlement, and through several cracks joining to create a long crack.

Two types of stress are exerted on a pipe: circumferential stress, around the pipe's circumference, and longitudinal or axial stress, along the axis of the pipe. Cracks are always perpendicular to the direction of the stress. Therefore, longitudinal (axial) cracks are found in areas of high circumferential stress, and circumferential (transverse) cracks are found in areas of high axial stress, as shown in Fig. 3-9.

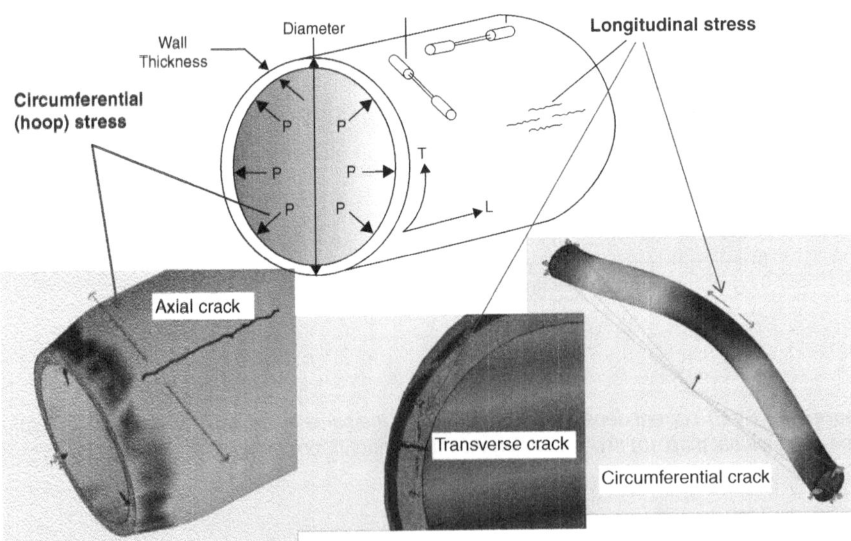

Figure 3-9 Schematic diagram of stresses exerted on a buried pipe.

Circumferential Stresses. Circumferential stress in a pipe has a variety of sources, including hoop stress due to internal operating pressure, which is usually the highest stress component in a pipe; residual stress created during pipe manufacturing; bending stresses generated when an oval or out-of round pipe is subjected to internal pressure; local stresses at the edge of double-submerged arc welds or associated with mechanical gouges and corrosion pits; secondary stresses that may cause a pipe to go out of round, such as soil settlement or land sliding; and stresses due to temperature differences through the thickness of a pipe wall. It has been reported [National Energy Board, 1996a] that of the 22 service failures in Canadian pipelines that have been investigated, 16 (73%) involve axial cracks, indicating that circumferential stresses control the failure.

The *operating pressure* of pipelines is defined as a percentage of the minimum yield strength (SMYS) specified for a pipe steel. However, the yield strength of the pipe is usually higher than the SMYS. The actual yield strength can be 10 to 30% higher than the SMYS. Therefore, a pipeline that is operating at 72% SMYS may reach only 60% of the actual yield strength of the line pipe steel. In particular, the Canadian Standards Association (CSA) Z662 sets the maximum allowable hoop stress depending on where the pipeline is located and the surrounding dwelling density, and four class locations have been set up based on the dwelling density [Canadian Standards Association, 2011]. On natural gas pipelines, the maximum allowable hoop or operating stress ranges from 80 to 44% SMYS, depending on the density of dwellings. For example, 80% SMYS is the maximum hoop stress allowed in a class 1 location, a sparsely inhabited region or rural area. The maximum allowable hoop stress governs the wall thickness of the pipe for a given grade (SMYS). Currently, regulations in many other countries limit the maximum hoop stress to 72% SMYS [Institute of Gas Engineers, 1993; U.S. Department of Transportation, 1994]. According to NEB's statistics [National Energy Board, 1996a], at the time of pipeline failures caused by SCC, the hoop stresses varied between 46 and 77% of a pipe's SMYS. Apparently, simply following the SMYS standard will not avoid SCC.

The internal pressure in a pipeline changes continually or fluctuates along its entire length of service. In a gas line it is affected by the rate at which gas is injected into the system and is withdrawn by downstream deliveries. In liquid pipelines the pressure fluctuates more widely since it is affected by the turning on and off of pumps and any changes in the density of the fluid being pumped. To fully characterize the operating pressure of a pipeline, three factors have to be considered: maximum operating pressure applied, the range of the pressure fluctuation, and the rate of the pressure changes [National Energy Board, 1996a].

When flat steel plates are formed into pipes, residual stresses are introduced into the pipe. The level of residual stress depends on the manufacturing process. It has been proposed [National Energy Board, 1996b] that operating pipelines can contain residual stresses that are least as about 25% of the yield strength of the material, although they may influence cracking behavior.

Studies of the effects of residual stress on pipeline SCC were limited before 2000. Canadian researchers have investigated this problem in recent years and have studied systematically the role of residual stress in pipeline SCC in both high- and nearly neutral–pH solutions. Li and Cheng [2008] found that deformation-induced residual stress, if not sufficiently high, actually has an inhibitory effect on pipeline corrosion, pitting occurrence, and crack initiation under high-pH conditions. This is attributed to enhanced generation of iron carbonate corrosion products and the resulting surface block effect at stressed areas. Tensile and compressive residual stresses have an identical inhibitive effect. Corrosion pits are easy to initiate around the neutral axis of a U-bend specimen, where the deformation and the resulting stress are ignorable. In nearly neutral–pH environments it was found [Van Boven et al., 2007] that the formation of micropits occurs preferentially in areas where the tensile residual stresses are highest (approximately 300 MPa), while SCC initiation occurs with a 71% normalized frequency in areas where the surface residual stress is in the range 150 to 200 MPa. The difference between residual stress levels occurring at SCC locations and those occurring at pitting locations results from the residual stress gradient in the depth direction. Furthermore, SCC can readily be blunted, due to plastic deformation and/or extensive anodic dissolution. As a result, a high positive tensile residual stress gradient is necessary for continued growth of SCC in pipeline steels exposed to a neutral-pH environment [Chen et al., 2007]. The tensile residual stress represents a large mechanical driving force for crack nucleation and short crack growth. Active cracks may become dormant as the near-surface residual stress gradient changes, due to self-equilibration, from highly tensile to a lower tensile state or to a compressive state. The change in residual stress level can occur within 1 mm of the surface, resulting in a large proportion of dormant SCC.

Furthermore, the residual stress generated during welding (i.e., welding stress) affects the corrosion reaction of the steel locally, especially at the HAZ, and thus enhances the propagation of cracks once they are initiated in a nearly neutral–pH solution. Moreover, the total synergistic effect of a hydrogen charging of 10 mA/cm^2 and the applied stress of 550 MPa on anodic dissolution of a welded X70 steel in a nearly neutral–pH solution is determined to be within the range 5.7 to 6.5, with a maximum value encountered at the HAZ [Zhang and Cheng, 2009].

Any irregularity on the surface of pipe can be a source of stress concentration. At points of surface damage, dents, or corrosion pits, stress levels in the circumferential and axial directions are higher than those on the rest of the pipe. Figure 3-10 shows the stress distribution of an X100 steel pipe containing a corrosion defect simulated by a finite element model under the synergistic effect of various internal pressures and soil strains [Xu and Cheng, 2012]. It is seen that the stress distribution around the corrosion defect is nonuniform. The high-stress region, up to 700 to 900 MPa, is concentrated at the bottom of the defect and propagates along both longitudinal and hoop directions with increased internal pressure and soil prestrain.

Longitudinal (Axial) Stresses. In addition to circumferential stress, operating pipelines experience stress that acts in the axial direction. The internal pressure of pipelines also causes stress in the axial direction of a pipeline which is a percentage of the hoop stress. For example, when a pipeline is completely buried and restrained from longitudinal movement by the soil, the axial stress is 28% of the hoop stress. When a pipeline is not completely restrained against longitudinal movement,

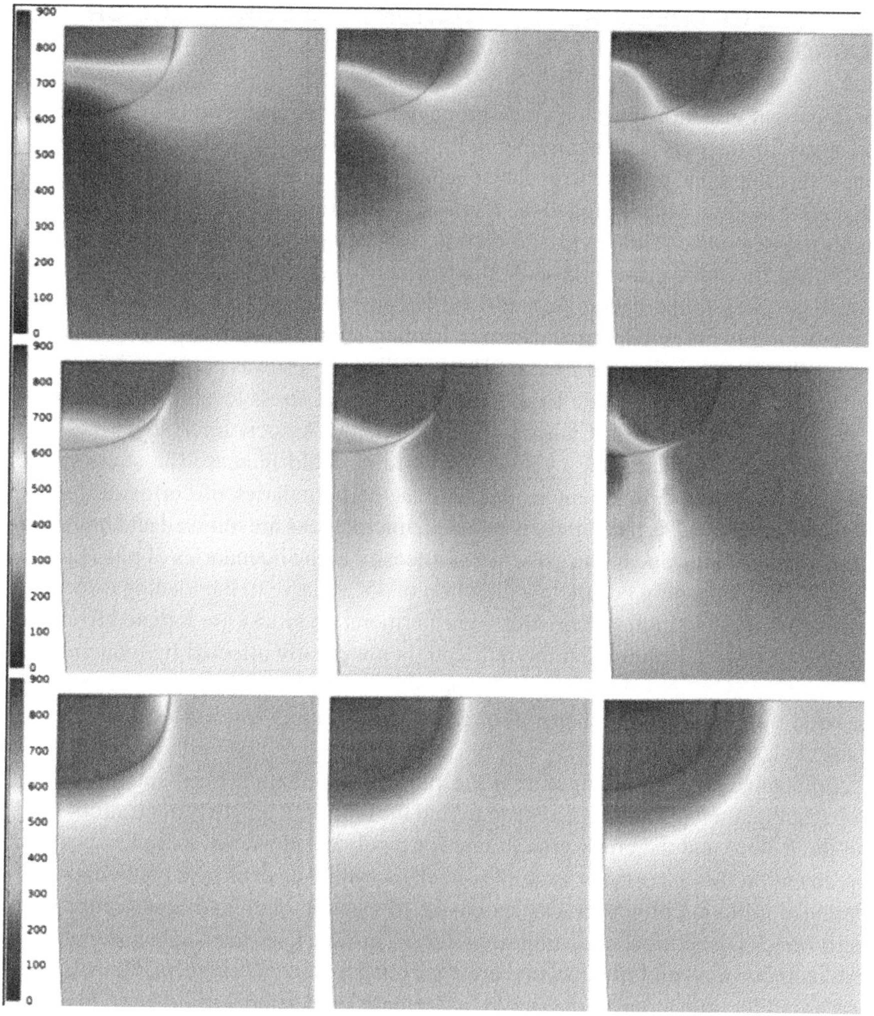

Figure 3-10 *Distribution of stress on a steel pipe with a corrosion effect (top: no soil strain, middle: tensile soil strain, bottom: compressive soil strain. From left to right: operating pressures of 15, 20, and 25 MPa). (From Xu and Cheng [2012].)*

the axial stress can be as high as 50% of the hoop stress [National Energy Board, 1996a].

Furthermore, longitudinal stress is most commonly caused by ground movement, such as landslides and settlement, or by the physical weight of the soil or loading above the pipe (overburden) (i.e., surface loading of pipelines). The level of these stresses is generally difficult to predict but could contribute to pipeline SCC. Accurate determination of soil-induced stress could be achieved through complex geotechnical engineering calculations.

3.5 ROLE OF PRESSURE FLUCTUATION IN PIPELINES: SCC OR CORROSION FATIGUE?

An important effect of stress on crack propagation is dependent on both stress level and stress fluctuations [Canadian Energy Pipeline Association, 1996e]. Internal pressure in pipelines, and thus the stress level in pipelines, fluctuates frequently. The stress fluctuation, commonly referred to the *R-ratio*, is the ratio of the minimum to the maximum stresses in a circumferential direction. It is believed that pressure fluctuations are critical for crack growth, based on experimental tests that no evidence of crack growth has been found under constant-load or constant-displacement conditions and that cracking is observed where dynamic loading conditions are applied [Canadian Energy Pipeline Association, 1996e]. When a full-scale test was performed, no crack growth was found under static load conditions, even at stress levels as high as 80% of the actual yield strength [Canadian Energy Pipeline Association, 1996f].

Furthermore, reducing the cyclic stress *R*-ratio could increase the stress range and subsequently the stress intensity range at the boundaries of corrosion pits or mechanical defects. As the *R*-ratio is reduced, microcracks are initiated and branched at the mouth of pits, due to the great stress intensity at the boundaries of pits [Eslami et al., 2011]. Branching occurs in a direction of 45° relative to the loading direction: the direction of maximum shear stress. Furthermore, no cracks are initiated from the bottoms of pits, indicating that the pit depth is not greatly affected by reducing the *R*-ratio. However, it is not impossible that the pit-to-crack transition occurs from the bottom of pit as electrochemical or geometrical factors reach critical conditions [Fang et al., 2010].

Although it is acknowledged that the pressure fluctuation is essential to crack propagation, attempts have been made to find the correlation between the stress level and the *R*-ratio and the crack growth rate for pipelines. However, there has been no agreement on this aspect. For example, crack growth can occur by a combination of continued initiation of new cracks, extension of existing ones, and coalescence with nearby cracks. At normal operating stress levels, growth rates seem to be independent of maximum stress and *R*-value, and are controlled by the environment. However, an increase in the maximum stress and/or a decrease in *R* value will be more likely to promote the formation of deeper cracks [Parkins, 1995]. Moreover, reducing pressure or pressure fluctuations in order to reduce stress in a pipeline with SCC will not stop cracking, but it may marginally reduce the crack growth rate. Industrial analysis of excavations indicated that crack depth increases at higher stress levels, which is attributed to higher loading rates rather than higher stress levels at locations near compressor stations [TransCanada Pipelines, 1996].

The relationship between the stress level at an individual *R* value and the crack growth rate has been tested in near neutrally–pH solutions. For X65 pipeline steel, there is a direct relationship between maximum stress and increasing crack growth rate at low *R* ratios [i.e., 0.5 (typical of oil pipeline operation) and 0.7]. When the *R*-ratio is up to 0.85, which is typical of gas pipeline operation, there is no effect of the stress level on the crack growth rates measured [Canadian Energy

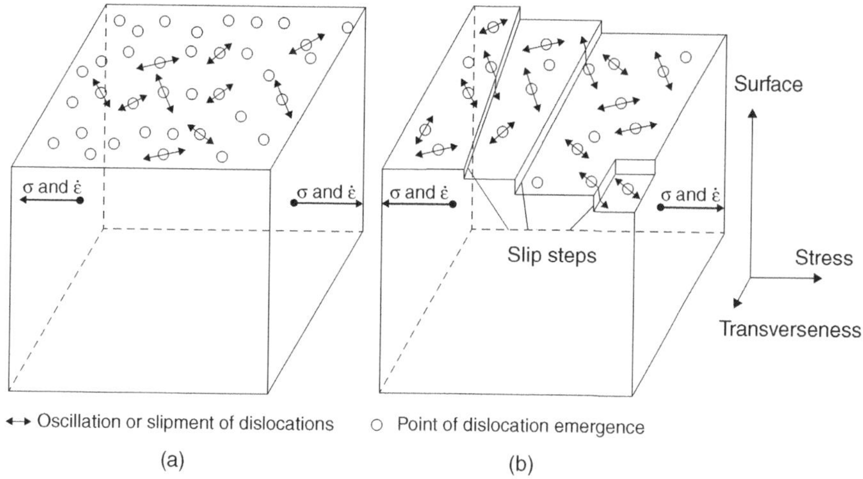

Figure 3-11 *Generation of dislocation emergence points (a) and slip steps (b) on the surface of a stressed or strained steel. (From Liu et al. [2009b].)*

Pipeline Association, 1996b]. Apparently, pressure fluctuations that are similar to cyclic stressing in pipelines would contribute to the cracking process.

It is also expected that there exist effects of the rate of pressure change or strain rate on crack initiation and growth. Generally, the slower frequencies give faster per-cycle growth rates [Canadian Energy Pipeline Association, 1996b]. At low frequencies, crack growth per cycle is higher since there is more time for the environment to interact with the crack, increasing the growth rate. Here, a *local additional potential* (LAP) *model* is introduced to illustrate the critical role of strain rate in the SCC of pipeline steels [Liu et al., 2009b].

It is acknowledged [Hertzberg, 1996] that an application of stress or strain will activate the dislocation mobility in steels. In the elastic region, some dislocations oscillate near their equilibrium positions, as shown in Fig. 3-11a. When the dislocation slip is generated locally, a stress concentration would be created at sites such as inclusions. Once the local stress approaches or exceeds the yield strength of the steel, significant dislocation movement occurs, resulting in the formation of dislocation emergence points and slip steps on the steel surface, as indicated in Fig. 3-11b.

Since an edge dislocation is easier to slip than a screw dislocation, the former is thus assumed to be the primary movable dislocation in steel. Dislocation emergence points and slip steps introduce local active sites on a steel surface, and electrons would flow and concentrate at these sites, resulting in a local charging effect. As a result, additional potential is generated locally when a steel specimen is exposed in solution. Moreover, the LAP is altered with the polarizing condition. For example, an anodic LAP is generated during anodic polarization, as shown in Fig. 3-12a and c, and a cathodic LAP is generated during cathodic polarization (Fig. 3-12b and c).

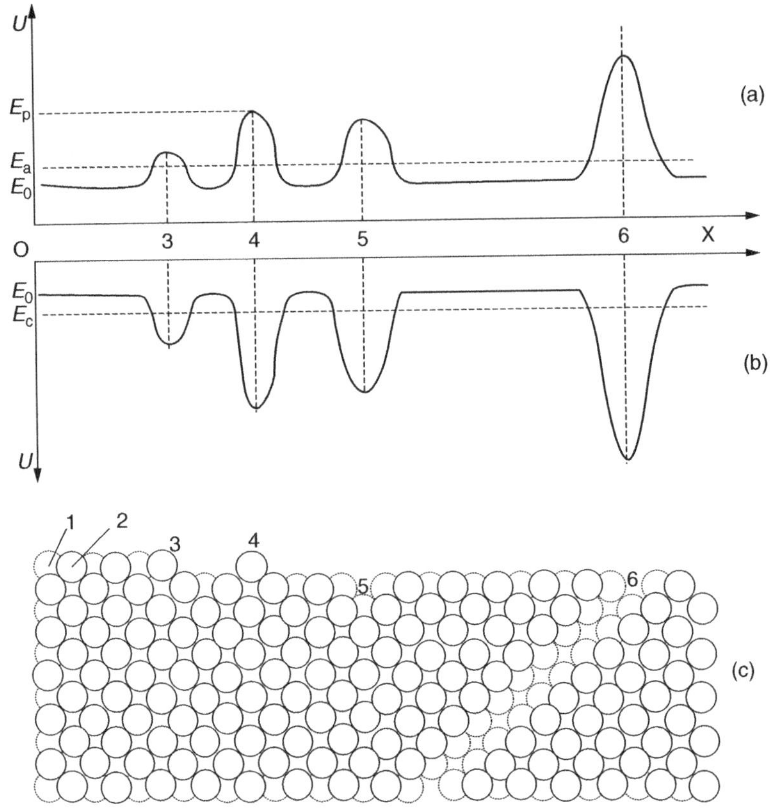

1 – Subsurface atom plane; 2 – (110) crystal plane; 3 – Slip steps;

4 – Insular atom on surface; 5 – Vacancy on surface; 6 – Dislocation emergence point

Figure 3-12 *Schematic diagram of local potential distribution at (a) anodic polarization and (b) cathodic polarization as well as (c) the corresponding surface defect distribution. (From Liu et al. [2009b].)*

The LAP is defined as a point-charge potential, and the total potential at the local active site, E_{pi}, can be expressed as

$$E_{\text{pi}} = E_0 + \frac{kq_i}{4\pi\varepsilon_r r} \tag{3-1}$$

where E_0 is the electrode potential at an intact site, q_{i} the charge of electrons, ε_r the dielectric constant of water, r the distance from the local charged point on the steel surface to the solution layer, and k is a constant. Thus, the potential at a local active site on a steel surface depends strongly on the amount of local charge (i.e., the number of electrons) at the steel–solution interface. Consequently, the LAP introduced by local

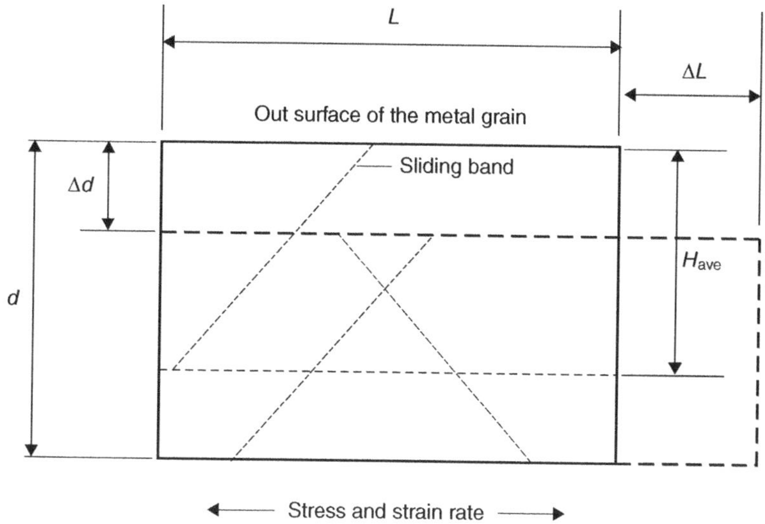

Figure 3-13 *Schematic diagram of a surface crystalline grain before and after deformation. (From Liu et al. [2009b].)*

electron concentration would affect remarkably the electrochemical behavior of steel in solution.

To establish the relationship between the LAP and the strain rate, the number of slip dislocations is derived. Figure 3-13 is a simplified view of a crystalline grain on a steel surface, where the upper surface of the grain is assumed to be exposed to the solution. The solid rectangular lines define the initial boundary of the grain, where L and d are the initial size and depth of the grain, respectively. The dashed lines constitute a new grain boundary after the deformation of ΔL and Δd. A two-dimensional diagram was used for simplification. W is the thickness of the grain, and its deformation is ΔW. It is assumed that the time period for L changing to $L + \Delta L$ is t, the average depth of the source dislocation is H_{ave}, the dislocation density is ρ, the generating frequency of new dislocations is f, and the shear stress is the only driving force to activate dislocations. Moreover, v_{th} is assumed to be the minimum mobility velocity of dislocations on the steel surface, and dislocation with a mobility velocity higher than v_{th} will reach the steel surface within time t.

The strain, ε, is defined as

$$\varepsilon = \frac{\Delta L}{L} = \frac{\Delta d}{d} = \frac{\Delta W}{W} \tag{3-2}$$

The strain rate, $\dot{\varepsilon}$, is thus

$$\dot{\varepsilon} = \frac{\varepsilon}{t} = \frac{\Delta L}{Lt} = \frac{\Delta d}{dt} = \frac{\Delta W}{Wt} \tag{3-3}$$

The average time for a dislocation leaving from its source to reach the grain surface is

$$t_0 = \frac{\sqrt{2}\,H_{\text{ave}}}{v_{\text{th}}} \tag{3-4}$$

The number of dislocations reaching the grain surface is

$$n = \rho f \frac{t}{t_0} H_{\text{ave}}(L + d + W) \tag{3-5}$$

The number of dislocations reaching the upper surface of the grain is then

$$n_{\text{upper}} = \rho f \frac{t}{t_0} H_{\text{ave}} L \tag{3-6}$$

Combining Eqs. (3-4) and (3-6), the number of dislocations reaching the upper surface of grain in unit time, n_0, is thus

$$n_0 = \frac{n_{\text{upper}}}{t} = \frac{\sqrt{2}}{2} \rho f L v_{\text{th}} \tag{3-7}$$

Furthermore, according to the geometric change of a grain after deformation under a tensile stress, the relationship between the released dislocation number, n, and strain is

$$n = \frac{(\Delta L/2) + L}{r_0} \frac{\Delta d}{r_0} + \frac{(\Delta L/2) + L}{r_0} \frac{\Delta W}{r_0} + \left(\frac{W - \Delta W}{2r_0} + \frac{d - \Delta d}{2r_0} \right) \frac{\Delta L}{r_0} \approx \frac{3}{2} \frac{Ld + LW}{r_0^2} \varepsilon \tag{3-8}$$

where r_0 is the atomic radius. Therefore, the number of dislocations reaching the upper surface of the grain is written as

$$n_{\text{upper}} = \frac{(\Delta L/2) + L}{r_0} \frac{\Delta d}{r_0} \approx \frac{Ld}{r_0^2} \varepsilon \tag{3-9}$$

The density, n_0, is then determined as

$$n_0 = \frac{n}{t} = \frac{S}{r_0^2} \frac{\varepsilon}{t} = \frac{S}{r_0^2} \dot{\varepsilon} \tag{3-10}$$

where $S = Ld$. Comparing Eqs. (3-7) and (3-10), we have

$$v_{\text{th}} = \frac{\sqrt{2} S}{\rho f L r_0^2} \dot{\varepsilon} \tag{3-11}$$

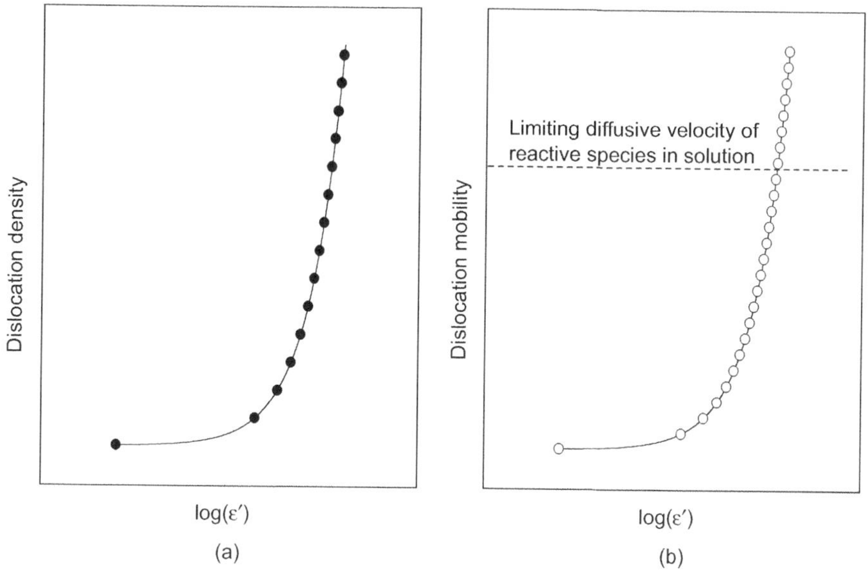

Figure 3-14 *Schematic view of density (a) and mobility (b) of dislocation emergence points as a function of strain rate. (From Liu et al. [2009b].)*

Apparently, both the density, n_0, and mobility, v_{th}, of dislocations on the steel surface (i.e., dislocation emergence points) depend linearly on the strain rate, as demonstrated in Eqs. (3-10) and (3-11), respectively.

Figure 3-14 shows the normalized dependence of the mobility and density of dislocation emergence points on the logarithm of strain rate. It is seen that as the strain rate increases, the density of the dislocation emergence points increases, which would enhance the local electrochemical activity of the steel electrode due to the LAP effect. Simultaneously, the mobility of the dislocation emergence point increases with strain rate. When the strain rate is sufficiently high to result in a sufficiently high mobility of dislocation emergence points on the steel surface, the reactive species transporting from the solution do not have an opportunity to adsorb at these active sites for an electrochemical reaction. Therefore, there is a dual role for strain rate in electrochemical activity of the steel under straining. With respect to the effect of the strain rate on SCC susceptibility, it is expected that a maximum value phenomenon will be observed. Detailed experimental results about the effect of strain rate on the susceptibility of pipeline steels to SCC are included in Chapter 6.

While cyclic stress facilitates crack propagation, it was also found that nearly neutral–pH SCC can occur under constant loading conditions [Fang et al., 2007]. In constant-load tests, crack initiation was associated with pit nucleation and growth as well as the associated stress concentration. When the local strain rate around a pit is sufficiently high, which may lie in the susceptible strain rate range for nearly neutral–pH SCC if the pit dimension is appropriate, a crack can initiate and even propagate. As a result, nearly neutral–pH SCC would occur at an appropriate load and test time under constant-load tests.

Cyclic loading can also cause steel to exhibit local plastic deformation called the *cyclic softening phenomenon*. The steel may behave elastically for a number of cycles and then exhibit plastic deformation for a number of load cycles. This phenomenon has been found on pipe steels involved in nearly neutral–pH SCC failures [Leis, 1991]. To date, research has been incomplete with regard to the cyclic softening behavior of pipe steels.

REFERENCES

Asher, SL, Leis, B, Colwell, J, Singh, PM (2007) Investigating a mechanism for transgranular stress corrosion cracking on buried pipelines in near-neutral pH environments, *Corrosion* 63, 932–939.

Baker, M, Jr. (2005) *Final Report on Stress Corrosion Cracking Study*, Integrity Management Program Delivery Order DTRS56-02-D-70036, Office of Pipeline Safety, U.S. Department of Transportation, Washington, DC.

Bueno, AHS, Castro, BB, Ponciano, JAC (2004) Laboratory evaluation of soil stress corrosion cracking and hydrogen embrittlement of API grade steels, *Proc. 5th International Pipeline Conference*, Paper IPC04-0284, Calgary, Alberta, Canada.

Canadian Energy Pipeline Association (1996a) Submission to the National Energy Board, *Proceeding MH-2-95*, Vol. 2, CEPA, Calgary, Alberta, Canada, App. D, Tab. 1, p. 13.

Canadian Energy Pipeline Association (1996b) Submission to the National Energy Board, *Proceeding MH-2-95*, Vol. 2, CEPA, Calgary, Alberta, Canada, App. D, Tab. 9, pp. 9–10.

Canadian Energy Pipeline Association (1996c) Submission to the National Energy Board, *Proceeding MH-2-95*, Vol. 1, CEPA, Calgary, Alberta, Canada, Issue 1, Tab. 1.10.

Canadian Energy Pipeline Association (1996d) Submission to the National Energy Board, *Proceeding MH-2-95*, Vol. 2, CEPA, Calgary, Alberta, Canada, App. D, Tab. 6, pp. 5–6.

Canadian Energy Pipeline Association (1996e) Submission to the National Energy Board, *Proceeding MH-2-95*, Vol. 2, CEPA, Calgary, Alberta, Canada, App. D, Tab. 5, p. 7.

Canadian Energy Pipeline Association (1996f) Submission to the National Energy Board, *Proceeding MH-2-95*, Vol. 2, CEPA, Calgary, Alberta, Canada, App. D, Tab. 4, p. 8.

Canadian Energy Pipeline Association (1997) *Stress Corrosion Cracking Recommended Practices*, CEPA, Calgary, Alberta, Canada.

Canadian Energy Pipeline Association (2007) *Stress Corrosion Cracking Recommended Practices*, 2nd ed., CEPA, Calgary, Alberta, Canada.

Canadian Standards Association (2011) *Oil and Gas Pipeline Systems*, Z662-11, CSA, Reydale, Ontario, Canada.

Chen, W, Van Boven, G, Rogge, R (2007) The role of residual stress in neutral pH stress corrosion cracking of pipeline steels: Part II. Crack dormancy, *Acta Mater.* 55, 43–53.

Corrosion Doctors, online source, http://corrosion-doctors.org/Pipeline/Williams-explosion .htm.

Delanty, B, O'Beime, J (1992) *Oil Gas J.*, June, p. 39.

Eslami, A, Kania, R, Worthingham, B, Van Boven, G, Eadie, R, Chen, W (2011) Effect of CO_2 and R-ratio on near-neutral pH stress corrosion cracking initiation under a disbonded coating of pipeline steel, *Corros. Sci.* 53, 2318–2327.

Fang, BY, Atrens, A, Wang, JQ, Han, EH, Ke, W (2003) Review of stress corrosion cracking of pipeline steels in "low" and "high" pH solutions, *J. Mater. Sci.* 38, 127–132.

Fang, BY, Han, EH, Wang, JQ, Ke, W (2007) Stress corrosion cracking of X-70 pipeline steel in near neutral pH solution subjected to constant load and cyclic load testing, *Corros. Eng. Sci. Technol.* 42, 123–129.

Fang, B, Eadie, R, Chen, W, Elboujdaini, M (2010) Pit to crack transition in X-52 pipeline steel in near-neutral pH environment: 1. Formation of blunt cracks from pits under cyclic loading, *Corros. Eng. Sci. Technol.* 45, 302–312.

Fu, AQ, Cheng, YF (2009) Characterization of corrosion of X65 pipeline steel under disbonded coating by scanning Kelvin probe, *Corros. Sci.* 51, 914–920.

Grundmeier, G, Schmidt, W, Stratmann, M (2000) Corrosion protection by organic coatings: electrochemical mechanism and novel methods of investigation, *Electrochim. Acta* 45, 2515–2533.

Hertzberg, RW (1996) *Deformation and Fracture Mechanics of Engineering Materials*, 4th ed., Wiley, New York.

Institute of Gas Engineers (1993) *Steel Pipelines for High Pressure Gas Transmission*, 3rd ed., IGE, London.

Jack, TR, Erno, B, Krist, K, Fessler, RR (2000) Generation of near-neutral pH and high pH SCC environments of buried pipelines, *Corrosion 2000*, Paper 362, NACE, Houston, TX.

Jin, TY, Liu, ZY, Cheng, YF (2010) Effects of non-metallic inclusions on hydrogen-induced cracking of API5L X100 steel, *Int. J. Hydrogen Energy* 35, 8014–8021.

Keddam, M, Mattos, OR, Takenouti, H (1981) Reaction model for iron dissolution studied by impedance electrode, *J. Electrochem. Soc.* 128, 257–266.

Leis, BN (1990) Update on SCC life prediction models for pipelines, *First Pipeline Technology Conference*, Ostende, Belgium.

Leis, BN (1991) Some aspects of SCC analysis for gas transmission pipelines, *CIM International Symposium on Materials Performance and Maintenance*, Ottawa, Ontario, Canada.

Leis, BN (1995) Characteristic features of SCC colonies in gas pipelines: implications for NDI systems and related integrity analyses, *Pipeline Technol.* 1, 611–614.

Leis, BN, Eiber, RJ (1997) Stress corrosion cracking on gas transmission pipelines: history, causes and mitigation, *Proc. First International Business Conference on Onshore Pipelines*, Berlin, Germany.

Leis, BN, Bubenik, TA, Nestleroth, JB (1996) Stress-corrosion cracking in pipelines, *Pipeline Gas J.* 223, 42–49.

Leng, A, Streckel, H, Stratmann, M (1999) The delamination of polymeric coatings from steel: 1. Calibration of the Kelvin probe and basic delamination mechanism, *Corros. Sci.* 41, 547–578.

Li, MC, Cheng, YF (2008) Corrosion of the stressed pipe steel in carbonate–bicarbonate solution studied by scanning localized electrochemical impedance spectroscopy, *Electrochim. Acta* 53, 2831–2836.

Liu, ZY, Li, XG, Du, CW, Lu, L, Zhang, YR, Cheng, YF (2009a) Effect of inclusionson initiation of stress corrosion cracks in X70 pipeline steel in an acidic soil environment, *Corros. Sci.* 51, 895–900.

Liu, ZY, Li, XG, Du, CW, Cheng, YF (2009b) Local additional potential model for effect of strain rate on SCC of pipeline steel in an acidic soil solution, *Corros. Sci.* 51, 2863–2871.

Meng, GZ, Zhang, C, Cheng, YF (2008) Effects of corrosion product deposit on the subsequent cathodic and anodic reactions of X-70 steel in near-neutral pH solution, *Corros. Sci.* 50, 3116–3122.

Modiano, S, Carreño, JAV, Fugivara, CS, Torresi, RM, Vivier, V, Benedetti, AV, Mattos, OR (2008) Changes on iron electrode surface during hydrogen permeation in borate buffer solution, *Electrochim. Acta* 53, 3670–3679.

National Energy Board (1996a) *Stress Corrosion Cracking on Canadian Oil and Gas Pipelines, Report of the Inquiry*, MH-2-95, NEB, Calgary, Alberta, Canada.

National Energy Board (1996b) Notes from 12 January 1996 meeting between NEB SCC Inquiry Panel and Camrose Pipe Company Ltd., Exhibit A-58, NEB, Calgary, Alberta, Canada.

National Energy Board (2008) *Focus on Safety and Environment: A Comparative Analysis of Pipeline Performance, 2000–2008*, NEB, Calgary, Alberta, Canada.

Parkins, RN (1994) *Overview of Intergranular Stress Corrosion Cracking Research Activities*, AGA PRC Report PR-232-9401, PRCI, Falls Church, Virginia, p. 85.

Parkins, RN (1995) *Stress Corrosion Cracking of Pipelines in Contact with Near-Neutral pH Solutions*, AGA PRC Report 232-9501, PRCI, Falls Church, Virginia, p. 23.

Parkins, RN, Blanchard, WK, Delanty, BS (1994) Transgranular stress corrosion cracking of high pressure pipelines in contact with solutions of near-neutral pH, *Corrosion* 50, 394–408.

Revie, RW (2011) *Uhlig's Corrosion Handbook*, 3rd ed., Wiley, Hoboken, NJ. http://onlinelibrary.wiley.com/doi/10.1002/9780470872864.fmatter/summary.

TransCanada Pipelines (1996) *Response to National Energy Board Information Request 2 of Proceeding MH-2-5*, TCP, Calgary, Alberta, Canada, p. 3.

Transportation Safety Board of Canada (1999a) *Crude Oil Pipeline Rupture near Glenavon, Saskatchewan, 27 February 1996 Involving Interprovincial Pipe Line*, Report P96H0008, TSBC, Ottawa, Ontario, Canada.

Transportation Safety Board of Canada (1999b) *Pipeline Reports, 1996*, Report P96H0008, TSBC, Ottawa, Ontario, Canada.

U.S. Department of Transportation (1994) *Pipeline Safety Regulations*, CFR Parts 192 and 195, U.S. DOT, Washington, DC.

Van Boven, G, Chen, W, Rogge, R (2007) The role of residual stress in neutral pH stress corrosion cracking of pipeline steels: I. Pitting and cracking occurrence, *Acta Mater.* 55, 29–42.

Van der Sluys, WA, Wilmott, M, Krist, K (1998) Effect of sulfide on stress corrosion crack growth in gas transmission pipe lines, *First International Pipeline Conference*, ASME, New York.

Welch, C (2003) Natural-gas trunk line through state at risk of cracking, *The Seattle Times*, Dec. 20.

Wilmott, MJ, Jack, TR, Van Boven, G, Sutherby, RL (1996) Pipeline stress corrosion cracking: crack growth sensitivity studies under simulated field conditions, *Corrosion 1996*, Paper 242, NACE, Houston, TX.

Xu, LY, Cheng, YF (2012) Assessment of the complexity of stress/strain conditions of X100 steel pipeline and the effect on the steel corrosion and failure pressure prediction, *Proc. 9th International Pipeline Conference*, Paper IPC2012-90087, Calgary, Alberta, Canada.

Xue, HB, Cheng, YF (2011) Characterization of microstructure of X80 pipeline steel and its correlation with hydrogen-induced cracking, *Corros. Sci.* 53, 1201–1208.

Zhang, C, Cheng, YF (2010) Corrosion of welded X100 pipeline steel in a near-neutral pH solution, *J. Mater. Eng. Perf.* 19, 834–840.

Zhang, GA, Cheng, YF (2009) Micro-electrochemical characterization of corrosion of welded X70 pipeline steel in near-neutral pH solution, *Corros. Sci.* 51, 1714–1724.

4

Nearly Neutral–pH Stress Corrosion Cracking of Pipelines

4.1 INTRODUCTION

Nearly neutral–pH SCC of pipelines was not documented until the mid-1980s, and was first identified in Canadian pipelines. This type of SCC is so named because the pH of the trapped electrolyte at the crack location under disbonded coating is between 5.5 and 7.5. Although pipeline failures that occurred in 1985 and 1986 were considered to be the first evidence of nearly neutral–pH SCC in Canada, it was subsequently determined that such SCC has actually been detected on other pipelines in the 1970s [National Energy Board, 1996]. Furthermore, the occurrence of SCC on pipelines in the United States, like other countries in the world, is primarily of the high-pH type; however, some recent failures have been attributed to nearly neutral–pH SCC [National Energy Board, 1996].

4.2 PRIMARY CHARACTERISTICS

The cracking environment associated with nearly neutral–pH SCC of pipelines features anaerobic, diluted groundwater containing primarily bicarbonate ions and dissolved CO_2 typically from the decay of organic matter and geochemical reactions in soil. This type of solution chemistry is formed since CP is inaccessible to the pipe

Stress Corrosion Cracking of Pipelines, First Edition. Y. Frank Cheng.

surface over a sufficiently long period, due either to the shielding disbondment of the coating or to a highly resistive soil. Thus, the pipe steel at the SCC sites is at a free corrosion potential of about −760 and −790 mV(Cu/$CuSO_4$). Furthermore, although a distinct relationship between the nearly neutral–pH SCC occurrence and temperature has been lacking, cracking tends to occur in a cold climate where the CO_2 concentration in groundwater is high.

Furthermore, up to 1996, on NEB-regulated pipelines, 65% of nearly neutral–pH SCC occurrs between the compressor station and the first downstream block valve, with 12, 5, and 18% occurring between the first and second valves, between the second and third values, and downstream of the third valves, respectively [National Energy Board, 1996]. Moreover, nearly neutral–pH SCC is associated with specific terrain conditions, such as alternating wet–dry soils and soils that tend to disbond or damage coatings.

Nearly neutral–pH stress corrosion cracks are primarily transgranular (i.e., the cracks grow across the grains), as shown in Fig. 3-3. The cracks are usually wider than those formed under high-pH conditions. Moreover, the crack sides are covered with the remains of corrosion products.

Generally, colonies of multiple parallel cracks are observed perpendicular to the direction of the highest stress on the external pipe surface. These cracks can vary in depth and length and grow in two directions: along the pipe wall thickness direction and along the axial direction on the pipe surface. Approximately two-thirds of pipeline incidents due to nearly neutral–pH SCC are axial cracking, indicating that the most important source to initiate cracking is the internal pressure of pipelines. The remainders of nearly neutral–pH SCC are oriented transverse to the pipeline's axis, due to the presence of secondary stresses which are generated from a dent, local bending, and/or soil movement–induced axial loading. Cracks increase in depth and length and tend to coalesce, or link together, to form longer cracks. At some point these cracks may reach a critical depth and length combination that can result in a rupture. A leak will occur if a crack grows through the pipe wall before it reaches a critical length for rupture. Furthermore, a very small number of the cracks in some colonies continue to grow, although the majority of cracks and colonies become dormant at a depth of less than 1 mm [Bouaeshi et al., 2007].

Furthermore, the nearly neutral–pH stress corrosion cracks could be either linked or independent cracks [Chen et al., 2002]. The former are distinguished by multiple peaks in the depth direction, considered to be a result of crack coalescence, while the latter have a single peak and a smooth depth profile. The linked cracks are usually less than 500 μm in depth, with little increase in depth as the crack length extends beyond about 2 mm. In contrast, the independent cracks exhibit a linear relationship between crack length and crack depth. The linked cracks, although dormant in the depth direction, appear to be actively corroding in the width direction near the crack root and possibly at the overlapped region between two joined cracks. For the independent cracks, corrosion in the width direction speeds up as the crack deepens, implying that the growing cracks are evolving toward a pit.

4.3 CONTRIBUTING FACTORS

4.3.1 Coatings

Coating properties and performance have been acknowledged as one of the most important factors in pipeline SCC, including nearly neutral–pH SCC. If the coating is intact and does not disbond from the steel substrate, corrosion and SCC will not occur since the corrosive electrolyte is not accessible to the steel. Coatings could fail in the field by means of a number of mechanisms, including disbondment, formation of holidays, "missing" coating, and general degradation [Jack et al., 2000]. In particular, "shielding" disbondment is formed for coatings that are impermeable to CP, and thus the pipe steel beneath the coating is shielded from CP current. This is an essential condition in developing the solution chemistry and electrochemistry that initiates nearly neutral–pH SCC.

The occurrence of nearly neutral–pH SCC of pipelines is usually associated with PE tape coating [National Energy Board, 1996]. It is further categorized that 65% of axial nearly neutral–pH stress corrosion cracks occur under disbonded PE, and 100% of transverse cracks are associated with PE tape. PE tapes were used predominantly in the early 1960s to 1980s, and have poor resistance to disbondment. Generally, the tapes are wrapped spirally around the pipe with an overlap at the helix line. Disbondment usually occurs between the pipe surface and the tape along the ridge created by longitudinal, spiral, and girth welds, and at the overlap between the helix of the wrap [National Energy Board, 1996].

When tape disbonds from a pipeline, moisture can accumulate beneath the tape surface. Since the tape has quite high electrical insulation properties, it thus prevents CP current from reaching water trapped between the tape and the pipe surface. The pipe steel is near its corrosion potential in the electrolyte. This has been proved by long-term soil-box tests, where there is minimal CP penetration into the disbondment [Been et al., 2005]. Furthermore, CO_2 is generated due to the decay of vegetable and organic matter in the soil and dissolved in the trapped water, generating a diluted bicarbonate solution. Nearly neutral–pH solution chemistry is generated in the absence of CP and the dissolution of CO_2 in the water. SCC might still occur when the soil is so resistive that the CP current cannot reach the pipeline.

Asphalt and coal tar coatings may also disbond, especially due to poor surface preparation. However, since these coatings tend to become saturated with moisture, or, if brittle, may break into pieces, CP current is able to reach the pipe steel surface in the disbonded area.

It is generally accepted that fusion-bonded epoxy (FBE) coatings are an effective protection against SCC. FBE has excellent adhesion to steel substrate if the steel surface is properly treated. Moreover, FBE is permeable to CP. Figure 4-1 shows the electrochemical characterization of FBE-separated testing cells, as shown in Fig. 4-2. The corrosion potential of an X70 steel working electrode (WE) that is separated from a saturated calomel electrode (SCE) reference electrode (RE) by an FBE membrane in two cells drops dramatically with time, even approaching the free corrosion potential of a bare steel in a soil solution [i.e., −0.87 V(SCE)],

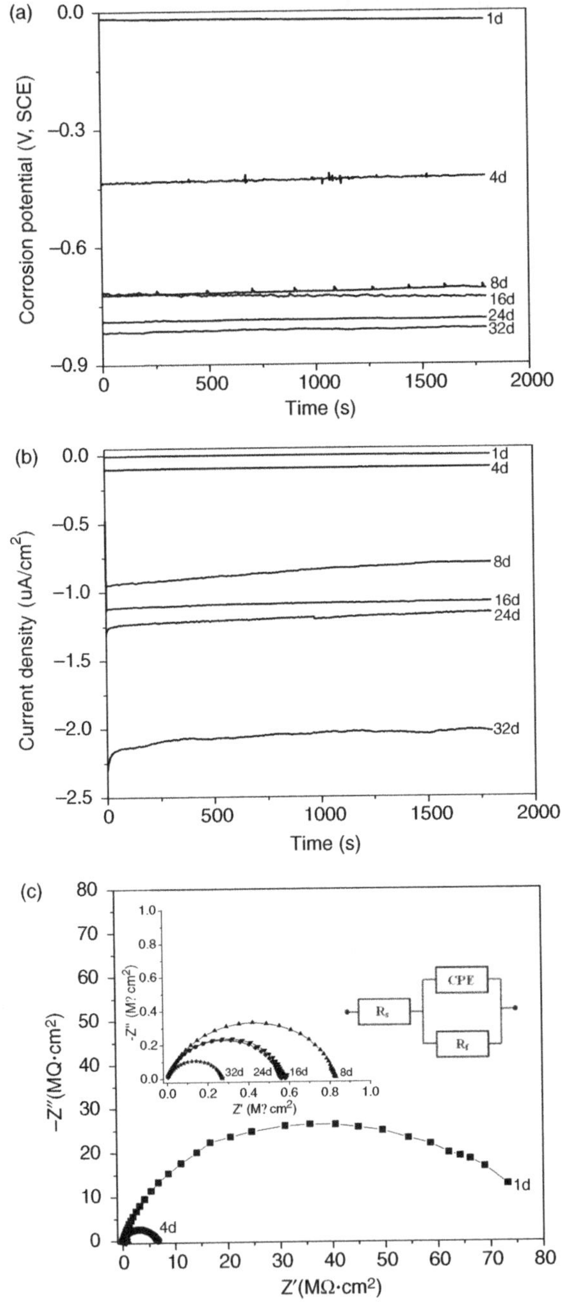

Figure 4-1 *Time dependence of electrochemical characterization measured on an X70 steel that is separated from RE and CE by an FBE membrane in two cells: (a) corrosion potential, (b) potentiostatic polarization current density at −1.5 V(SCE), and (c) Nyquist diagrams. (From Fu and Cheng [2011].)*

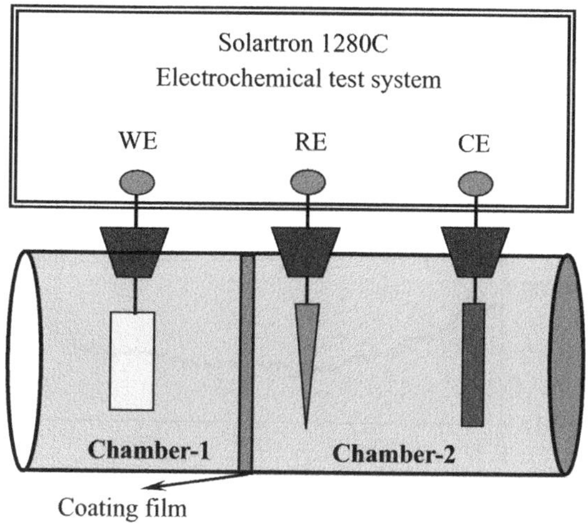

Figure 4-2 *Schematic diagram of the experimental setup to study the permeability of FBE coating to CP, where WE (working electrode) is an X70 pipeline steel, RE (reference electrode) is an SCE, and CE (counter electrode) is a platinum plate. The cells contain an identical soil solution. (From Fu and Cheng [2011].)*

after 32 days of immersion, as shown in Fig. 4-1a. The cathodic current densities obtained at a potentiostatic polarization of −1.5 V(SCE) shifts negatively with time (Fig. 4-1b). The Nyquist diagrams show regular semicircles, the size decreasing with time, indicating increasing hydrogen evolution at the cathodic polarization applied. Apparently, CP could permeate through FBE, as indicated by the electrochemical responses measured.

Furthermore, newly developed multiple-layer coatings, such as high-performance composite coating (HPCC) is not permeable to CP, as demonstrated in Fig. 4-3, where the corrosion potential, potentiostatic polarization current, and Nyquist diagrams are measured on an X70 steel electrode that is separated from RE and CE by the HPCC membrane in two cells. It is seen that although the corrosion potential of the steel decreased with time (Fig. 4-3a), it never reached the corrosion potential of bare steel [i.e., −0.87 V(SCE)], in the soil solution. Even at a potentiostatic polarization of −1.5 V(SCE), the anodic current density is obtained, and there is little effect of time on the current density measured (Fig. 4-3b). The Nyquist diagrams measured at various times are highlighted by the large semicircle. Due to the random data obtained in the low-frequency range, the semicircles are not complete. Moreover, there is little change in semicircle size with time, and the impedance behavior of HPCC shows a purely capacitive behavior. According to coating performance evaluation criteria developed previously [King et al., 2002], HPCC, behaving as an ideal capacitor, is capable of providing full protection of the steel. Apparently, HPCC is impermeable to the CP applied, due to its extremely high impedance.

In addition to the intrinsic property of coatings, the coating application practice can have a profound effect on the performance and disbondment of a coating in the

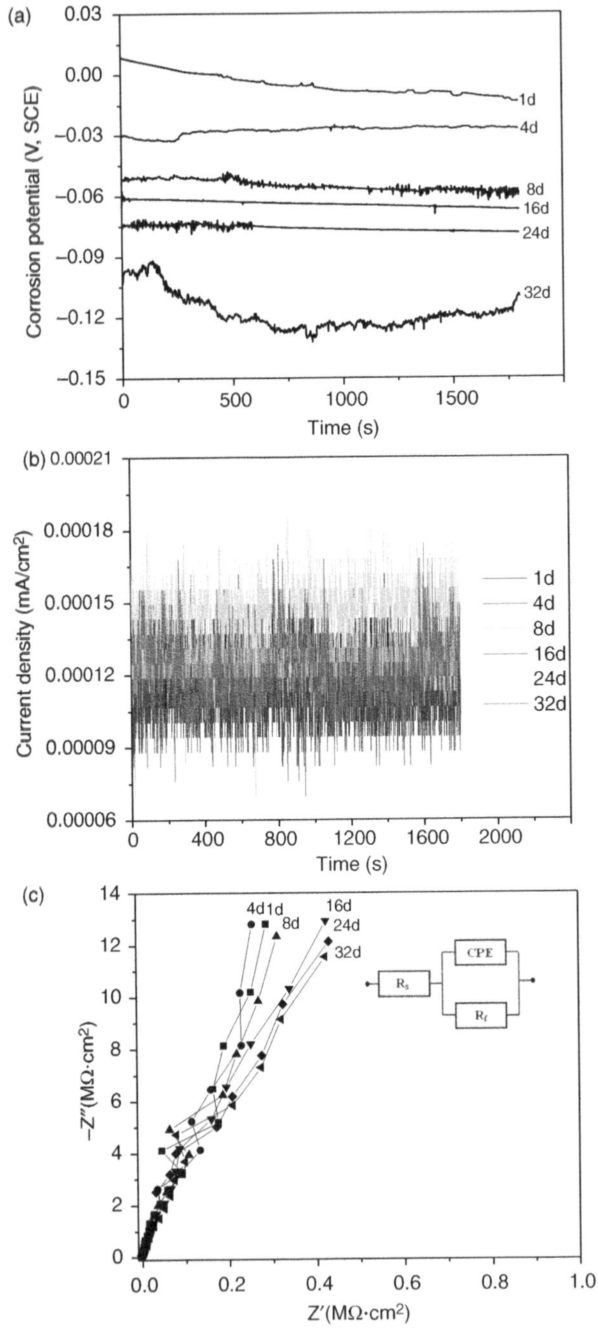

Figure 4-3 *Electrochemical characterization of an X70 steel electrode that is separated from RE and CE by a HPCC membrane as a function of time: (a) corrosion potential, (b) potentiostatic polarization current density at −1.5 V(SCE), and (c) Nyquist diagrams. (From Fu and Cheng [2011].)*

field. For example, poor surface preparation of line pipe steel would enhance the possibility of coating disbondment. Moreover, pipeline construction could damage the coating and generate holidays on the coating.

4.3.2 Cathodic Protection

Nearly neutral–pH SCC develops where CP current cannot penetrate the coating to reach the steel pipe. It has been established that there is limited permeation of CP through the coating disbondment from the open holiday. For example, for a disbonded coating membrane with a disbondment thickness of 0.9 mm, the CP potential applied cannot reach farther than a few centimeters from the coating holiday (3.5 cm × 1 cm) because of the shielding effects of the disbondment, as shown in Fig. 4-4. Although an increase in the CP current or potential applied could cause CP to permeate further into the disbonded area, this is impractical since the disbondment can be 1 m or more long. Moreover, a sufficiently negative CP potential or current density would enhance the hydrogen evolution, resulting in hydrogen-induced SCC of pipelines. As a result, a high CP level will not be effective to help prevent SCC. Furthermore, excessive CP current would increase the susceptibility of coating to cathodic disbondment.

The CP applied on a coating holiday is also partially shielded to reach the base of the holiday, due to the geometrical restraint, such as one with a diameter as small as 0.2 mm [Dong et al., 2008]. Figure 4-5 shows Nyquist diagrams measured by LEIS with a microprobe positioned at the 0.2-mm defect on an HPCC-coated X70 steel specimen under various cathodic potentials in a carbonate–bicarbonate solution. For a comparison, the EIS diagrams were also measured on bare X70 steel which was under cathodic potentials identical to those in Fig. 4-5, and the results are shown in Fig. 4-6.

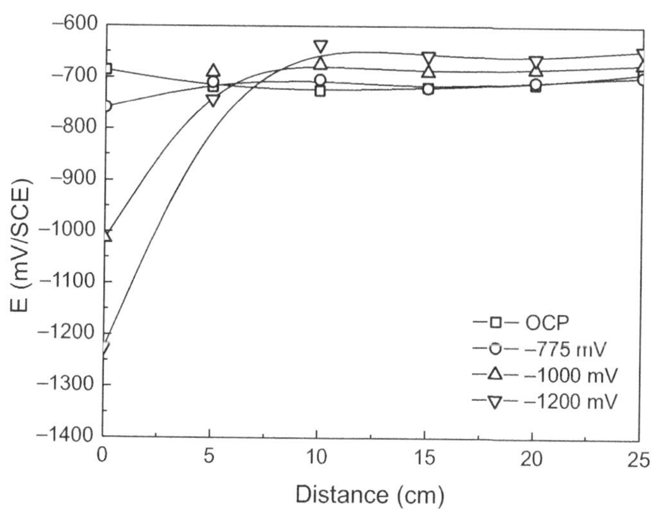

Figure 4-4 Potential profile within a coating disbondment after 0.5 h of test: application of various cathodic potentials at the coating holiday located at 0 cm. (From Chen et al. [2009].)

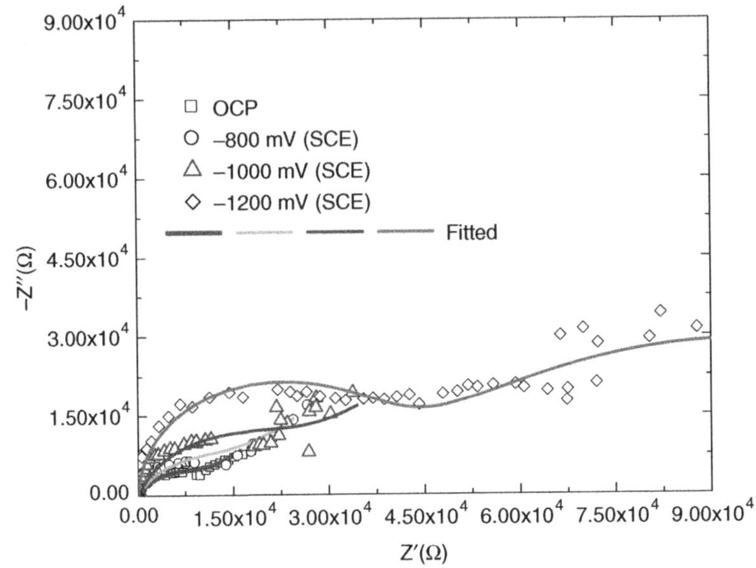

Figure 4-5 *Nyquist diagrams measured by LEIS with a microprobe positioned at the defect (0.2 mm in diameter) on HPCC-coated X65 steel under various cathodic potentials. (From Dong et al. [2008].)*

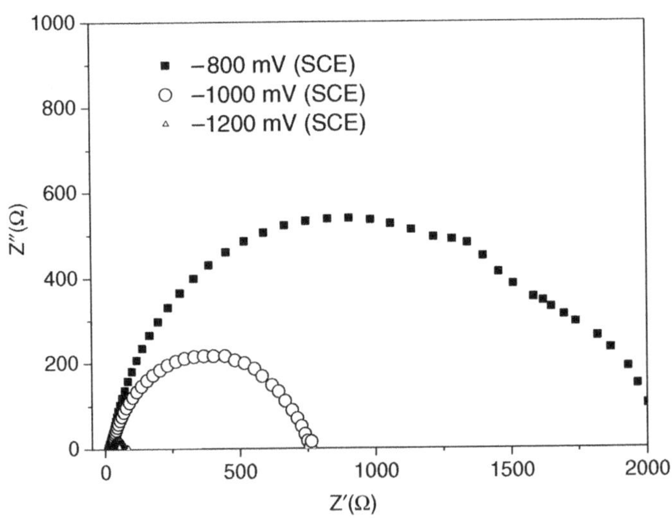

Figure 4-6 *Nyquist diagrams measured on a bare X70 steel specimen under various cathodic potentials in a carbonate–bicarbonate solution. (From Dong et al. [2008].)*

It is apparent that the presence of the defect (0.2 mm in diameter) would affect the permeation of the cathodic potential to the steel at the base of the defect. In particular, in addition to the high-frequency semicircle that is attributed to the interfacial charge-transfer reaction occurring at the defect base, a low-frequency straight line with an approximate 45° slope is also observed in Fig. 4-5. This indicates the mass-transfer process of a reactant, such as oxygen through a narrow, deep defect (a depth/width ratio of 5.5). Even at a very negative potential [e.g., −1200 mV(SCE)], the impedance measured is still associated with the diffusion-controlled charge-transfer reaction at the base of the defect rather than with the pure charge-transfer reaction shown in Fig. 4-6. This is attributed to the fact that the CP applied is partially shielded by the defect, with a narrow, deep geometry.

Preliminary studies by the NOVA Research and Technology Center (NRTC) on pulsed CP indicated that pulsing can penetrate a disbonded area more deeply than can conventional CP systems [Van Boven and Wilmott, 1995]. Depending on the size of the disbonded area, the pulsed CP technique may help to control nearly neutral–pH SCC.

4.3.3 Soil Characteristics

SCC occurs in a wide variety of soils, covering a range of color, texture, and pH. No single characteristic was found to be common to all the soil samples [Wenk, 1974]. A summary from more than 450 investigative excavations performed on TCPL systems in the mid- to late 1980s showed that in tape-coated portions of pipelines, nearly neutral–pH SCC was found in many types of terrains and soils (e.g., muskeg, clay, silt, sand, and bedrock). There was no apparent difference in the soil chemistry for SCC and non-SCC sites [Delanty and O'Beirne, 1991, 1992]. Moreover, nearly neutral–pH SCC was also found in asphalt-coated pipeline portions, of which over 83% were found in extremely dry terrains consisting of either sandy soils or a mixture of sand and bedrock. There was inadequate CP in these locations.

Based on the SCC severity rating model for nearly neutral–pH SCC for tape-coated TCPL pipelines [Delanty and Marr, 1992], the most aggressive soil types were lacustrine soil (formed by deposits in lakes), followed by organics over glaciofluvial soil (formed by deposits in streams fed by melting glaciers), and organics over lacustrine soil. The prevalence of SCC in glaciofluvial soils was about 13% of that in lacustrine soils and about 17% of that in soils with organics over glaciofluvial or lacustrine. Very poorly or poorly drained soils were found to be the most aggressive, whereas level-depressed soil was found to be the most aggressive topography.

Furthermore, nearly neutral–pH SCC may be associated with local topographical regions: for example, at the bottom of hills or streams, where the groundwater is either parallel to an pipeline or across it. Extensive water in the environment would be favorable to supply CO_2 to an electrolyte trapped in a disbonded area so that the nearly neutral–pH environment is maintained. To date, the majority of laboratory tests have been performed in NS4 solution, a diluted bicarbonate solution purged with 5% CO_2 balanced with N_2 gas to simulate the trapped electrolyte beneath disbonded PE coatings at locations where nearly neutral–pH SCC was found [National Energy

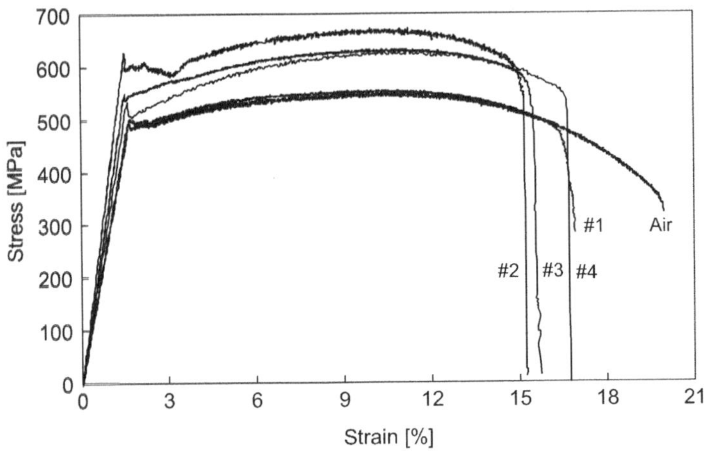

Figure 4-7 Stress–strain curves of X65 pipeline steel specimens in air and in various soil solutions. (From Cheng [2007a].)

Board, 1996]. The principal component of the electrolyte is sodium bicarbonate [Fessler et al., 1973].

As summarized, it seems that the soil chemistry is not related directly to the occurrence of pipeline SCC. Instead, an electrolyte trapped under disbonded coating results in pipeline SCC. However, tests have shown that the high susceptibility of pipeline steels to SCC in soil-extracted solutions is always associated with a high hydrogen permeation current in the solution [Cheng, 2007a]. This indicates clearly that the soil chemistry would contribute to the occurrence of pipeline SCC and even change its mechanism either by hydrogen evolution and permeation processes or by other potential processes, such as the microorganism culture.

Furthermore, the SCC behavior of pipeline steels has been investigated in several soil-extracted solutions that have compositions different from that of NS4 solution, but they all have nearly neutral pH with values of 6.6 to 7.0. Figure 4-7 shows the stress–strain curves measured on an X70 pipeline steel specimen in various soil solutions and in air. It is seen that the strengths of the steel are higher in solutions than in air, accompanying the reduction in elongation. The SCC susceptibility that is evaluated by the fracture strain shows that the susceptibility of the steel to SCC in soil solutions is ranked as $2 > 3 > 4 > 1$. Moreover, the electrochemical hydrogen permeation current through the X70 steel specimen (1 mm in thickness) is measured in various soil solutions, and the subsurface hydrogen concentration measured in each soil solution is shown in Fig. 4-8. The subsurface hydrogen concentration ranges from 0.2 to 0.5 mol H/m^3. The No. 2 soil solution generates the highest hydrogen concentration, while solution 1 has the lowest subsurface hydrogen concentration. The ranking of the soil solutions to generate the subsurface hydrogen concentration is identical to the SCC susceptibility. It has actually been suggested [Been et al., 2005] that there are differences in the nature of the trapped water environment for different soils, which would be expected to result in different SCC behavior.

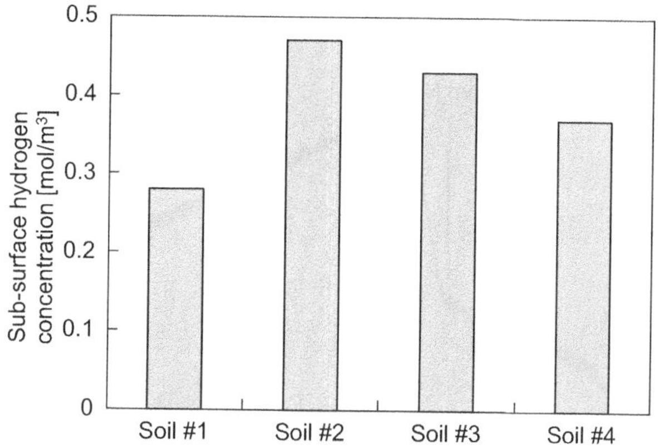

Figure 4-8 Subsurface hydrogen concentration measured in various soil solutions. (From Cheng [2007a].)

Soil chemistry has a profound effect on crack growth rate, especially in the regime characterized by a low stress intensity factor [Chen and Sutherby, 2004]. In a regime of high stress intensity the crack growth rate is insensitive to the soil chemistry. However, in a low-ΔK regime, the growth rate is shown to have a strong dependence on the testing environments. Generally, soil solutions with a low general corrosion rate are associated with a blunt crack tip and a wide crack crevice, which would result in lower stress intensity at the crack tip and a weaker crack closure effect.

4.3.4 Microorganisms

It has been proposed that there exists the possibility of the involvement of a great variety of microorganisms in the initiation and propagation of stress corrosion cracks in pipe steel [International Science and Technology Center, 1992]. Moreover, the participation of some microorganisms in pipeline SCC follows the mechanism of hydrogen embrittlement [Parkins, 2000]. In particular, an anaerobic soil is believed to be necessary to create a nearly neutral–pH SCC environment [Canadian Energy Pipeline Association, 1996a; National Energy Board, 1996]. Under anaerobic soil conditions, SRB may form in soil electrolyte, including clays, loams, and sandy loams. SRB is capable of reducing sulfate in the soil to sulfide. When a corrosion reaction occurs in a line pipe steel, hydrogen atoms are generated. Since the sulfide could serve as a "poison" that prevents atomic hydrogen from forming molecular hydrogen, hydrogen atoms would penetrate the pipe steel. The permeated hydrogen could diffuse toward the stress concentration zones, such as the tip of a crack, and embrittle the steel locally, making it easier for a crack to propagate [Hirth, 1980].

There has been a lot of important evidence to support the involvement of SRB in pipeline SCC, including the crack colonies located on the steel only in areas with sulfur deposits, steel decarbonization along the crack face due to carbon biochemical transformations, and specific sediments exhibiting calcium abundance as

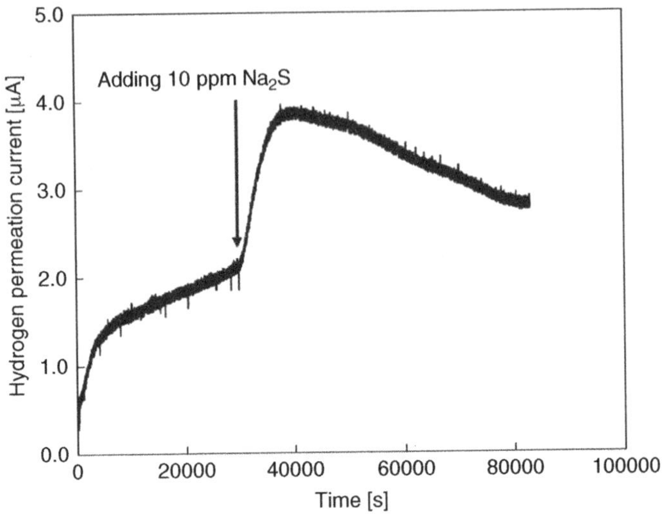

Figure 4-9 *Hydrogen permeation current curve measured on X65 steel in a 10 mM $NaHCO_3$ solution with the addition of 10 ppm Na_2S. (From Cheng [2007b].)*

an organogenic formation of calcium carbonate or as a product of microorganism activity and their remains [Ott, 1998]. Moreover, all cases of SCC of buried pipelines investigated showed that soil microorganism activity, including SRB, heterotrophic, and acid-forming bacteria colonies, was 10 times higher than that of soil in an average state [National Energy Board, 1996].

There have been investigations using sodium sulfide to simulate the presence of SRB in soils to measure the electrochemical hydrogen permeation current through an X65 pipeline steel membrane (1 mm in thickness) [Cheng, 2007b]. A hydrogen permeation current curve measured in 0.01 M $NaHCO_3$ solution with the addition of 10 ppm Na_2S is shown in Fig. 4-9. Apparently, adding Na_2S increases the hydrogen permeation current remarkably. However, no steady-state value of hydrogen current is achieved. After a maximum value, the hydrogen permeation current decreases with time. Numerical analysis of the hydrogen permeation current curve concludes that there is little difference in hydrogen diffusivity before and after Na_2S addition in 0.01 M $NaHCO_3$ solution, which are 9.49×10^{-7} and 8.13×10^{-7} cm^2/s, respectively. However, the subsurface hydrogen concentration increases from 0.20 mol H/m^3 to 0.54 mol H/m^3 upon the addition of sulfide ions [Cheng, 2007b]. Both sulfide ions and SRB have been recognized as promoters for hydrogen entry into steel [Iyer et al., 1990]. Since sulfide ions enhance the subsurface hydrogen concentration but do not affect the hydrogen diffusivity, it is reasonable to assume that sulfide ions promote hydrogen permeation primarily by inhibiting the hydrogen recombination and increasing the hydrogen entry. The maximum-value phenomenon in the hydrogen permeation current curve measured in sulfide-containing solution is attributed to the activation of hydrogen trapping sites and to the generation of microcracks on the steel surface and inside the specimen with the rapid rise in hydrogen concentration to exceed a critical value. When the surface microcracks fissure, the number of hydrogen diffusion paths is reduced and the hydrogen permeation current drops, resulting in a hump in the permeation current curve. Generally, when the charging solution contains "poisons" for hydrogen permeation, such as sulfide, the hydrogen concentration inside the steel would probably reach and exceed a critical value, and consequently, the maximum value of hydrogen permeation current is achieved.

4.3.5 Temperature

The effect of temperature on nearly neutral–pH SCC has not been well established. Moreover, field experiences suggest that temperature probably is not a significant factor for that form of SCC [Baker, 2005]. Compared to high-pH SCC, the role of temperature is not a critical factor affecting the nearly neutral–pH SCC of pipelines. Statistically, 50% of nearly neutral–pH SCC failures on TCPL line 2 occurred within 10 miles downstream of compressor stations versus 90% for high-pH SCC in the same range of distance from a station [Delanty and O'Beirne, 1991]. Moreover, it has been demonstrated in the lab that there is no dependence of cracking on temperature in the range 5 to 45°C under nearly neutral–pH conditions, which is in conformity with field observations [Parkins et al., 1994].

4.3.6 Stress

Buried pipelines usually experience quite complex stress conditions. In addition to the hoop stress generated due to internal pressure, longitudinal stress would be exerted on the pipe due to ground movement [Xu and Cheng, 2012a]. Moreover, various residual stresses, including welding stress and bending stress, would not be eliminated completely and contribute to the cracking process. For example, it was found [Van Boven et al., 2007] that for an X65 pipeline steel containing various levels of tensile and comprehensive residual stresses in a neutral-pH soil-extracted electrolyte, micropitting occurs preferentially in areas where the tensile residual stresses were the highest, while SCC initiation occurred with a 71% normalized frequency in areas where the surface residual stress was below the maximum residual stress. The difference between residual stress levels occurring at SCC locations versus pitting locations resulted from the change in residual stress when cyclic stress is applied during SCC testing. Furthermore, an X65 steel sample in nearly neutral–pH environments in the absence of stress did not undergo SCC [Asher and Singh, 2009]. Furthermore, smooth steel samples held at a stress of 85% of the yield strength of the steel in the environment did not show any higher susceptibility to cracking. However, when stress concentration raisers such as surface scratches are contained, the steel exhibits quasi-cleavage fracture when samples are held at 80% yield stress. Thus, a stress concentration is required for hydrogen to accumulate (hydrogen is believed to participate the nearly neutral–pH SCC of pipelines) and cause quasi-cleavage fracture on steel samples in nearly neutral–pH environments.

The *threshold stress* is the stress level below which there is no crack to initiate. Research on the crack initiation of steels is primarily to find the threshold stress level below which cracking will not occur. Unfortunately, the threshold stress for SCC is difficult to define. To date, a threshold stress for nearly neutral–pH SCC in pipelines has not been established. The attempt to establish a threshold stress value for pipeline SCC by laboratory testing is not reliable because pipelines in the field encounter complex, multiaxial stresses resulting from various sources, whereas laboratory testing is generally uniaxial tension applied by a test machine [Beavers, 1993]. Despite this fact, some laboratory data have been useful in inferring threshold stress levels for nearly neutral–pH SCC initiation [Canadian Energy Pipeline Association, 1996b]. For example, tests on a sample of grade 448-MPa pipeline steel could grow cracks at a stress level of 69% SMYS but not at lower stress levels. It is also pointed out that the data are limited and need to be replicated to have validity. More important, the attempt to apply a single threshold stress value over the entire length of a pipeline is unwise.

Hoop stress resulting from internal pressure can be controlled appropriately. However, residual stresses from manufacturing and welding, surface loading from heavily loaded trucks, longitudinal stress from ground movement, and stress concentration at dents or gouges or corrosion pits can enhance the stress level exerted on pipelines remarkably, but they are beyond the control of pipeline operators. However, for deformation-induced stress or other residual stresses that are elastic in nature, there is no effect on electrochemical corrosion of steel [Xu and Cheng, 2012b]. The critical failure strain of corrosion scale formed on a steel surface is calculated to be 0.00357 to 0.00417, which is higher than the maximum strain of 0.0029 under an applied stress of 582 MPa on the specimen, as shown in Fig. 4-10. Thus, the corrosion scale formed does not fracture during elastic loading. There is no difference in the steady-state corrosion potential detected upon application of an elastic tensile or compressive

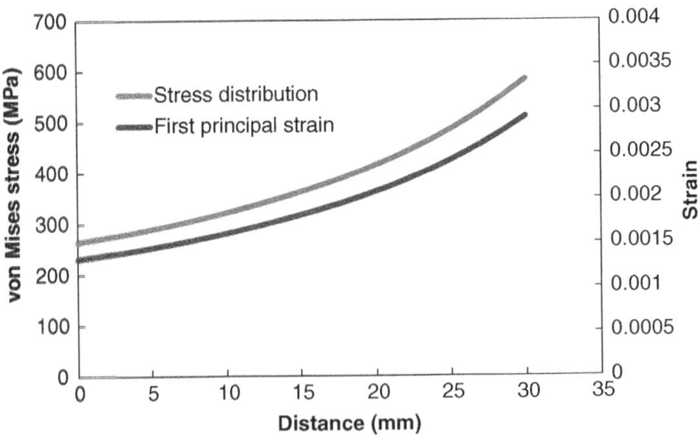

Figure 4-10 Von Mises stress and strain distribution along an X100 steel specimen that is under a load of 200 N. (From Xu and Cheng [2012b].)

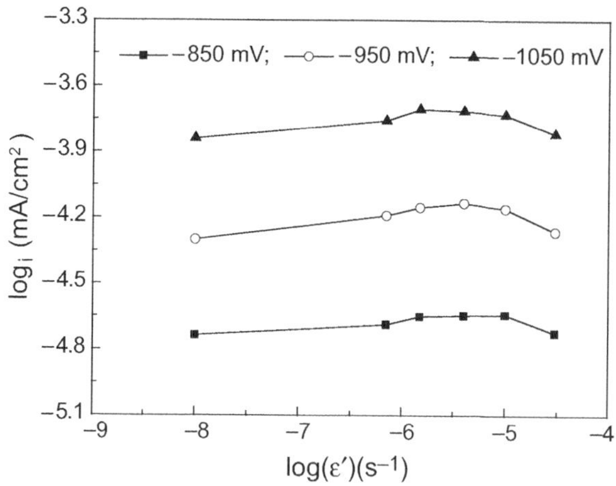

Figure 4-11 *Cathodic current density as a function of strain rate measured on an X70 pipeline steel specimen in a nearly neutral–pH solution. (From Liu et al. [2009].)*

stress. If the residual stress results in plastic strain of the pipe steel, the effect on steel corrosion would be significant [Xu and Cheng, 2012c].

Since hydrogen is involved in nearly neutral–pH SCC, any factors affecting hydrogen permeation and absorption would affect the SCC process. Tests showed that an applied stress that is equivalent to the gross hoop stress in a pipe wall under an operating internal pressure of 70 bars significantly accelerates hydrogen absorption to X52, X70, and X100 steels [Capelle et al., 2010]. The difference between subsurface hydrogen concentrations in unloaded and stressed steel can exceed several-fold. Thus, stressed pipelines would be able to self-accelerate hydrogen permeation, contributing to the occurrence of cracking.

Furthermore, the role of strain rate in the susceptibility of steels to SCC, including nearly neutral–pH SCC, should not be ignored. In addition to the anodic process, the strain rate affects the cathodic reaction kinetics and thus the SCC mechanism. Figure 4-11 shows the effect of the strain rate on the cathodic current densities measured on X70 pipeline steel at an individual potential in NS4 solution [Liu et al., 2009]. It is seen that the cathodic current density increases with strain rate. After reaching a maximum at a strain rate of approximately 5×10^{-6} s^{-1}, the cathodic current density decreases. According to the LAP model in Chapter 3, with an increase in strain rate, the number of dislocation emergence points on a steel surface increases, resulting in a negative shift of LAP. As a result, the cathodic reaction is favorable and the cathodic current density increases. When the strain rate is sufficiently high and up to 5×10^{-6} s^{-1}, a maximum of cathodic current density is observed. The mobility of the dislocation emergence points is so fast that the reactive species do not have a chance to adsorb at these active sites for reductive reaction. Thus, the cathodic current density decreases. Since the cathodic reaction occurring in deoxygenated, nearly neutral–pH solutions is associated with hydrogen evolution, the effect of the strain rate on cathodic reaction kinetics would be expected to alter the hydrogen-involved SCC susceptibility of the steel. Furthermore, the dependence of the cathodic current density on the strain rate becomes less significant when the potential is shifted more negatively, such as to −1050 mV(SCE), where the cathodic reactions are thermodynamically feasible. The strain-induced LAP effect becomes less important than the cathodic potential applied on the steel.

4.3.7 Steel Metallurgy

The chemical composition and microstructure of steels are dependent on the metallurgical processing. However, there is a small effect of the microstructure and microchemistry of pipeline steels on SCC occurrence. SCC has been found on the conventional low grades of steels that have been used for over 50 years, including those from X25 to X65 steels. Furthermore, the vast majority of line pipes for gas transmission are produced by seam-welded processes and contain varying amounts of ferrite, pearlite, and bainite with wide variations in the crystalline grain size. However, there is no strong evidence that any of the items above either promote or inhibit SCC [Baker, 2005].

Since hydrogen is actively involved in nearly neutral–pH SCC processes, the metallurgical features of steels affecting hydrogen absorption and permeation would affect the SCC behavior. In a nearly neutral–pH bicarbonate solution, the resistance to hydrogen absorption decreases with decreasing steel strength (yield stress), due to the different microstructures in the steels [Capelle et al., 2010]. For example, the resistance to hydrogen trapping efficiency is significantly lower for a ferritic, microstructure, which is typical for low-grade pipeline steels such as X52 and X65 than for a ferritic–bainitic microstructure, typical in X70 and X80 steels. Moreover, hydrogen diffusion is maximum in steel specimens of highest relaxation, such as normalizing. Hydrogen diffuses preferentially through ferrite grains [Asher and Singh, 2008]. Furthermore, steel that has ferrite grains alternating with a pearlite microstructure shows a higher susceptibility to SCC than that of the water-spray structure of incomplete transformation of pearlite in a recrystallized ferrite matrix with fewer precipitates. Quenched steel containing a typical structure of martensite is most susceptible to SCC in diluted bicarbonate solutions [Torres-Islas et al., 2008].

Recently, the strength and toughness of pipeline steels has been increased by microalloying treatment, by controlled rolling and cooling processes, or both. The fine crystalline grain and hard microphases, including acicular ferrite and ferritic bainite are generated in the steels. There have been a number of studies investigating the correlation between steel composition and microstructure and hydrogen permeability, and thus susceptibility to SCC. For X80 steel, quantitative determination of the hydrogen permeating rate, diffusivity that represents the apparent lattice diffusivity of the dissolved and reversibly trapped hydrogen, and the apparent hydrogen solubility corresponding to the hydrogen in the lattice and in reversible traps is 5.2×10^{-10} mol H/m · s, 2.0×10^{-11} m^2/s, and 26 mol H/m^3, respectively [Xue and Cheng, 2011]. Moreover, the number of hydrogen-trapping sites per unit volume in the steel is calculated to be 3.3×10^{27} m^{-3}, indicating that significant amounts of metallurgical irregularities, such as dislocations, grain boundaries, inclusions, and precipitates, are contained in the steel, serving as hydrogen traps and affecting the hydrogen diffusivity. In particular, irreversible hydrogen traps such as nonmetallic inclusions would effectively trap hydrogen and initiate stress corrosion cracks locally. Chapter 9 includes a detailed analysis of the correlation of high-strength steel metallurgy with hydrogen permeation and HIC of the steel.

4.4 INITIATION OF STRESS CORROSION CRACKS FROM CORROSION PITS

Typically, SCC colonies are initiated at locations on the outside surface where there is already pitting or general corrosion [Parkins et al., 1994]. In a typical nearly neutral–pH condition, pitting occurs on pipeline steels due primarily to the galvanic effect at metallurgical discontinuities such as grain boundaries [Chu et al., 2004; Liu et al., 2010], inclusions [Elboujdaini et al., 2000; Jin and Cheng, 2011], or phase interfaces [Kushida et al., 2001].

Corrosion pits are formed on pipeline steels when they experience fluctuations of cathodic potential, even in the absence of external stress. Figure 4-12 shows the LEIS maps measured on X70 steel electrode that is under various times of square-wave polarization (SWP) from −950 to −800 mV(SCE) in NS4 solution [Liu et al., 2012]. It is seen that the local impedance distribution over the scanning area of the electrode depends on the SWP time. After 48 h of SWP (Fig. 4-12a), the impedance distribution is uniform, with some sites, such as points A, B, and C, having lower impedance than the adjacent area. However, after 96 h of SWP (Fig. 4-12b), the impedance fluctuation becomes more apparent, and the marked points A, B, C, and D show drops in the local impedance. After 144 h of SWP (Fig. 4-12c), the LEIS map indicates further decrease of impedance at points A, B, C, and D. These sites represent active locations on the steel surface on SWP that simulate CP fluctuations in the field. The surface morphology of the steel electrode after 144 h of SWP between −950 and −850 mV(SCE) shows apparently that corrosion pits are initiated on the electrode surface, as seen in Fig. 4-13. Moreover, most pits are observed along the grounding lines. As a comparison, the surface images of the steel electrode potentiostatically at −950 and −850 mV(SCE) are shown in Fig. 4-14, where no corrosion or pitting corrosion is observed since the steel is under full cathodic protection.

Apparently, upon cathodic potential fluctuations, corrosion pits can be initiated on pipelines and become potential sites of crack incubation. An electrochemical state conversion model (ESCM) has been developed to illustrate mechanistically the occurrence of pitting corrosion on cathodically polarized pipelines in nearly neutral–pH solutions [Liu et al., 2011]. Various microdefects existing on the steel surface would result in cathodic local additional potentials (LAPs), enhancing the cathodic reactions locally [Liu et al., 2009]. The effect of LAP on the local current density is defined as

$$i_D = \frac{\Delta E}{\beta_c} + i_N \tag{4-1}$$

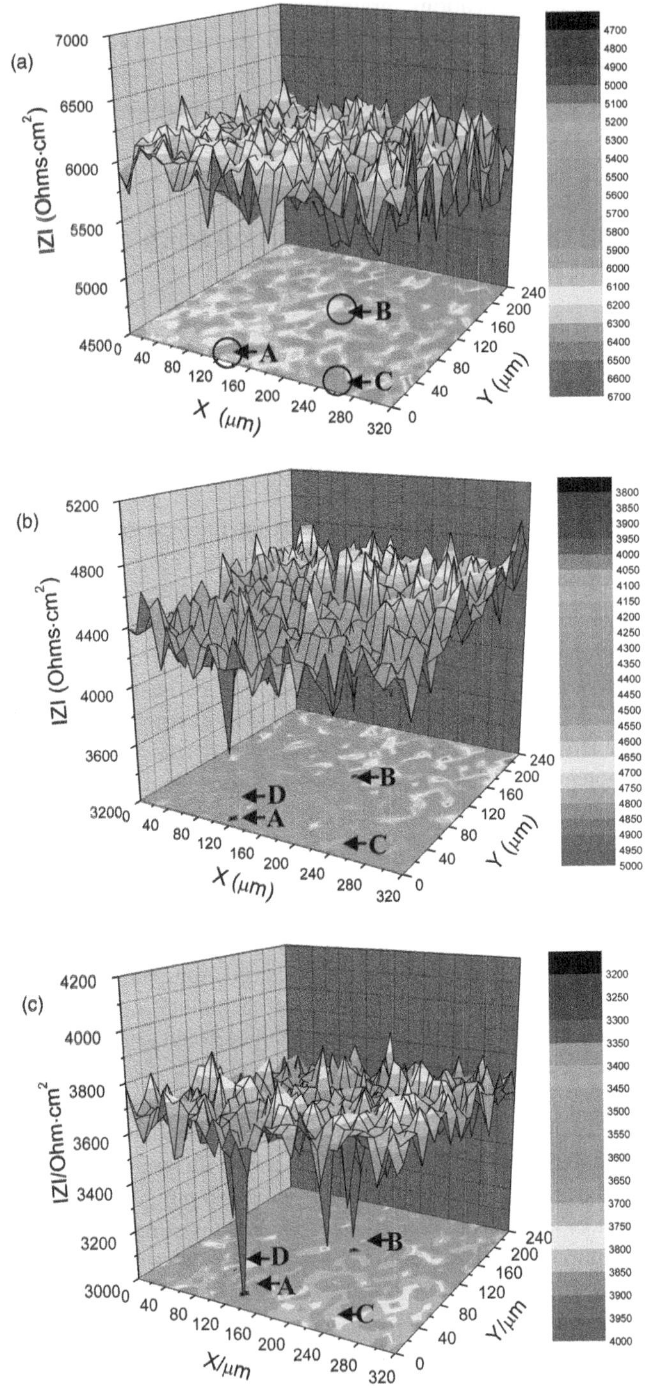

Figure 4-12 *LEIS maps measured on the steel electrode under SWP from −950 to −800 mV(SCE) for (a) 48 h, (b) 96 h, and (c) 144 h. (From Liu et al. [2012].)*

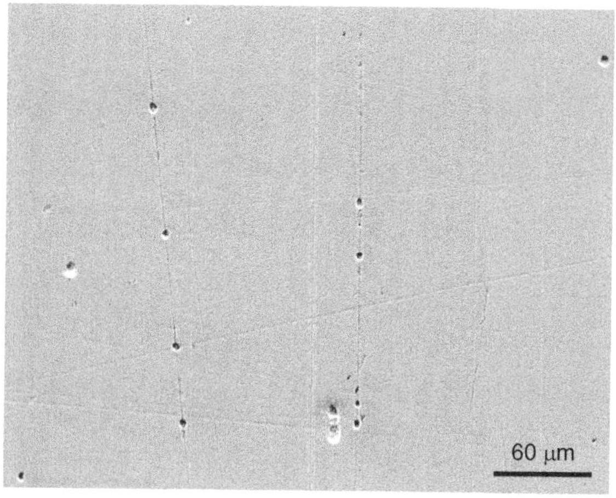

Figure 4-13 Pit nucleating positions on an X70 steel electrode after 144 h of SWP between −950 and −850 mV(SCE) in an NS4 solution. (From Liu et al. [2012].)

where i_D is the current density at defect, ΔE the LAP at defect, β_c the cathodic Tafel slope, and i_N the current density at a nondefect area (i.e., an intact area). If cathodic reaction is controlled by a mass-transport step, cathodic reactants such as H^+, HCO_3^-, and H_2CO_3 will be consumed more rapidly at defects than in an intact area. Figure 4-15a is a schematic view of the double-charge layer on a steel surface under a constant cathodic potential field, where both the defect and nondefect areas are associated with a uniform double-charge layer. All cathodic reactions are in equilibrium with the mass-transport process of reactants, and the steel is under full CP and does not corrode.

When a potential fluctuation is generated, the double-charge layer structure is disturbed, where reactants are consumed more rapidly at microdefects than in a nondefect area, as shown in Fig. 4-15b. Consequently, the defect undergoes a temporary anodic potential field, resulting in anodic dissolution locally to nucleate pits, while the nondefect area is still protected by the cathodic potential. If the potential fluctuation is sufficiently high so that the cathodic reactants are consumed rapidly over the entire surface of the electrode (Fig. 4-15c), the double-charge layer becomes anodized on the entire electrode. The entire electrode thus encounters anodic dissolution, with a high rate at defects. If the potential fluctuation exists repeatedly for a long time, the pit growth will become self-catalytic.

In an anaerobic, nearly neutral–pH solution, the potential cathodic reactants contain H_2CO_3, HCO_3^-, and H_2O, which will be reduced at different cathodic potentials [Zhang and Cheng, 2009a]:

$$2H_2CO_3 + 2e \rightarrow H_2 + 2HCO_3^- \quad (4\text{-}2)$$

$$2HCO_3^- + 2e \rightarrow H_2 + 2CO_3^{2-} \quad (4\text{-}3)$$

$$2H_2O + 2e \rightarrow H_2 + 2OH^- \quad (4\text{-}4)$$

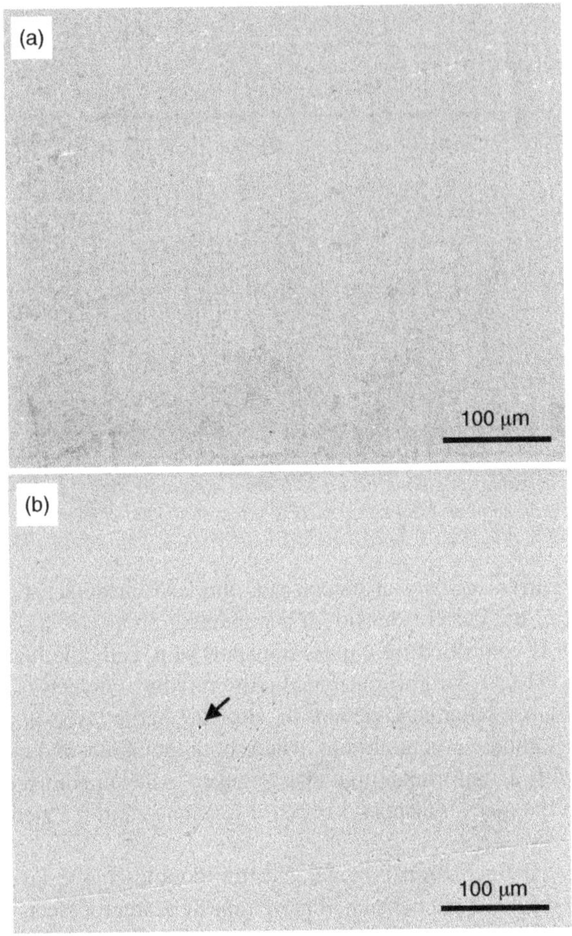

Figure 4-14 Surface morphologies of an X70 steel electrode after 144 h of potentiostatic polarization at (a) −950 mV(SCE) and (b) −850 mV(SCE) in an NS4 solution. (From Liu et al. [2012].)

Assume that the solution pH is 7 and the ionic concentrations are those contained in a nearly neutral–pH solution (i.e., an NS4 solution) and that the equilibrium electrode potentials for reactions (4-2) to (4-4) are $\varphi_{H_2CO_3/H_2}$ of -0.620 V_{SCE}, $\varphi_{HCO_3^-/H_2}$ of -0.849 V_{SCE}, and φ_{H_2O/H_2} of -0.658 V_{SCE}, respectively. If the applied potential is sufficiently negative, the cathodic potential field covers the entire metal surface, including both defects and nondefective areas, protecting the steel from corrosion. However, if a potential fluctuation occurs, altering the steel from a cathodic potential to a less cathodic value, the electrode equilibrium is disturbed. The anodic dissolution occurs locally at defects due to the presence of a temporary anodic potential field, resulting in the initiation of "corrosion pits."

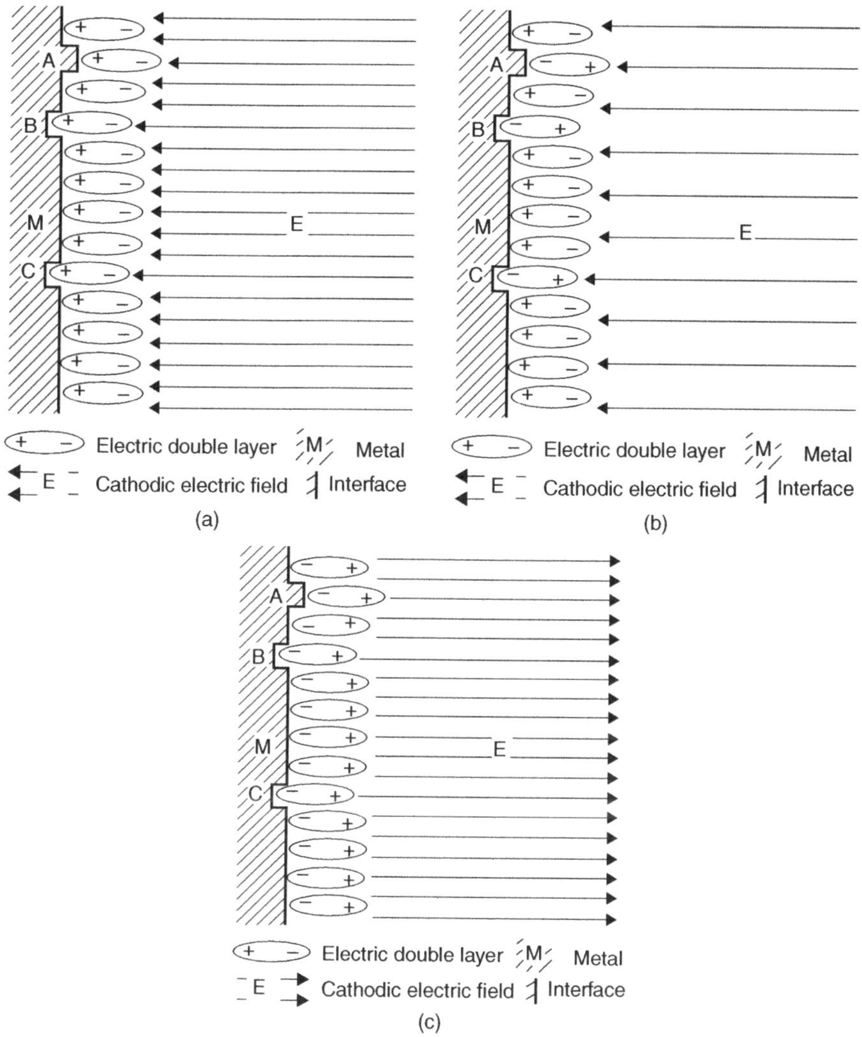

Figure 4-15 *Schematic view of the double-charge layer under (a) a constant cathodic potential, (b) a potential fluctuation resulting in a temporary anodic state at defects, and (c) a potential fluctuation, causing an anodic reaction over the entire electrode surface, where points A, B, and C refer to microdefects. (From Liu et al. [2011].)*

The anodic reactions of steel in a deoxygenated, nearly neutral–pH solution are [Heuer and Stubbins, 1999]

$$Fe + 2OH^- \rightarrow Fe(OH)_2 + 2e \quad (4\text{-}5)$$

$$Fe + CO_3^{2-} \rightarrow FeCO_3 + 2e \quad (4\text{-}6)$$

The equilibrium electrode potentials of reactions (4-5) and (4-6) are calculated by

$$\varphi_{Fe(OH)_2/Fe} = \varphi^{\theta}_{Fe(OH)_2/Fe} + \frac{2.303RT}{F}(14 - pH) \tag{4-7}$$

$$\varphi_{FeCO_3/Fe} = \varphi^{\theta}_{FeCO_3/Fe} - \frac{2.303RT}{2F}\log\left(C_{CO_3^{2-}}\right) \tag{4-8}$$

where $\varphi^{\theta}_{Fe(OH)_2/Fe}$ and $\varphi^{\theta}_{FeCO_3/Fe}$ are the standard equilibrium electrode potentials of reactions (4-5) and (4-6), F is the Faraday constant, R is the ideal gas constant, and T is temperature. The value of $\varphi_{Fe(OH)_2/Fe}$ is −0.664 V(standard hydrogen electrode, SHE) and that of $\varphi_{FeCO_3/Fe}$ is −0.805 V(SHE) at pH 7, assuming that the concentration of CO_3^{2-} in the solution is uniform.

Figure 4-16 shows the potentiodynamic polarization curves of an X70 line pipe steel in NS4 solution measured at various potential sweep rates. It is seen that, apparently, with a change in potential sweep rate, the corrosion potential is shifted. The corrosion potential divides the potential range into three zones. In zone I, at both slow and fast scannings, only an anodic reaction is observed. In zone II, the corrosion potential shifts negatively with an increase in potential sweep rate and remains constant when the sweep rate is faster than 50 mV/s. The anodic reaction is possible, at least temporarily, under cathodic polarization. In zone III, cathodic reactions are dominant. Therefore, it is expected that the local anodic dissolution (pitting) of steel may take place under unstable cathodic polarization. Pitting may occur when the potential fluctuates among zones I, II, and III, whereas no pitting takes place when the potential stays within zone III.

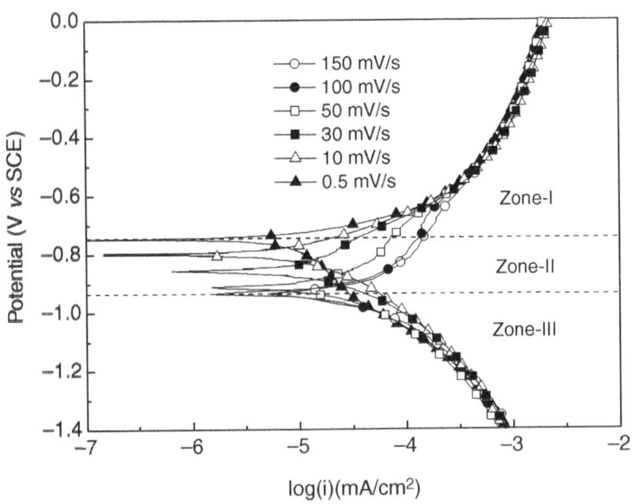

Figure 4-16 *Polarization curves measured on an X70 steel electrode at various potential sweep rates in a nearly neutral–pH soil solution. (From Liu et al. [2011].)*

This finding is significant both scientifically and technically. Since corrosion pits are the primary sites for initiating stress corrosion cracks on pipelines, stress concentration would develop at the pit base to enhance local anodic dissolution and/or to accumulate hydrogen. Stress concentration usually develops on pipelines by localized corrosion attack at or near inclusions to form corrosion pits. It was reported [Asher and Singh, 2009] that transgranular stress corrosion cracks develop on X65 steel at a stress around the yield strength of the steel, while cracklike fractures due to direct removal of inclusions can occur at a stress of only 30% of yield strength.

Furthermore, the presence of mill scale on a steel surface would facilitate the formation of corrosion pits. When mill-scaled steel is immersed in nearly neutral–pH solutions, three stages of change are experienced on both mill scale and the steel: that is, the initial breakdown of the scale, coupling of the dissolution of the underlying steel to the reductive dissolution of mill scale, and active corrosion of the steel at the base of pores in the scale [Qin et al., 2004]. The porous structure of mill scale increases the possibility of local separation of anodic and cathodic sites, promoting localized corrosion at the base of pores. These stress-raising pits could eventually act as the precursor sites for the initiation of stress corrosion cracks. Furthermore, the pits could grow in both the surface and the depth directions [Eslami et al., 2010]. Cyclic loading, combined with the local dissolution at pits, could facilitate pit-to-crack transition in regions with the highest stress concentration.

It has been well accepted that pipeline integrity is maintained by both coating and CP. The CP current permeates the coating or reaches the steel directly at the coating holidays, polarizing the pipeline steel cathodically. The CP current could fluctuate for a number of reasons, such as coating permeability and soil resistivity. Moreover, the stray current released from nearby high-voltage power transmission lines would also disturb the CP performance [Fu and Cheng, 2010]. All of the factors would cause potential fluctuations on the steel, resulting in pitting corrosion. Furthermore, the pressure or pressure fluctuations inside the pipeline generate a significant hoop stress or cyclic stress on the steel, enhancing pit initiation and growth [Liu et al., 2011]. Thus, for pipelines under CP application, the occurrence of corrosion pitss is inevitable, which could result further in the initiation of stress corrosion cracks. Actually, locations with high levels of cyclic stress, old pipes, and insufficient (but not totally lacking) CP are more prone to SCC initiation [Eslami et al., 2010]. It is worth pointing out that it will actually take a long time for corrosion pits to turn into well-defined stress corrosion cracks in a lab. It may take several years to observe SCC cracks on pipelines in the field [National Energy Board, 1996].

Furthermore, stress corrosion cracks could be initiated on pipeline steels in neutral-pH environments by hydrogen-enhanced local microplastic deformation at surface discontinuities, such as dislocation slip bands [Lu et al., 2010]. When the accumulated microplastic deformation at critical points reaches a threshold value, a crack can nucleate or re-nucleate. The former occurs at a free surface exposed to the corrosive medium, and the latter occurs at a crack tip. Moreover, both hydrogen and dissolution can increase the surface plasticity of steel, contributing to crack initiation and propagation. However, this model does not give quantitative information about the plastic increase due to hydrogen and anodic dissolution, and more important, the synergism of hydrogen and dissolution on the cracking process are not addressed. Actually, hydrogen- and dissolution-enhanced plasticity seems to be the other version of previously developed hydrogen- and stress-enhanced local dissolution at a crack tip [Cheng, 2007c].

4.5 STRESS CORROSION CRACK PROPAGATION MECHANISM

4.5.1 Role of Hydrogen in Enhanced Corrosion of Steels

During SCC, the condition at a crack tip has always been close to that of fresh, bare steel, which is in a nonequilibrious electrochemical reaction state, whereas the crack wall is covered with a corrosion product layer under which an electrochemical reaction takes place in a quasiequilibrium state. Consequently, the different electrochemical status between the crack tip and crack sides results in a combined effect of hydrogen embrittlement and anodic dissolution on the SCC of pipeline steels under nearly neutral–pH conditions [Liu et al., 2012]. There has been considerable evidence to support this mechanism [Parkins, 2000; Bueno et al., 2004; Asher and Singh, 2009; Lu et al., 2010]. In particular, the high susceptibility of pipeline steels to SCC is always associated with a high hydrogen permeation current measured in soil-extracted solutions [Cheng, 2007a].

A nearly neutral–pH environment is capable of generating a catalytic surface effect on hydrogen evolution in pipeline steels [Cheng and Niu, 2007]. Figures 4-17

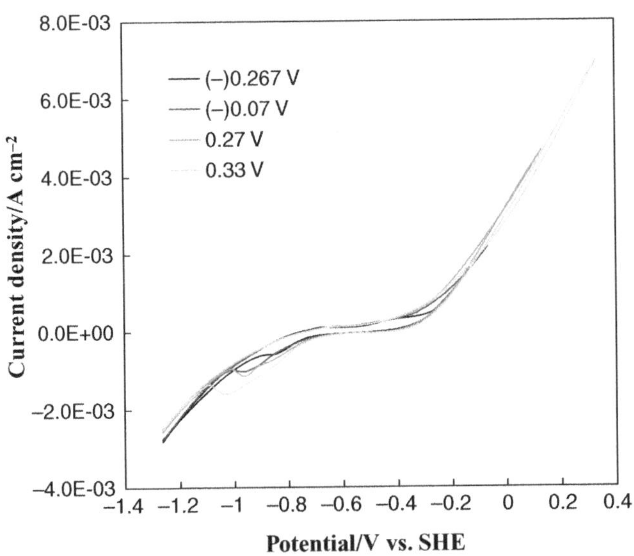

Figure 4-17 *Cyclic voltammograms measured on an X70 steel with the various upper switch potentials at a potential sweep rate of 50 mV/s in 0.01 M $NaHCO_3$ solution. (From Cheng and Niu [2007].)*

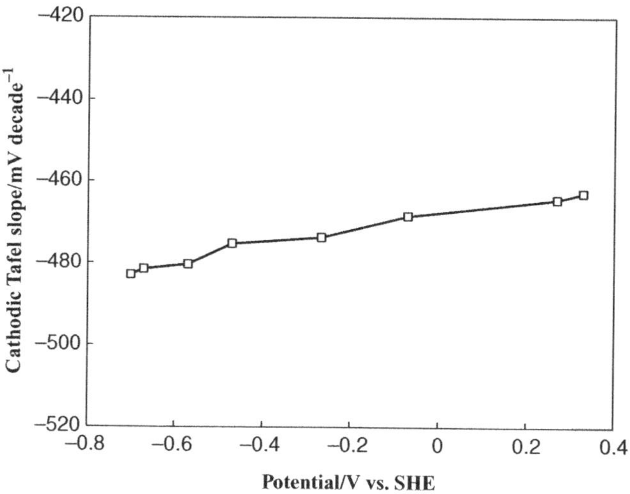

Figure 4-18 *Cathodic Tafel slope as a function of the upper switch potential during CV measurements in Fig. 4-17. (From Cheng and Niu [2007].)*

and 4-18 show cyclic voltammograms measured on X70 pipeline steel with various upper switching potentials at a potential sweep rate of 50 mV/s in 0.01 M $NaHCO_3$ solution, and the cathodic Tafel slope as a function of the upper switch potential, respectively. It is seen that the cathodic Tafel slope becomes less negative with an increase in upper switch potential; that is, the absolute value of the cathodic Tafel slope decreases with increasing upper potential limit. The cathodic Tafel slope for hydrogen evolution reaction, b_c, is related to the charge-transfer coefficient, α, by

$$|b_c| = \frac{2.303RT}{\alpha nF} \tag{4-9}$$

where n is the number of electrons exchanged in the electrode reaction. A decrease in the absolute value of b_c causes an increase in α, which refers to the proportion of energy used to overcome the total energy barrier for hydrogen evolution under the cathodic polarization applied. Therefore, a greater proportion of cathodic polarization is used for hydrogen evolution when the upper potential limit is shifted positively, resulting in enhanced hydrogen evolution kinetics. Thus, an appropriate oxidation of steel to generate an oxide or mill scale on its surface is capable of enhancing the hydrogen evolution.

There are remarkable effects of the oxidation state of steels on hydrogen evolution [Flis and Zakroczymski, 1992; King et al., 2000]. The presence of mill scale or corrosion product that is formed at potentials slightly positive to the corrosion potential would catalyze the hydrogen evolution [King et al., 2000; Qin et al., 2004; Meng et al., 2008]. These deposited layers are generally characterized by a porous structure that could increase the probability of the local separation of anodic and cathodic reaction sites. The occurrence of local dissolution underneath the deposit would generate an acidified electrolyte with low pH. Therefore, the cathodic hydrogen evolution is enhanced during a negative sweep, as indicated by the decreased absolute value of a cathodic Tafel slope at about −0.4 to −0.5 V(SHE), which is around the free corrosion potential of steel in nearly neutral–pH environments. Actually, it has been found [Capelle et al., 2010] that both low- and high-grade steels, including X52, X70, and X100, demonstrate sensitivity to hydrogenating in a deoxygenated, nearly neutral–pH NS4 solution at the corrosion potential or under relatively weak cathodic polarization. In particular, for X52 pipeline steel, a critical hydrogen concentration level exists, which is about 4.3×10^{-6} mol/cm^3, resulting in a significant loss of local fracture resistance of the steel [Capelle et al., 2008]. This value is suggested as an important parameter for reliability assessment of the exploited pipelines. Moreover, for an assessment of local strength at notches, dents, and pits in the presence of hydrogen, a diagram of the work of local fracture versus hydrogen concentration is effective and can be used for assessment of the susceptibility of pipelines to hydrogen embrittlement under assigned testing conditions.

The hydrogen-enhanced corrosion effect has been attributed to the alteration of chemical potential and exchange current density of the steel in nearly neutral–pH environments [Li and Cheng, 2007]. The anodic dissolution of uncharged and hydrogen-charged steels could be expressed as:

$$\text{Uncharged steel:} \quad Fe \rightarrow Fe^{2+} + 2e \tag{4-10}$$

$$\text{Charged steel:} \quad Fe_{(H)} \rightarrow Fe^{2+} + 2e \tag{4-11}$$

The equilibrium potential, E^{eq}, for reaction (4-10) and $E^{eq}_{(H)}$ for reaction (4-11) can be written as

$$E^{eq} = \frac{\mu_{Fe^{2+}} - \mu_{Fe}}{2F} \tag{4-12}$$

$$E^{eq}_{(H)} = \frac{\mu_{Fe^{2+}} - \mu_{Fe(H)}}{2F} \tag{4-13}$$

where $\mu_{Fe(H)}$, μ_{Fe}, and $\mu_{Fe^{2+}}$ are chemical potentials of charged and uncharged steels and iron ions, respectively. The anodic current density, i_a, of uncharged steel at the anodic potential, E_a, which is away from the equilibrium potential, is

$$i_a = i^0 \exp\left(\frac{2\beta F(E_a - E^{eq})}{RT}\right) \tag{4-14}$$

where i^0 is the exchange current density and β is the charge-transfer coefficient. Similarly, the anodic current density of the charged steel, $i_{a(H)}$, at E_a is

$$i_{a(H)} = i^0_H \exp\left(\frac{2\beta_H F(E_a - E^{eq}_{(H)})}{RT}\right) \tag{4-15}$$

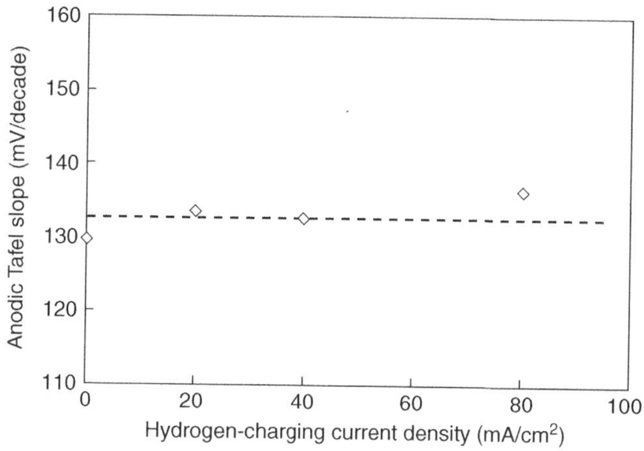

Figure 4-19 *Dependence of the anodic Tafel slope of X70 steel in a nearly neutral–pH NS4 solution as a function of the hydrogen-charging current density. (From Li and Cheng [2007].)*

where β_H is the charge-transfer coefficient of the charged steel. The anodic Tafel slopes, b_a, can be determined by fitting the measured anodic polarization curves on steel with various hydrogen-charging current densities, and the results are shown in Fig. 4-19. It is seen that the values of b_a are approximately independent of the charging current density. It is thus reasonable to assume that β is approximately equal to β_H. Reorganize Eqs. (4-13) to (4-15):

$$i_{a(H)} = i_a \left[\frac{i_H^0}{i^0} \exp\left(\frac{\beta \Delta\mu}{RT} \right) \right] = k_H i_a \tag{4-16}$$

where $\Delta\mu$ is the difference in the chemical potentials of the steel in the presence and absence of hydrogen charging (i.e., $\Delta\mu = \mu_{Fe(H)} - \mu_{Fe}$), and k_H represents the effect of hydrogen on the anodic dissolution rate of the steel. Therefore, the hydrogen-enhanced anodic dissolution is attributed to the hydrogen-induced alteration of the chemical potential of the steel and the exchange current density, both of which contribute to the enhanced corrosion of steel upon hydrogen charging.

4.5.2 Potential-Dependent Nearly Neutral–pH SCC of Pipelines

Pipeline SCC in nearly neutral pH environments is potential-dependent. Figure 4-20 shows potentiodynamic polarization curves measured on an X70 steel electrode at both slow and fast potential sweep rates in NS4 solution and SCC susceptibility as a function of potential. When the potential applied is more positive than a null-current potential of about −730 mV(SCE), which is close to the corrosion potential of the steel under quasiequilibrium conditions in nearly neutral–pH environments, the polarization curves measured at both slow and fast scanning rates are within the anodic polarization range, indicating that cracking under these conditions is

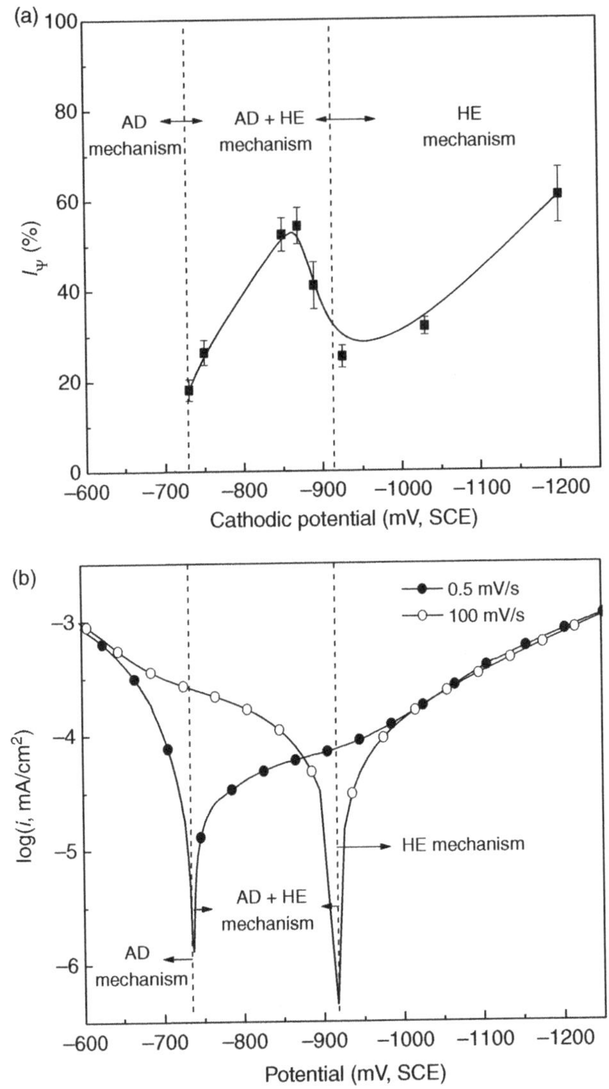

Figure 4-20 *SCC susceptibility of an X70 steel in an NS4 solution as a function of the cathodic potential (a) and the corresponding polarization curves (b) measured at slow (0.5 mV/s) and fast (100 mV/s) potential scanning rates. (From Liu et al. [2012].)*

controlled by anodic reaction and the SCC mechanism is anodic dissolution (AD)–based. When the potential applied is more negative than the null-current potential measured at a fast scanning rate [i.e., more negative than −920 mV(SCE)], the steel is in the cathodic polarization region, and the cathodic reaction charges hydrogen to the steel. Thus, the SCC mechanism is under a hydrogen embrittlement mechanism. When the polarized potential is between the two null-current potentials, the steel is in cathodic polarization at a slow potential scanning rate (e.g., 0.5 mV/s), and in anodic polarization if the scanning rate is fast (e.g., 100 mV/s). In other words, the steel is in a nonequilibrium state between the two null-current potentials. Dissolution could occur under a cathodic polarization potential to contribute to the cracking process. This, combined with the hydrogen effect, would accelerate SCC. Therefore, SCC of the steel is under the combined effect of AD and HE between potentials of −730 and −920 mV(SCE).

Apparently, the SCC of pipelines in nearly neutral–pH environments depends on the potential of the steel, which is affected by the applied CP, status of coating, soil properties, and other factors. In reality, CP is used to prevent corrosion in pipelines. The CP would penetrate the coating to reach the steel surface. However, the CP effectiveness (i.e., the cathodic potential of the steel under CP) depends on the coating permeability [Howell and Cheng, 2007; Chen et al., 2009]. Moreover, the CP could pass completely or partially through coating defects to reach the pipeline [Dong et al., 2008]. Thus, pipeline steel could be in a variety of cathodic potentials. This model demonstrates that AD would occur to contribute to SCC even when the pipeline is under CP.

Furthermore, in an AD + HE region, with the negative shift in applied potential, the cathodic current density measured at the slow scanning rate increases, while the anodic current density at the fast scanning rate decreases. A maximum of SCC susceptibility is observed. In the relatively weak cathodic polarization region [i.e., from −730 to −860 mV(SCE)], the SCC susceptibility increases with the negative shift in potential, as indicated by the increasing I_Ψ, the ratio of the reduction in area obtained in solution to that in air (Fig. 4-20a). Since hydrogen evolution is enhanced with cathodic current density, it is believed that the HE effect dominates SCC growth, and dissolution would accelerate the cracking process. When the potential is more negative than −860 mV(SCE), the SCC susceptibility is decreased, which is due primarily to the significant reduction of dissolution at sufficiently cathodic potentials. Apparently, the AD effect is not ignorable and could affect SCC susceptibility at a cathodic potential. Moreover, hydrogen affects primarily the anodic process, as the cathodic polarization potential is more positive than −950 mV(SCE). When the cathodic potential is more negative, HIC would occur directly.

Furthermore, it has been demonstrated [Asher et al., 2007; Torres-Islas et al., 2008] that a change in applied potential will change the environmental conditions under which the pipeline steel is exposed, resulting in a change in the pipeline SCC between the intergranular and transgranular modes.

4.5.3 Pipeline Steels in Nearly Neutral–pH Solutions: Always Active Dissolution?

It has been well accepted that pipeline steels are in an active dissolution state in nearly neutral–pH solutions and that conventional film rupture–repassivation theory does not apply for pipeline nearly neutral–pH SCC [Parkins, 2000]. A potentiodynamic polarization curve measured on pipeline steels in a bulk NS4 solution does confirm that the steel cannot be passivated in the solution [Fu et al., 2009]. Relatively subtle changes in potential, pH, concentration, and temperature could determine whether dissolution and passivation are favored or SCC follows a hydrogen-based mechanism [King et al., 2000]. Higher pH, more concentrated solutions, less negative potential, and higher temperature favor dissolution and passivation of the steel surface, whereas the opposite conditions favor a hydrogen mechanism. Actually, these conclusive remarks were drawn based on electrochemical corrosion measurements conducted in bulk solutions.

Generally, an electrolyte trapped under a disbonded coating is very thin, especially at the early stage of coating disbondment, when an electrochemical environment for corrosion reaction develops. Corrosion of steel in a thin electrolyte is distinctly different from that in a bulk solution. For example, the small ohmic potential drop and nonuniform current distribution in a thin electrolyte layer are expected to affect significantly the electrochemical mechanism of corrosion reaction in the steel [Nishikata et al., 1995]. Figure 4-21 shows polarization curves measured on X70 pipeline steel in thin layers of NS4 solution by a scanning Kelvin probe after 1 and 24 h of immersion. Since the current density is quite low, the difference between passive and active behavior measured at different electrolyte layer thicknesses is not very apparent but is still distinguishable. When the steel electrode is in the 60-μm solution layer, it could be passivated with a stable passive potential range from approximately 0 to 0.5 V(SHE). This feature is not observed in polarization curves measured at 90- and 140-μm solution layers. With a solution layer thickness of 90 μm, passivity of the steel cannot be maintained after 24 h of immersion. Under the 140-μm solution layer, the steel cannot be passivated for both 1 and 24 h of immersion, and shows an active dissolution state, just like the polarization behavior measured in bulk NS4 solutions.

Apparently, under a thin layer of nearly neutral–pH solution, such as 60 or 90 μm, passivity can be developed in pipeline steel, which is an important finding. The important fact is that in contrast to conventional comments on the electrochemical corrosion behavior of pipeline steels in a nearly neutral–pH solution, steel passivity is feasible. This is attributed to the fact that the Fe^{2+} concentration in an aqueous phase could reach a saturation state in a thin solution layer. Although the amount of CO_3^{2-} in a diluted NS4 solution is not high, a solubility level can be reached that will cause $FeCO_3$ to deposit out of solution as a precipitate once the Fe^{2+} concentration achieves saturation or supersaturation.

With an increase in solution layer thickness, it becomes more difficult for Fe^{2+} to reach saturation. It is apparent from Fig. 4-21 that when the solution layer increases to 140 μm, steel can no longer be passivated. Furthermore, in the presence of chloride ions in bicarbonate solutions, even at a small amount of 0.005 M, the steel passivity is not maintained [Liu and Mao, 1995]. The disappearance of passivity would be a result of the effect of chloride ions.

Therefore, when a thin layer of nearly neutral–pH electrolyte forms on a steel surface, it is feasible to develop stable passivity on the steel. During coating disbondment, especially at an early stage, the trapped electrolyte is usually very thin. Moreover, the space under the coating disbondment is often very limited. The steel would be in a passive state rather than in an active dissolution state. Apparently, the conventional assumption that pipeline steel is in an active state under a nearly

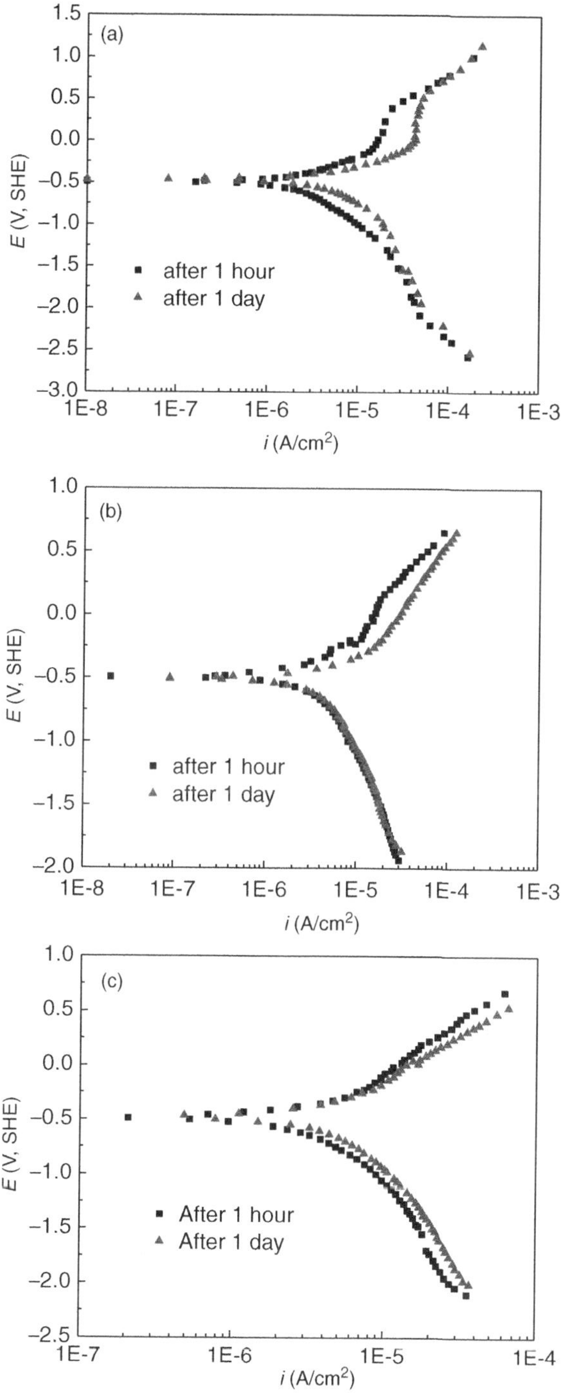

Figure 4-21 *Polarization curves of X70 steel measured in a thin NS4 solution layer after immersions of 1 and 24 h: (a) 60 μm; (b) 90 μm; (c) 140 μm. (From Fu et al. [2009].)*

neutral–pH environment is questionable. Furthermore, once cracks are initiated on steel that is under a thin electrolyte, the role of hydrogen is actually less important, due to the potential inhibition of the hydrogen permeation into the steel by the surface film [Song et al., 1990]. As a consequence, the well-accepted proposition regarding hydrogen involvement in pipeline SCC should be rechecked for its reliability.

In summary, the reality of pipeline SCC occurring in nearly neutral–pH environments is that at an early stage of coating disbondment, the electrolyte trapped under the coating is sufficiently thin and the steel is in passivity. At this stage the cracking process follows a film rupture–repassivation mechanism. With the increase in time and solution layer thickness, the passivity of the steel cannot be maintained, and the steel transits toward an active dissolution state. Due to the surface self-catalytic effect on hydrogen evolution, hydrogen is involved in the cracking process, accompanying the anodic dissolution occurring at the crack walls. Apparently, pipeline SCC occurring in nearly neutral–pH solution experiences a mechanistic change with the increase in solution layer thickness.

This finding is consistent with the statistic analysis taken from the field of colonies of surface cracks from a pipeline exhibiting nearly neutral–pH SCC [Bouaeshi et al., 2007], where two different types of crack characteristics are assumed to imply the different mechanisms associated with the cracking process under a nearly neutral–pH environment. Type 1 cracks grow to a depth of about 0.5 mm and then tend to stop growing in the depth direction while continuing to widen and possibly get longer. The dormancy results from a reduction in the corrosion driving force at the crack tip, and there is a significant amount of corrosion product in the crack. Type 2 cracks show little corrosion on the crack flanks and seem to be growing actively, particularly in the depth direction. The cracks remain sharp and the growth in depth increases relative to the growth in length. It is thus reasonable to assume that the first type of crack, developed under nearly neutral–pH conditions, is associated with the corrosion-passivity mechanism, where the film rupture–repassivation theory applies, whereas the second type of crack is due to the increasing importance of mechanical factors to support the role of hydrogen in the cracking process.

4.6 MODELS FOR PREDICTION OF NEARLY NEUTRAL–PH SCC PROPAGATION

A number of models have been developed to predict quantitatively the growth rate of stress corrosion cracks in pipelines under nearly neutral–pH environmental conditions. In addition to dissolution reactions, hydrogen is involved in the cracking process. Generally, the electrochemical reactions of pipeline steel without applied stress in deoxygenated, nearly neutral–pH solutions include an anodic oxidation of Fe and the cathodic reduction of water. Hydrogen atoms generated and adsorbed on a steel surface may either combine into hydrogen molecules chemically or electrochemically, or penetrate the steel and become absorbed hydrogen atoms. Thus, only some of the hydrogen atoms generated from the cathodic reaction will permeate the steel.

It has been established [Daft et al., 1979] that in nearly neutral–pH solutions in the absence of hydrogen-permeation "poisons," the electrochemical reaction dominates the hydrogen recombination process. Therefore, the entire electrode reaction can be described as

$$Fe(xH) + 2H_2O \rightarrow Fe^{2+} + 2OH^- + \frac{2(1-x)}{2-x}H_2 + \frac{2x}{2-x}H_{abs} \quad (4\text{-}17)$$

where x is the number of hydrogen atoms permeating the steel, and Fe(xH) represents a steel specimen containing x hydrogen atoms. For a stressed steel specimen containing cracks, hydrogen may penetrate the steel and diffuse toward the crack tip, where a high triaxial stress concentration exists [Hirth, 1980]. Therefore, there will be more hydrogen diffusing into the stressed steel specimen, preferably accumulating in sites with a high stress concentration, such as a crack tip. The electrochemical reaction of the stressed steel in the presence of cracks in deoxygenated, nearly neutral–pH solutions is described as

$$Fe(\sigma, yH) + 2H_2O \rightarrow Fe^{2+} + 2OH^- + \frac{2(1-y)}{2-y}H_2 + \frac{2y}{2-y}H_{abs} \quad (4\text{-}18)$$

where y is the number of hydrogen atoms permeating the stressed steel, and $y > x$. Fe(yH) represents a steel specimen containing y hydrogen atoms.

The electrochemical potential difference between reactions (4-17) and (4-18) is caused by the free-energy changes resulting from the stress status and amount of hydrogen in steels. If the free-energy changes and electrochemical potentials in reactions (4-17) and (4-18) are denoted as ΔG_1 and ΔG_2, and E_1 and E_2, respectively, then

$$\Delta G_1 = G_{Fe2+} + 2G_{OH^-} + \frac{2(1-x)}{2-x}G_{H2} + \frac{2x}{2-x}G_{Habs} - G_{Fe}(xH) - 2G_{H_2O} = -nFE_1 \quad (4\text{-}19)$$

$$\begin{aligned}\Delta G_2 &= G_{Fe2+} + 2G_{OH^-} + \frac{2(1-y)}{2-y}G_{H_2} + \frac{2y}{2-y}G_{Habs} - G_{Fe}(\sigma, yH) - 2G_{H_2O} \\ &= \Delta G_1 - \frac{2(y-x)}{(2-y)(2-x)}(G_{H2} - 2G_{Habs}) - \Delta G_{Fe}(\sigma, H) = -nFE_2\end{aligned} \quad (4\text{-}20)$$

where G represents the formation free energy of the individual species. $\Delta G_{Fe}(\sigma, H) = G_{Fe}(\sigma, yH) - G_{Fe}(xH)$ is due to the presence of stress and the different concentrations of hydrogen in the steels.

The anodic dissolution current density i_1 in reaction (4-17) is

$$i_1 = i^0 \exp\left(-\frac{\Delta G_1}{RT}\beta\right) = i^0 \exp\left(\frac{nFE_1}{RT}\beta\right) \quad (4\text{-}21)$$

The dissolution current density i_2 at the crack tip in reaction (4-18) is

$$i_2 = i_0\,(\sigma, yH)\exp\left(-\frac{\Delta G_2}{RT}\beta\right)$$
$$= i_0\,(\sigma, yH)\exp\left[-\frac{\Delta G_1 - \dfrac{2(y-x)}{(2-y)(2-x)}(G_{H2} - 2G_{Habs}) - \Delta G_{Fe}(\sigma, H)}{RT}\beta\right] \tag{4-22}$$

If it is assumed that the exchange current density keeps constant, then

$$i_2 = i_1 \exp\left[\frac{\dfrac{2(y-x)}{(2-y)(2-x)}(G_{H2} - 2G_{Habs}) + \Delta G_{Fe}(\sigma, H)}{RT}\beta\right] \tag{4-23}$$

When the steel is free of stress ($\sigma = 0$), it is expected that there is no stress corrosion crack in the steel, and the steel corrodes freely in nearly neutral–pH solution. The amount of hydrogen penetrating the steel, x, is approximately equal to y. Since there is no difference with respect of the free-energy change for equations (4-17) and (4-18), $i_2 = i_1$.

In the presence of stress, cracks are initiated in the steel. The synergistic effect of hydrogen and stress on anodic dissolution can be given as

$$i_2 = i_1 \exp\left[\frac{\dfrac{2(y-x)}{(2-y)(2-x)}(G_{H2} - 2G_{Habs})}{RT}\beta\right]\exp\left(\frac{\Delta U - T\Delta S}{RT}\beta\right)$$
$$\times \exp\left[\frac{W_m(\sigma_1^2 + \sigma_2^2 + \sigma_3^2)}{2E\rho RT}\beta\right] \times \exp\left(\frac{\sigma_h \bar{V}_H}{RT}\beta\right) = k_H k_\sigma k_{H\sigma} i_1 \tag{4-24}$$

where ΔU is the internal energy change; ΔS is the entropy change; W_m is the molar weight; σ_1, σ_2, and σ_3 are the principal stresses; ρ is the density, E is Young's modulus, σ_h is the volume stress, and $\bar{V}_H$ is the average volume of hydrogen in steel. In particular, k_H is the effect of hydrogen on the anodic dissolution rate in the absence of stress (i.e., the free corrosion rate of Fe in a deoxygenated, nearly neutral–pH solution), k_σ is the effect of stress on anodic dissolution in the absence of hydrogen (i.e., a pure anodic dissolution-based cracking mechanism), and $k_{H\sigma}$ is the synergistic effect of hydrogen and stress on the anodic dissolution at the crack tip.

According to the slip-oxidation model [Ford, 1996], the crack growth rate (CGR) by anodic dissolution mechanism can be represented as

$$\mathrm{CGR} = \frac{iW}{nF\rho} \tag{4-25}$$

Thus, the CGR in the presence of hydrogen and stress can be given as

$$\mathrm{CGR}(\sigma, \mathrm{H}) = k_{\mathrm{H}} k_{\sigma} k_{\mathrm{H}\sigma} i_1 \frac{W}{nF\rho} = k_{\mathrm{H}} k_{\sigma} k_{\mathrm{H}\sigma} \mathrm{CGR}(\sigma, \mathrm{H} = 0) \tag{4-26}$$

where $\mathrm{CGR}(\sigma, \mathrm{H} = 0)$ is the crack growth rate without the effect of hydrogen. Therefore, a quantitative prediction of the rate of crack growth in the pipeline steel in deoxygenated, nearly neutral–pH solutions depends on determination of the three affecting factors [Cheng, 2007c].

Cheng and co-workers used various microelectrochemical techniques, including LEIS and SVET, to measure the local dissolution current density at the crack tip in pipeline steels that are under a variety of hydrogen-charging current densities and applied stresses to simulate the potential environments to which the pipelines are exposed. It is determined that for a crack-free pipeline steel, the hydrogen effect factor, k_{H}, is approximately 1.1 at a charging current density of 20 mA/cm^2 [Li and Cheng, 2007]. Moreover, the stress effect factor, k_{σ}, was determined to be 1.3 at an applied stress of 525 MPa [Tang and Cheng, 2009]. The synergistic effect factor of hydrogen and stress, $k_{\mathrm{H}\sigma}$, is calculated from the electrochemical impedance measurements and the fitted charge-transfer resistances for the stressed steel under various charging current densities by

$$K_{\mathrm{H}\sigma} = \frac{\mathrm{R}^0_{\mathrm{ct},\sigma}}{R^{\mathrm{H}}_{\mathrm{ct},\sigma}} \tag{4-27}$$

where $R^0_{\mathrm{ct},\sigma}$ and $R^{\mathrm{H}}_{\mathrm{ct},\sigma}$ are the charge-transfer resistances of steel without and with hydrogen charging, respectively. At a charging current density of 20 mA/cm^2, the synergistic effect factor is 5.4. The total effect of hydrogen and stress on anodic dissolution of a smooth steel specimen under nearly neutral–pH conditions is

$$\text{total effect factor} = k_{\mathrm{H}} \times k_{\sigma} \times k_{\mathrm{H}\sigma} = 1.1 \times 1.3 \times 5.4 = 7.7 \tag{4-28}$$

When a steel specimen contains cracks, the stress concentration introduced by the cracks, or by cracklike flaws, would enhance remarkably the local dissolution rate of the steel. As shown in Fig. 4-22, there are higher current densities at the crack tip than in other regions. It was determined that the stress effect factor on the local dissolution rate of the steel at a crack tip is about 3.3 times that away from the crack. Thus, it is assumed that under an applied stress of 525 MPa, the local stress effect factor at a crack tip is approximately 4.3 (1.3 × 3.3, where the factor 1.3 is the stress-enhanced dissolution factor determined previously on a smooth steel specimen).

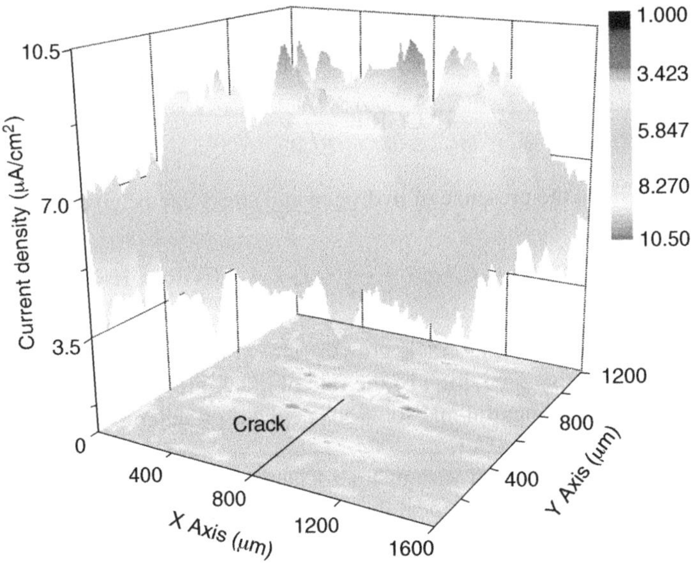

Figure 4-22 *SVET measurement on a precracked X70 steel specimen in an NS4 solution under a tensile force of 3000 N. (From Tang and Cheng [2009].)*

Furthermore, the hydrogen effect factor at a crack tip of unstressed steel at a charging current density of 20 mA/cm^2 is 1.3 [Zhang and Cheng, 2009b]. Similarly, the SVET current density maps are measured on a cracked steel specimen that is under loading and various hydrogen-charging current densities, as shown in Fig. 4-23. The synergistic effect factor, determined based on Eq. (4-27) and an extrapolation of the charge-transfer resistance at the crack tip to a charging current density of 20 mA/cm^2, is assessed to be approximately 4.0 [Tang and Cheng, 2011]. The total effect of hydrogen and stress on anodic dissolution of a precracked steel specimen under nearly neutral–pH conditions is determined as

$$\text{total effect factor} = k_{\mathrm{H}} \times k_{\sigma} \times k_{\mathrm{H}\sigma} = 1.3 \times 4.3 \times 4.0 = 22.4 \qquad (4\text{-}29)$$

A simple test using an electrochemical scratch electrode technique [King et al., 2000] showed that if the crack propagation follows a purely anodic dissolution mechanism, the crack growth rate is approximately 5×10^{-8} mm/s under nearly neutral–pH conditions. However, field measurement by ultrasound has demonstrated that an average of 10^{-6} mm/s of crack growth rate is experienced by pipeline steel in a nearly neutral–pH environment [Baker, 2005]. The 20-fold difference in crack growth rate has been assumed to be associated with hydrogen involvement, where the interaction of hydrogen and stress enhances SCC propagation.

According to Cheng's model [2007c], under an applied stress of 525 MPa and a hydrogen-charging current density of 20 mA/cm^2, the total effect factor for nearly neutral–pH SCC growth on a smooth pipeline steel specimen is only 7.7 much lower than the 20-fold difference compared to the dissolution-based SCC propagation rate determined by electrochemical scratch electrode testing. Therefore, in the absence of cracks or cracklike flaws, the stressing and hydrogen-charging conditions employed are incapable of incurring sufficient anodic dissolution on steel in a nearly neutral–pH environment. However, when cracks are present, the total effect factor determined is

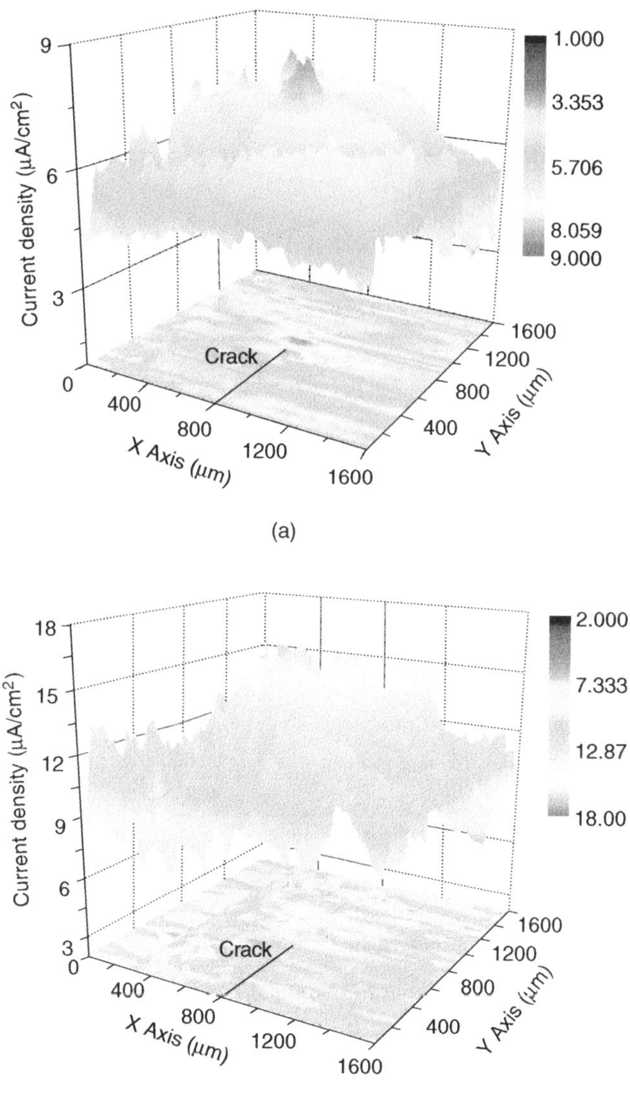

Figure 4-23 *SVET mapping on a precracked X70 steel specimen at a constant tensile force of 3000 N without and with 2 h of hydrogen charging at various current densities: (a) no charging; (b) 0.5 mA/cm²; (c) 2 mA/cm²; (d) 8 mA/cm². (From Tang and Cheng [2011].)*

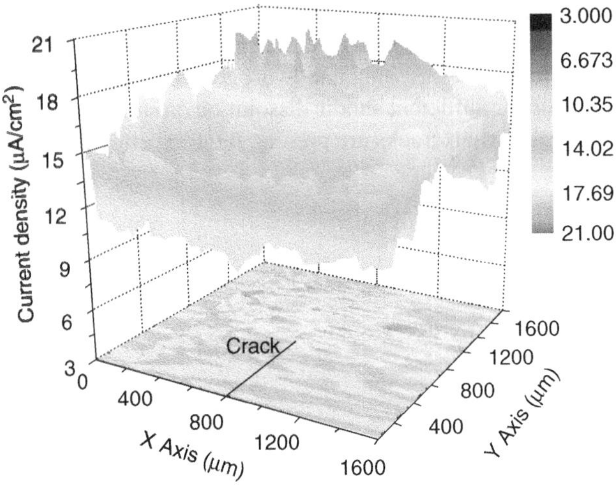

(c)

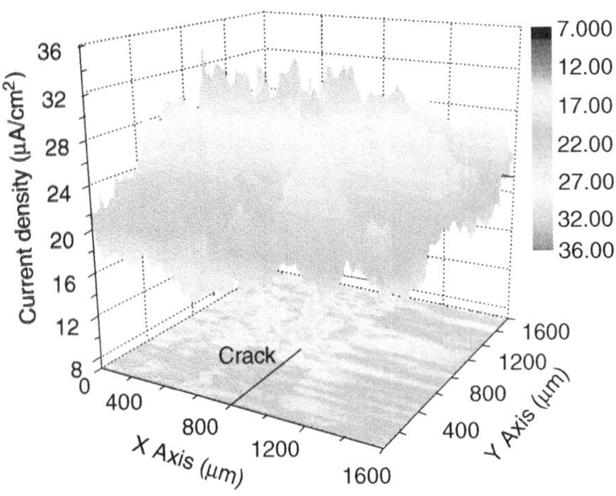

(d)

Figure 4-23 *(Continued)*

up to 22.4, which is very close to the 20-fold difference in crack growth rates in the absence and presence of hydrogen involvement. Apparently, the presence of microcracks and/or cracklike flaws is critical to developing a local dissolution rate at a crack tip, resulting in a crack propagation rate similar to that measured in the field. It is further deduced that stress corrosion cracks would be preferential to propagate around metallurgical "weak" sites, such as inclusions, high-angle grain boundaries, corrosion pits, and mechanical dents. Therefore, the model developed considers simultaneously the importance of metallurgical, environmental, and mechanical factors to the propagation of stress corrosion cracks.

Furthermore, the role of cyclic stress that results from the pressure fluctuations in pipelines during crack propagation is not negligible. Crack growth behavior is sensitive to cyclic loading conditions, such as cyclic frequency [Kim et al., 2004]. Crack growth has been observed under cyclic loading only, and no crack growth was detected in a specimen subjected to monotonic loading at the highest stress intensity factor used [Chen and Sutherby, 2004]. Therefore, the crack growth was categorized as corrosion fatigue instead of SCC. Actually, fracture mechanics concepts have been used to quantify crack growth in pipelines. For example, the loading rate for each crack condition was calculated in order to correlate the crack growth rate with the loading rate, and the *J*-integral was then calculated through nonlinear finite element analyses for semielliptical cracks based on the stress–strain relationships obtained from tensile tests on a pipeline steel specimen [Kim et al., 2004]. A linear relationship has been obtained between the crack growth rate and the loading rate, thereby demonstrating the possibility of predicting crack growth rates for various loading conditions in the field. Furthermore, a crack growth model based on pressure fluctuation considerations was proposed [Chen, 2006], where the crack growth rate da/dN was defined as $f(\Delta K^2\ K_{max}\, f^{0.1})$, where a is the crack length, N the number of stress cycles, ΔK the stress intensity factor, K_{max} the maximum stress intensity factor, and f the frequency. The model is modified further as

$$\frac{da}{dN}(\text{total}) = \frac{da}{dN}(\text{fatigue}) + \frac{1}{f}\frac{da}{dt}(\text{SCC}) \qquad (4\text{-}30)$$

The total crack growth rate combines the contributions from corrosion fatigue and SCC. The predictive possibility does not apply to the start stage of crack growth featuring shallow cracks. For stable crack growth, the cracks are deep, with a large stress intensity factor. The predictive model works well to fit the cracking rate obtained from field data.

REFERENCES

Asher, S, Singh, PM (2008) Hydrogen production and permeation in near-neutral pH environments, *Corrosion 2008*, Paper 08411, NACE, Houston, TX.

Asher, SL, Singh, PM (2009) Role of stress in transgranular stress corrosion cracking of transmission pipelines in near-neutral pH environments, *Corrosion* 65, 79–87.

Asher, SL, Leis, B, Colwell, J, Singh, PM (2007) Investigating a mechanism for transgranular stress corrosion cracking on buried pipelines in near-neutral pH environments, *Corrosion* 63, 932–939.

Baker, M (2005) *Final Report on Stress Corrosion Cracking Study*, Integrity Management Program Delivery Order DTRS56-02-D-70036, Office of Pipeline Safety, U.S. Department of Transportation, Washington, DC.

Beavers, JA (1993) On the mechanism of stress corrosion cracking of natural gas pipelines, *Proc. 8th Line Pipe Symposium*, Paper 17, PRCI, Falls Church, VA.

Been, J, King, F, Yang, L, Song, F, Sridhar, N (2005) The role of coatings in the generation of high- and near-neutral pH environments that promote environmentally assisted cracking, *Corrosion 2005*, Paper 05167, NACE, Houston, TX.

Bouaeshi, W, Ironside, S, Eadie, R (2007) Research and cracking implications from an assessment of two variants of near-neutral pH crack colonies in liquid pipelines, *Corrosion* 63, 648–660.

Bueno, AHS, Castro, BB, Ponciano, JAC (2004) Laboratory evaluation of soil stress corrosion cracking and hydrogen embrittlement of API grade steels, *Proc. 5th International Pipeline Conference*, Paper IPC04-0284, Calgary, Alberta, Canada.

Canadian Energy Pipeline Association (1996a) Submission to the National Energy Board, *Proceeding MH-2-95*, Vol. 1, CEPA, Calgary, Alberta, Canada, Issue 2, p. 9.

Canadian Energy Pipeline Association (1996b) Submission to the National Energy Board, *Proceeding MH-2-95*, Vol. 2, CEPA, Calgary, Alberta, Canada, App. D, Tab. 6, p. 8.

Capelle, J, Gilgert, J, Dmytrakh, I, Pluvinage, G (2008) Sensitivity of pipelines with steel API X52 to hydrogen embrittlement, *Int. J. Hydrogen Energy* 33, 7630–7641.

Capelle, J, Dmytrakh, I, Pluvinage, G (2010) Comparative assessment of electrochemical hydrogen absorption by pipeline steels with different strength, *Corros. Sci.* 52, 1554–1559.

Chen, W (2006) SCC or corrosion fatigue: some thoughts on integrity management and crack control, *Workshop on Pipeline SCC*, TransCanada Pipelines, Calgary, Alberta, Canada.

Chen, W, Sutherby, R (2004) Environmental effect on crack growth rate of pipeline steel in near-neutral pH soil environments, *Proc. 5th International Pipeline Conference*, Paper IPC04-0449, Calgary, Alberta, Canada.

Chen, W, King, F, Vokes, E (2002) Characteristics of near neutral pH stress corrosion cracks in an X65 pipeline, *Corrosion* 58, 267–275.

Chen, X, Li, XG, Du, CW, Cheng, YF (2009) Effect of cathodic protection on corrosion of pipeline steel under disbonded coating, *Corros. Sci.* 51, 2242–2245.

Cheng, YF (2007a) Fundamentals of hydrogen evolution reaction and its implications on near-neutral pH stress corrosion cracking of pipelines, *Electrochim. Acta* 52, 2661–2667.

Cheng, YF (2007b) Analysis of electrochemical hydrogen permeation through X-65 pipeline steel and its implications on pipeline stress corrosion cracking, *Int. J. Hydrogen Energy* 32, 1269–1276.

Cheng, YF (2007c), Thermodynamically modeling the interactions of hydrogen, stress and anodic dissolution at crack-tip during near neutral pH SCC in pipelines, *J. Mater. Sci.* 42, 2701–2705.

Cheng, YF, Niu, L (2007) Mechanism for hydrogen evolution reaction on pipeline steel in near-neutral pH solution, *Electrochem. Commun.* 9, 558–562.

Chu, R, Chen, W, Wang, SH, King, F, Jack, TR, FesslerRR (2004) Microstructure dependence of stress corrosion cracking initiation in X-65 pipeline steel exposed to a near-neutral pH soil environment, *Corrosion* 60, 275–283.

Daft, EG, Bohnenkamp, K, Engell, HJ (1979) Investigations of the hydrogen evolution kinetics and hydrogen absorption by iron electrodes during cathodic polarization, *Corros. Sci.* 19, 591–612.

Delanty, BS, Marr, JE (1992) Stress corrosion cracking severity rating model, *Proc. International Conference on Pipeline Reliability*, CANMET, Ottawa, Ontario, Canada.

Delanty, BS, O'Beirne, J (1991) Low-pH stress corrosion cracking, *Proc. 6th Symposium on Line Pipe Research*, Paper L30175, PRCI, Falls Church, VA.

Delanty, BS, O'Beirne, J (1992) Major field study compares pipeline SCC with coatings, *Oil Gas J.* 90, 24.

Dong, CF, Fu, AQ, Li, XG, Cheng, YF (2008) Localized EIS characterization of corrosion of steel at coating defect under cathodic protection, *Electrochim. Acta* 54, 628–633.

Elboujdaini, M, Wang, YZ, Revie, RW, Parkins, RN, Shehata, MT (2000) Stress corrosion crack initiation processes: pitting and microcrack coalescence, *Corrosion 2000*, Paper 00379, NACE, Houston, TX.

Eslami, A, Fang, B, Kania, R, Worthingham, B, Been, J, Eadie, R, Chen, W (2010) Stress corrosion cracking initiation under the disbonded coating of pipeline steel in near-neutral pH environment, *Corros. Sci.* 52, 3750–3756.

Fessler, RR, Groeneveld, T, Elsea, A (1973) Stress-corrosion and hydrogen-stress cracking in buried pipelines, *International Conference on Stress Corrosion Cracking and Hydrogen Embrittlement of Iron Base Alloys*, Firminy, France.

Flis, J, Zakroczymski, T (1992) Enhanced hydrogen entry in iron at low anodic and low cathodic polarizations in neutral and alkaline solutions, *Corrosion* 48, 530–539.

Ford, FP (1996) Quantitative prediction of environmentally assisted cracking, *Corrosion* 52, 375–395.

Fu, AQ, Cheng, YF (2010) Effects of alternating current on corrosion of a coated pipeline steel in a chloride-containing carbonate/bicarbonate solution, *Corros. Sci.* 52, 612–619.

Fu, AQ, Cheng, YF (2011) Characterization of the permeability of a high performance composite coating to cathodic protection and its implications on pipeline integrity, *Prog. Org. Coat.* 72, 423–428.

Fu, AQ, Tang, X, Cheng, YF (2009) Characterization of corrosion of X70 pipeline steel in thin electrolyte layer under disbonded coating by scanning Kelvin probe, *Corros. Sci.* 51, 186–190.

Heuer, JK, Stubbins, JF (1999) An XPS characterization of $FeCO_3$ films from CO_2 corrosion, *Corros. Sci.* 41, 1231–1243.

Hirth, JP (1980) Effects of hydrogen on the properties of iron and steel, *Metall. Trans. A* 11, 861–890.

Howell, GR, Cheng, YF (2007) Characterization of high performance composite coating for the northern pipeline application, *Prog. Org. Coat.* 60, 148–152.

International Science and Technology Center (1992) Project 1344-D, ISTC, Moscow, Russia.

Iyer, RN, Takeuchi, I, Zamanzadeh, M, Pickering, HW (1990) Hydrogen sulfide effect on hydrogen entry into iron: a mechanistic study, *Corrosion* 46, 460–468.

Jack, TR, Erno, B, Krist, K, Fessler, R (2000) Generation of near-neutral pH and high-pH SCC environments on buried pipelines, *Corrosion 2000*, Paper 362, NACE, Houston, TX.

Jin, TY, Cheng, YF (2011) In-situ characterization by localized electrochemical impedance spectroscopy of the electrochemical activity of microscopic inclusions in an X100 steel, *Corros. Sci.* 53, 850–853.

Kim, BA, Zheng, W, Williams, G, Laronde, M, Oguchi, N, Hosokawa, Y (2004) Experimental study on SCC susceptibility of X60 steel using full pipe sections in near-neutral pH environment, *Proc. 5th International Pipeline Conference*, Paper IPC2004-0280, Calgary, Alberta, Canada.

King, F, Jack, T, Chen, W, Wilmott, M, Fessler, RR, Krist, K (2000) Mechanistic studies of initiation and early stage crack growth for near-neutral pH SCC on pipelines, *Corrosion 2000*, Paper 361, NACE, Houston, TX.

King, F, Cheng, YF, Gray, L, Drader, B, Sutherby, R (2002) *Proc. 4th International Pipeline Conference*, Paper 27100, Calgary, Alberta, Canada.

Kushida, T, Nose, K, Asahi, H, Kimura, M, Yamane, Y, Endo, S, Kawano, H (2001) Effects of metallurgical factors and test conditions on near-neutral pH SCC of pipeline steels, *Corrosion 2001*, Paper 01213, NACE, Houston, TX.

Li, MC, Cheng, YF (2007) Mechanistic investigation of hydrogen-enhanced anodic dissolution of X-70 pipe steel and its implication on near-neutral pH SCC of pipelines, *Electrochim. Acta* 52, 8111–8117.

Liu, X, Mao, X(1995) Electrochemical polarization and stress corrosion cracking behavior of a pipeline steel in diluted bicarbonate solution with chloride ions, *Scr. Mater.* 33, 145–150.

Liu, ZY, Li, XG, Du, CW, Cheng, YF (2009) Local additional potential model for effect of strain rate on SCC of pipeline steel in an acidic soil solution, *Corros. Sci.* 51, 2863–2871.

Liu, ZY, Li, XG, Cheng, YF (2010) In-situ characterization of the electrochemistry of grain and grain boundary of an X70 steel in a near-neutral pH solution, *Electrochem. Commun.* 12, 936–938.

Liu, ZY, Li, XG, Cheng, YF (2011) Electrochemical state conversion model for occurrence of pitting corrosion on a cathodically polarized carbon steel in a near-neutral pH solution, *Electrochim. Acta* 56, 4167–4175.

Liu, ZY, Li, XG, Cheng, YF (2012) Mechanistic aspect of near-neutral pH stress corrosion cracking of pipelines under cathodic polarization, *Corros. Sci.* 55, 54–60.

Lu, BT, Luo, JL, Norton, PR (2010) Environmentally assisted cracking mechanism of pipeline steel in near-neutral pH groundwater, *Corros. Sci.* 52, 1787–1795.

Meng, GZ, Zhang, C, Cheng, YF (2008) Effects of corrosion product deposit on the subsequent cathodic and anodic reactions of X-70 steel in near-neutral pH solution, *Corros. Sci.* 50, 3116–6122.

National Energy Board (1996) *Stress Corrosion Cracking on Canadian Oil and Gas Pipelines, Report of the Inquiry*, MH-2-95, NEB, Calgary, Alberta, Canada.

Nishikata, A, Ichihara, Y, Tsuru, T (1995) An application of electrochemical impedance spectroscopy to atmospheric corrosion study, *Corros. Sci.* 37, 897–911.

Ott, KF (1998) *Stress Corrosion on Gas Pipelines: Hypothesis, Arguments and Facts, Summarized Data*, Information and Advertising Center, GasProm, Moscow, p. 73.

Parkins, RN (2000) A review of stress corrosion cracking of high pressure gas pipelines, *Corrosion 2000*, Paper 363, NACE, Houston, TX.

Parkins, RN, Blanchard, WK, Delanty, BS (1994) Transgranular stress corrosion cracking of high pressure pipelines in contact with solutions of near-neutral pH, *Corrosion* 50, 394–408.

Qin, Z, Demko, B, Noel, J, Shoesmith, D, King, F, Worthingham, R, Keith, K (2004) Localized dissolution of mill scale-covered pipeline steel surfaces, *Corrosion* 60, 906–914.

Song, RH, Pyun, SI, Oriani, RA (1990) Hydrogen permeation through the passive film on iron by time-lag method, *J. Electrochem. Soc.* 137, 1703–1706.

Tang, X, Cheng, YF (2009) Micro-electrochemical characterization of the effect of applied stress on local anodic dissolution behavior of pipeline steel under near-neutral pH condition, *Electrochim. Acta* 54, 1499–1505.

Tang, X, Cheng, YF (2011) Quantitative characterization by micro-electrochemical measurements of the synergism of hydrogen, stress and dissolution on near-neutral pH stress corrosion cracking of pipelines, *Corros. Sci.* 53, 2927–2933.

Torres-Islas, A, Gonzalez-Rodriguez, JG, Uruchurtu, J, Serna, S (2008) Stress corrosion cracking study of microalloyed pipeline steels in dilute $NaHCO_3$ solutions, *Corros. Sci.* 50, 2831–2839.

Van Boven, G, Wilmott, M (1995) *Pulsed Cathodic Protection 1995: An Investigation of Current Distribution Under Disbonded Pipeline Coatings*, Report 01118, NRTC Calgary, Alberta, Canada.

Van Boven, G, Chen, W, Rogge, R (2007) The role of residual stress in neutral pH stress corrosion cracking of pipeline steels: I. Pitting and cracking occurrence, *Acta Mater.* 55, 29–42.

Wenk, RL (1974) Field investigation of stress corrosion cracking, *Proc. 5th Symposium on Line Pipe Research*, Paper L30174, PRCI, Falls Church, VA.

Xu, LY, Cheng, YF (2012a) Reliability and failure pressure prediction of various grades of pipeline steel in the presence of corrosion defects and pre-strain, *Int. J. Press. Vessels Piping* 89, 75–84.

Xu, LY, Cheng, YF (2012b) An experimental investigation of corrosion of X100 pipeline steel under uniaxial elastic stress in a near neutral pH solution, *Corros. Sci.* 59, 103–109.

Xu, LY, Cheng, YF (2012c) Assessment of the complexity of stress/strain conditions of X100 steel pipeline and the effect on the steel corrosion and failure pressure prediction, *Proc. 9th International Pipeline Conference*, Paper IPC2012-90087, Calgary, Alberta, Canada.

Xue, HB, Cheng, YF (2011) Characterization of inclusions of X80 pipeline steel and its correlation with hydrogen-induced cracking, *Corros. Sci.* 53, 1201–1208.

Zhang, GA, Cheng, YF (2009a) On the fundamentals of electrochemical corrosion of X65 steel in CO_2-containing formation water in the presence of acetic acid in petroleum production, *Corros. Sci.* 51, 87–94.

Zhang, GA, Cheng, YF (2009b) Micro-electrochemical characterization of corrosion of welded X70 pipeline steel in near-neutral pH solution, *Corros. Sci.* 51, 1714–1724.

5

High-pH Stress Corrosion Cracking of Pipelines

5.1 INTRODUCTION

High-pH stress corrosion cracking of pipelines is a classical SCC, noted originally in gas transmission pipelines. Actually, the first documented case of SCC resulting in a gas release, explosion, and fire with several fatalities, which occurred in Louisiana in the mid-1960s, was caused by high-pH SCC [National Energy Board, 1996]. Since then, research on high-pH SCC in pipelines has been ongoing. In the late 1960s, a concentrated carbonate–bicarbonate solution was identified as the most likely environment to in high-pH SCC, and evidence of this solution at the pipe surface was found in a limited number of cases [Fessler, 1969]. To date, high-pH SCC has been recognized on pipelines in the United States, Australia, Iran, Iraq, Italy, Pakistan, Saudi Arabia, and many other countries, thus becoming a worldwide problem affecting the safe operation of pipelines.

5.2 PRIMARY CHARACTERISTICS

Pipeline integrity is maintained by coating and CP as well as by a comprehensive pipeline safety maintenance program. The concentrated carbonate–bicarbonate solution identified as the most probable environment responsible for high-pH SCC may develop as a result of the interaction between hydroxyl ions produced by CP-driven cathode reactions, such as water reduction, and carbon dioxide in the soil generated

Stress Corrosion Cracking of Pipelines, First Edition. Y. Frank Cheng.

by the decay of organic matter. Cathodic reactions cause the pH of the electrolyte trapped under disbonded coatings to increase, and carbon dioxide dissolves readily in the elevated-pH electrolyte, resulting in the generation of concentrated carbonate–bicarbonate electrolyte. The pH of this electrolyte, depending on the relative concentration of carbonate and bicarbonate, is generally between 9 and 11. Apparently, the high-pH electrolyte causing pipeline SCC does not refer to the nearby soil chemistry.

High-pH SCC normally occurs in a relatively narrow cathodic potential range [−600 to −750 mV($Cu/CuSO_4$)] in a concentrated carbonate–bicarbonate environment. Coatings are usually permeable, permitting CP and oxygen to move toward the steel surface. High-pH SCC is temperature sensitive, and temperatures greater than 40°C are necessary for high-pH SCC susceptibility. Thus, there is a greater likelihood of this type of SCC immediately downstream of compressor stations, where the operating temperature might reach 65°C [National Energy Board, 1996]. Thus, the number of failures decreases with increased distance from compressor or pump and lower pipe temperature. Furthermore, evaporation of soil water at a relatively warm pipe surface might lead to generation of a high-pH SCC electrolyte on the steel surface exposed by missing or permeable coatings [Jack et al., 2000]. Cracking would occur only where the CO_2 content in equilibrium with evaporating solution buffers the pH in the cracking range, such as a pH around 8.5, and the potential of the steel surface remains in the cracking range.

The fracture mode tends to be intergranular, often with small branches. The cracks are usually narrow and tight, with almost no evidence of secondary corrosion of crack walls. Moreover, there exists extensive secondary cracking between grains as a result of the branched nature of this failure. High-pH stress corrosion cracks are primarily axial cracks, with very few transverse cracking cases [National Energy Board, 1996].

Furthermore, pipeline SCC can occur over a wide potential range from −500 to −1100 mV(SCE), but with different mechanisms and fracture modes. Intergranular SCC occurs at more anodic potentials than transgranular cracking, which starts as a potential of −890 mV(SCE) is approached. At more cathodic potentials, transgraular hydrogen embrittlement occurs, with little secondary cracking [Mustapha et al., 2012].

5.3 CONTRIBUTING FACTORS

5.3.1 Coatings

Properties of coatings are critical to the type of pipeline SCC. Generally, the majority of high-pH SCC failures have been associated with bituminous coatings such as coal tar or asphalt. Since CP is essential to the generation of a high-pH electrolyte to support high-pH SCC, for a high-pH electrolyte to be generated the coating must be permeable to CP. The potential range for high-pH SCC lies between the corrosion potential of steel in soils and the CP potential of −850 mV($Cu/CuSO_4$). When CP is fully accessible to the line pipe steel, the electrochemical condition for high-pH SCC is not met. Once CP is partially shielded by coatings where insoluble salts, such as cadmium carbonate, are deposited, the potential of the pipeline would relax and possible lie in the potential range for cracking.

Moreover, the resistance of a coating to disbondment and the type of surface preparation used with the coating are two additional factors affecting the resistance of pipelines to high-pH SCC [Beavers, 1992; Beavers et al., 1993a,b]. The ability of a coating to resist disbonding is a primary performance property and affects almost all forms of external corrosion of pipelines, including SCC. Coatings with good adhesion properties are generally resistant to the mechanical action of soils caused by wet–dry and freeze–thaw cycles. They are also able to resist water transmission and cathodic disbondment.

The development of high-pH environments under disbonded permeable coatings is determined by the relative rates of OH^- generation during reduction of water driven by the CP applied and the rate of OH^- transport away from the steel surface through the coating. In addition to an increase in trapped-water pH, additional processes, such as evaporation or rapid CO_2 permeation through coating, are required to generate a high-pH SCC environment [Been et al., 2005].

The solution pH has a significant effect on the cracking potential, with pH values varying with the different concentrations of CO_2, CO_3^{2-}, and HCO_3^- ions. The potential range for intergranular cracking, with solution pH ranging from 6.7 to 11.0, is about 100 mV wide at ambient temperature, with the middle of the potential range falling from about −0.25 to about −0.7 V(SCE) as the pH is increased. The crack growth rates are reduced by an increase in the pH of the solution. In low-pH solutions, the different morphologies of quasi-cleavage can be observed at different cracking potential regimes [Parkins and Zhou, 1997].

With respect to the surface conditions susceptible to SCC, grit-blasted surfaces are generally more resistant than mill-scaled surfaces to high-pH SCC initiation. Grit blasting could impart a compressive residual stress at the pipe surface to inhibit the cracking occurrence and remove mill scale [Beavers et al., 1993a]. Generally, the majority of single-layer FBE coatings are applied on grit-blasted pipe surfaces that have been prepared to a white [National Association of Corrosion Engineers, 1999a] or nearly white [National Association of Corrosion Engineers, 1999b] surface finish. Years ago, conventional bituminous coatings were frequently applied to mill-scaled surfaces. More recently, bituminous coatings have been employed in mills using commercial blast cleaning. Actually, the low-quality grit blast that is commonly used with plant-applied bituminous coatings decreases the SCC resistance compared to that found with a mill-scaled surface, primarily by creating stress raisers at embedded mill-scale particles.

5.3.2 Cathodic Protection

Occurrence of high-pH SCC on pipelines is dependent directly on the permeation of CP to line pipe steel. A concentrated carbonate–bicarbonate electrolyte to support high-pH SCC is generated due to the CP-driven cathodic reduction of water and the generation of hydroxyl ions, as well as the dissolution of carbon dioxide in the solution. Thus, if the pipeline steel is not maintained at a CP potential, a high-pH environment would not be created.

Furthermore, CP contributes to moving the potential of pipeline steel into the cracking range, which generally lies between the corrosion potential of the steel and the CP potential [e.g., –850 mV($Cu/CuSO_4$)] [Parkins, 1974]. Seasonal fluctuations of CP potential could increase the possibility that the pipeline falls into the cracking potential range. Thus, an accurate determination of this sensitive range is critical to avoiding high-pH SCC.

The use of CP would affect the passivity of steel and the passive film stability, and thus the SCC susceptibility of the steel in high-pH solutions. Figure 5-1 shows the polarization curves measured on X70 pipeline steel in low, intermediate, and high concentrations of carbonate–bicarbonate solution with 30 μm in thickness in the presence of a CP of −850 mV($Cu/CuSO_4$). It is seen that the preapplication of CP causes a negative shift of corrosion potential and a significant increase in anodic current density in all solutions. Apparently, the CP application history would enhance the corrosion activity of steel in carbonate–bicarbonate solutions, degrading the passivity of the steel developed in the solutions.

During full CP application on pipe steel, hydrogen evolution, in addition to oxygen reduction, may not be a negligible cathodic process at the CP potential. The hydrogen atoms generated may penetrate the steel and accumulate at local irregularities such as inclusions, microvoids, defects, and/or crack tips. Upon hydrogen entry, the corrosion potential of the steel in a concentrated carbonate–bicarbonate solution is shifted negatively, as shown in Fig. 5-2, and the passive current density increases (Fig. 5-3), indicating hydrogen-enhanced corrosion activity and degradation of the passivity of the steel.

To further illustrate mechanistically the effect of hydrogen on the corrosion of steel, the corrosion potentials of hydrogen-charged and uncharged steels in the solution are expressed as

$$E_{corr}^{uncharged} = E_{Fe^{2+}/Fe}^{0} + \frac{2.303\ RT}{2F} \log \frac{[Fe^{2+}][H^+]^{1/2}}{[H_{ad}^0]^{1/2}} - E_{ref} \qquad (5\text{-}1)$$

$$E_{corr}^{charged} = E_{Fe^{2+}/Fe}^{0} + \frac{2.303\ RT}{2F} \log \frac{[Fe^{2+}][H^+]^{1/2}}{[H_{ad}]^{1/2}} - E_{ref} \qquad (5\text{-}2)$$

where $[H^+]$ is the concentration of hydrogen ions in solution, $[H_{ads}^0]$ the subsurface concentration of the adsorbed hydrogen atoms in uncharged steel, and $[H_{ads}]$ the subsurface concentration of the adsorbed hydrogen atoms in charged steel. The difference of the corrosion potentials is

$$\Delta E = E_{corr}^{charged} - E_{corr}^{uncharged} = \frac{2.303RT}{2F} \log \frac{[H_{ad}^0]^{1/2}}{[H_{ad}]^{1/2}} \qquad (5\text{-}3)$$

Since $[H_{ads}^0] < [H_{ads}]$, the corrosion potential of charged steel is more negative than that of uncharged steel.

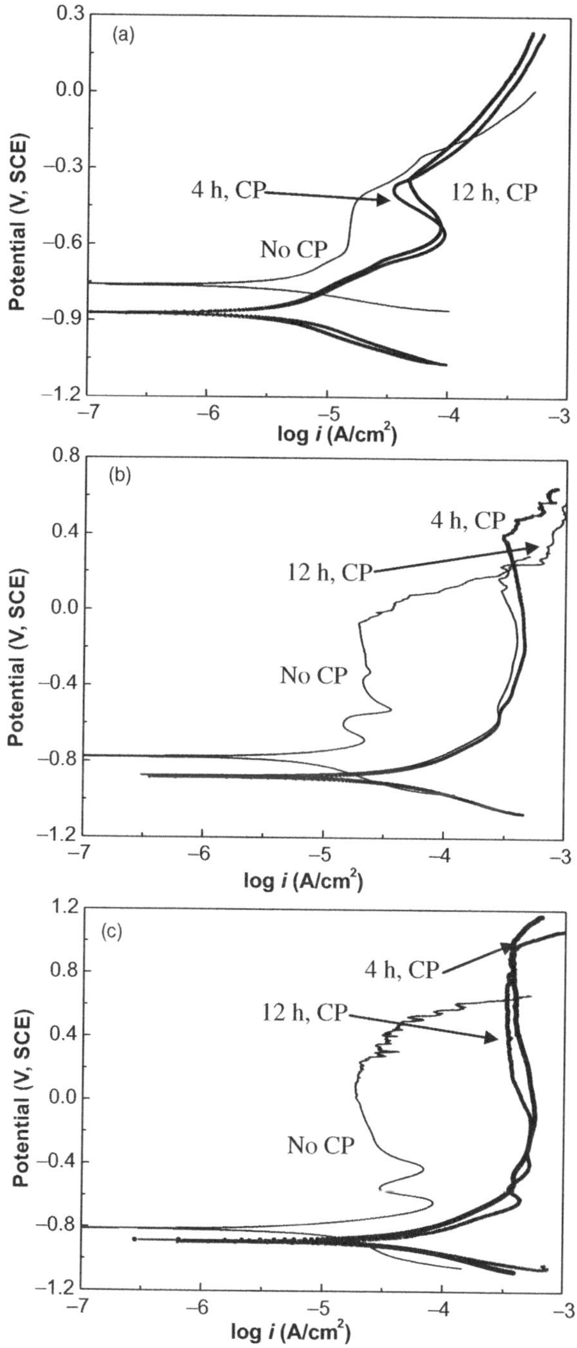

Figure 5-1 *Polarization curves of X70 pipeline steel measured in low (a), intermediate (b), and high (c) concentrations of carbonate–bicarbonate solutions with 30 μm in thickness as a function of CP duration. (From Fu and Cheng [2010].)*

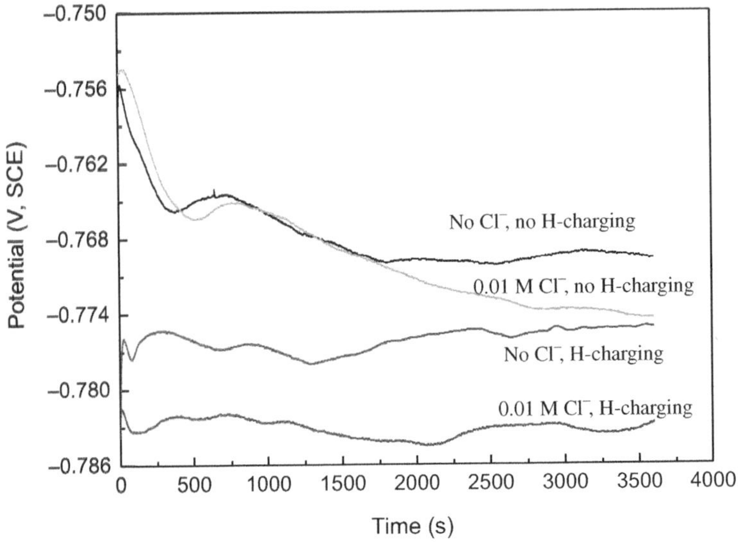

Figure 5-2 *Time dependence of corrosion potentials measured on uncharged and precharged X80 steel electrodes in concentrated carbonate–bicarbonate solutions without and with 0.01 M NaCl. (From Xue and Cheng [2010].)*

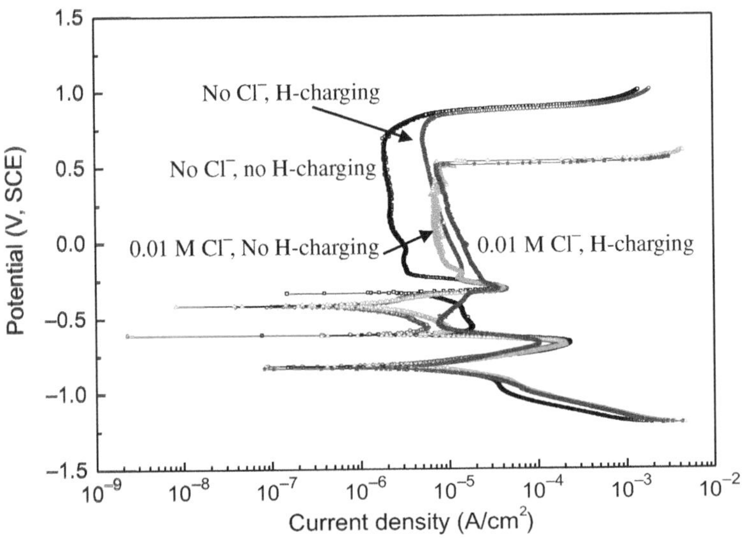

Figure 5-3 *Polarization curves measured on uncharged and precharged X80 steel electrodes in concentrated carbonate–bicarbonate solutions without and with 0.01 M NaCl. (From Xue and Cheng [2010].)*

Similarly, upon application of CP, the solution pH would increase, due to the generation of OH^-. The solution pH is dependent on the CP potential by

$$pH = -\frac{E_c + 0.241}{0.0592} \tag{5-4}$$

As the CP potential shifts negatively, the pH of the solution increases and the concentration of hydrogen ions decreases. The corrosion potentials of the steel in the absence and presence of CP can be expressed as

$$E_{corr} = E^0_{Fe^{2+}/Fe} + \frac{2.303\,RT}{2F} \log \frac{[Fe^{2+}][H^+]^{1/2}}{[H^0_{ad}]^{1/2}} - E_{ref} \tag{5-5}$$

$$E^{CP}_{corr} = E^0_{Fe^{2+}/Fe} + \frac{2.303\,RT}{2F} \log \frac{[Fe^{2+}][H^+_{CP}]^{1/2}}{[H^0_{ad}]^{1/2}} - E_{ref} \tag{5-6}$$

where $[H^+]$ and $[H^+_{CP}]$ are the concentrations of hydrogen ions in the solution in the absence and presence of CP, respectively. Since $[H^+] > [H^+_{CP}]$, the corrosion potential of steel under a CP potential is more negative than that without CP.

Moreover, when charged steel is subjected to anodic polarization, the anodic current density is greater than that measured on uncharged steel. Oxidization of the hydrogen atoms adsorbed contributes to increasing anodic current density. Furthermore, the H^+ produced reduces the solution pH under disbonded coating, which would also increase the anodic current density of the steel. Similar results have been observed by other researchers [Yu et al., 2001; Ningshen et al., 2006].

5.3.3 Soil Characteristics

Identical to soils associated with nearly neutral–pH SCC, high-pH SCC of pipelines occurs in a wide variety of soils. Investigations of soil and water extracts from soil taken from high-pH SCC locations [Wenk, 1974] showed that both the compositions of water extracts and the physical properties of soils do not show a certain relationship with SCC occurrence. However, the primary components of the electrolytes analyzed contain sodium carbonate and bicarbonate. The presence of concentrated carbonate and bicarbonate electrolytes has been identified as the typical solution chemistry that results in high-pH SCC [National Energy Board, 1996].

Pipeline SCC occurs in electrolyte trapped under disbonded coating, and the solution chemistry and electrochemistry depend on the synergistic effect of CP and the coating performance and properties. It is thus reasonable to assume that soil chemistry would not affect the pipeline SCC directly, including high-pH SCC. Actually, any consistent effect of soil chemistry on the high-pH SCC of pipelines has not been found based on extensive analyses of soil data from both the UK and the United States [Mercer, 1979]. However, the moisture content of soils, the ability of soils to cause coating damage, and localized variations in the level of CP are the primary soil-related factors affecting SCC occurrence. In particular, the chloride content in soils would affect corrosion and SCC of pipelines once chloride ions have access to the trapped electrolyte by diffusion or migration.

Inclusion of chloride in solution would shift the corrosion potential of the steel negatively, as shown in Fig. 5-2. Moreover, chloride ions are able to increase the passive current density and shift the pitting potential negatively, thus decreasing the passive range of steels in carbonate–bicarbonate solutions (Fig. 5-3). Apparently, a high content of chloride ions in soils would have the potential to increase corrosion activity, especially the pitting corrosion susceptibility, of pipeline steels once chloride ions permeate the electrolyte. Since SCC usually starts from pits, the susceptibility of crack initiation increases in chloride-containing solutions.

Furthermore, introduction of a trace amount of oxygen in the nominally deaerated solution could switch the initially anaerobic corrosion of steels to aerobic corrosion [Sherar et al., 2010]. The properties of film formed on the steel surface would experience alterations due to the diffusion of oxygen into faults of the siderite film that formed in anaerobic conditions, leading to a local formation of Fe^{3+} by oxidation of the Fe^{2+}. The hydrolysis of Fe^{3+} produces protons and leads to local acidification. The accompanying process could be the coupling of the steel dissolution at the base of the acidified location and the reduction of Fe_3O_4. Once pore opening is achieved, the separation of anodes and cathodes could be stabilized. The corrosion potential achieved is within the range to support high-pH SCC, −680 to −710 mV(SCE), in concentrated carbonate–bicarbonate solutions. Thus, the oxygen-driven transition from anaerobic to aerobic corrosion and the subsequent local anode and cathode separation could play an essential role in the initiation or early-stage growth of high-pH stress corrosion cracks.

The soil seasonal change is one of the major factors that contribute to high-pH SCC of pipelines [Wenk, 1974; Parkins, 1994]. A mathematic model has been developed to predict the temporal and spatial variations of the solution chemistry (pH; concentrations of Na^+, CO_3^{2-}, and HCO_3^-), steel potentials, and corrosion rate in a disbonded region of a coating when the holiday potential is set to vary as a result of the effect of soil seasonal wet–dry cycles [Song, 2010]. According to the model, in the wet season, CP at a pipe coating holiday could raise the local pH and the concentrations of Na^+, CO_3^{2-}, and HCO_3^- in the coating disbonded region, making them much greater than in the soil groundwater. This causes a SCC-susceptible condition: a high-pH, concentrated CO_3^{2-} and HCO_3^- solution, to form under the disbonded coating. With CP, CO_3^{2-} and HCO_3^- ions tend to displace out the more diffusive and free ions, such as Cl^- from coating disbondment. They then reside and combine with the soluble alkaline ions such as Na^+ to form high-pH and a more concentrated carbonate–bicarbonate solution. Conversely, during a dry season, the high-pH solution may not be possible to form without CP, even though a concentrated solution could form by water evaporation. It is questionable how long such a concentrated solution can endure before it dries out. A threshold thickness of the solution layer is critical to support pipeline corrosion and SCC [Fu and Cheng, 2010], and thus any SCC-susceptible condition will cease to exist with drying out of the solution.

5.3.4 Microorganisms

High-pH SCC of pipelines usually occurs in aerobic soils. Thus, microorganisms that depend on oxygen for their metabolism could be found in soils where high-pH SCC is detected. So far, no conclusion has been drawn as to the certain relationship between the type of aerobic microorganism and the occurrence of high-pH SCC.

The SRB microorganisms may participate in, and enhance, pipeline pitting corrosion and crack initiation [Abedi et al., 2007]. The consumption of oxygen during cathodic reaction and the presence of rich organic carbon sources in the soil provide suitable conditions for SRB growth and activities. In the presence of SRB, the reaction product is usually a sulfur-containing compound such as FeS. This could be the evidence that SRB is involved in the cracking process if the sulfide is found at the crack tip. Generally, SRB-affected crack tips have been rounded (blunted) with a baglike shape [Abedi et al., 2007].

5.3.5 Temperature

Temperature has a remarkable effect on high-pH SCC in pipelines. For example, the number of SCC-induced failures falls markedly with increased distance from a compressor or pump and lower pipe temperature [National Energy Board, 1996]. Moreover, the crack growth rate increases exponentially with temperature increase. Thus, the pipeline operators have attempted to control operating temperature by installing cooling towers to help control high-pH SCC. Temperature also has an effect on the cracking potential range in concentrated carbonate–bicarbonate solutions [Parkins and Zhou, 1997]. The cracking range usually moves to a more negative potential range as the temperature rises.

5.3.6 Stress

The threshold stress for high-pH SCC is on the order of 60 to 100% of the yield strength of pipeline steels based on laboratory results [Baker, 2005]. However, it is worth pointing out that the correlation of lab-determined threshold stress with the actual stresses in service is challenging. The lab tests determine the minimum threshold stress intensity for SCC (K_{ISCC}) of 25 MPa·m$^{1/2}$.

Like nearly neutral–pH SCC, actual service stresses contributing to high-pH SCC include primarily internal operating pressure causing hoop stress; stresses from external sources, such as ground movement and surface loading; and various residual stresses, such as welding stress and bending stress. The residual stress generated due to pipe bending, even below the threshold stress, would affect the SCC processes by altering the corrosion reaction of steels. It has been found [Li and Cheng, 2008] that deformation-induced residual stress, if not sufficiently high, has an inhibitive effect on steel corrosion, especially pitting corrosion and crack initiation under a high-pH carbonate–bicarbonate condition. The inhibitive effect is due to enhanced generation and deposition of carbonate corrosion products and the resulting surface block effect in the stressed zones. For example, local strains at points $x = 0$ mm and $x = 3.6$ mm

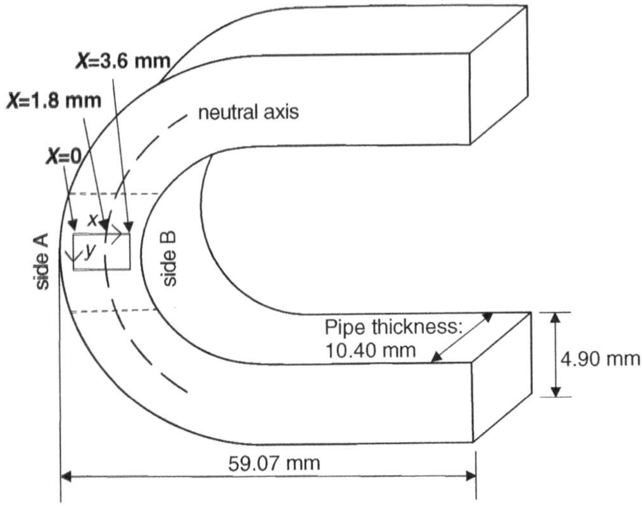

Figure 5-4 Schematic diagram of a U-bend X70 steel specimen and a LEIS scanning area. (From Li and Cheng [2008].)

in Fig. 5-4 were calculated to be approximately 17.1% and 22.8%, respectively. The localized EIS measurements in Fig. 5-5 at these points showed that there was the smallest semicircle size at the point near the neutral axis of the U-bend specimen, that is, $x = 1.8$ mm, while a semicircle measured at the tensile stress area ($x = 0$ mm) is larger than that at the compressive stress area ($x = 3.6$ mm). The tensile and compressive stresses have an identical effect on the corrosion inhibition. Tensile stress enhances steel dissolution more significantly than does compressive stress and thus generates more carbonate corrosion products, resulting in higher surface block effect and impedance.

Furthermore, it was found that it is common for pits to occur around the neutral axis of a U-bend specimen, where deformation of steel and the resulting stress are ignorable, as shown in Fig. 5-6. Deformation-induced stress is not sufficient to break corrosion product film formed at a stressed zone, which is thus more resistant to corrosion than is an unstressed zone (i.e., the zone around the neutral axis). The more active, unstressed zone would be the preferential site to initiate pits. This is an important finding because it is demonstrated that a small stress applied on steel is actually beneficial to maintaining the stability of steel, due to the deposit of corrosion product and the resulting surface block effect. On the contrary, an unstressed zone will experience significant dissolution and pitting corrosion since there is no apparent corrosion product deposit.

It has also been acknowledged [Baker, 2005] that when the stress level in pipelines is not sufficiently high, existing cracks will not propagate, since the slight applied stress inhibits cracking. The further growth of cracks depends on the presence of two types of more aggressive stress conditions: periodic cyclic stress due to pressure fluctuations of the high-pressure natural gas inside pipelines, and high-level stress concentration. These stress conditions meet the requirement to break the

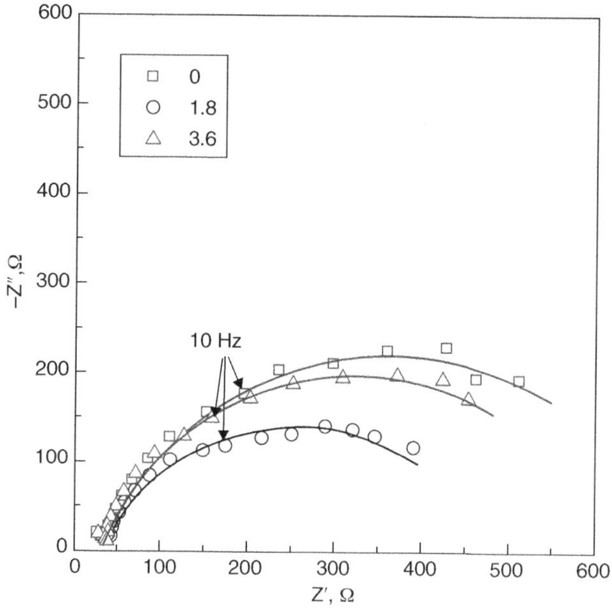

Figure 5-5 *LEIS plots measured at three points across the various stressed areas marked in Fig. 5-4, where* x = *0, 1.8, and 3.6 mm, respectively, with a constant* y = *1.35 mm in a concentrated carbonate–bicarbonate solution. (From Li and Cheng [2008].)*

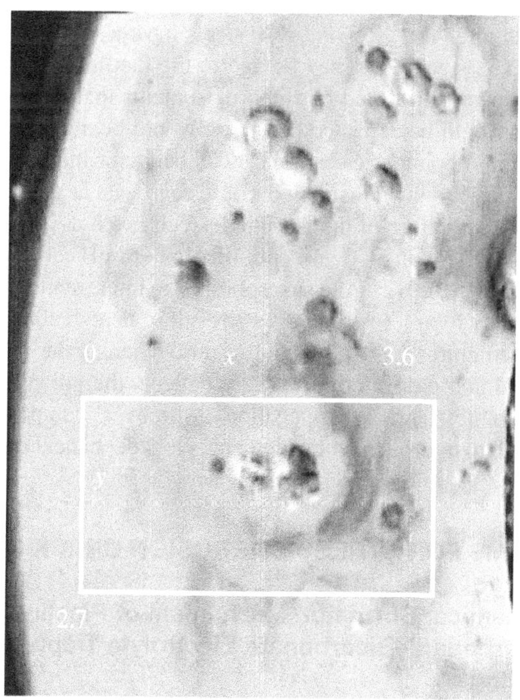

Figure 5-6 *Optical image of the morphology of a U-bend X70 steel specimen after testing in a carbonate–bicarbonate solution. (From Li and Cheng [2008].)*

carbonate–oxide film that has been generated, initiating pits and cracks. This is supportive evidence collected from pipeline operations to demonstrate the role of stress in SCC processes.

5.3.7 Metallurgies

Identical to the effect on nearly neutral–pH SCC, there has been no certain relationship between the composition and structure of steels and the high-pH SCC occurrence. Actually, high-pH SCC has been found on pipelines made of grade X25 to X65 steels. Investigations have been conducted about the high-pH SCC behavior of six different heat treatments of X52 steel, as well as three heat treatments of X65, X70, and X80 steels [Danielson et al., 2001]. It was concluded that in general, the microstructure and microchemistry of steels had a small effect on SCC behavior. Furthermore, the microstructure and microchemistry of X52 steel do not have a major effect on the K_{ISCC} value or the steady crack growth rate [Danielson et al., 2000]. However, other characteristics of steels, such as the creep response to cyclic loading, may be important to SCC susceptibility [Beavers and Harper, 2004]. Since high-pH stress corrosion cracks follow intergranularly along ferrite–ferrite and ferrite–pearlite boundaries, it is thus assumed that knowledge of the composition and activity of grain boundaries would be essential to understanding the cracking mechanism.

Moreover, in comparison to the SCC behavior of several different grades of pipeline steels, including X52, X65, and X70 steels, no difference is found in the steady crack growth rate. The X65 and X70 steels have a significantly lower K_{ISCC} value than that of X52 steel. This may be due to the fact that the former steels are much cleaner (i.e., lower sulfur and phosphorus contents than the latter) [Danielson et al., 2000], and it would be expected that there would be a lower level of S and P segregation at the grain boundaries. Similar SCC behavior indicates the possibility that some of the microalloying constituents could segregate at grain boundaries.

Recently, the microstructure of high-strength steels, such as X100 steel, has been studied in terms of its effect on SCC susceptibility in high-pH carbonate–bicarbonate solutions [Mustapha et al., 2012]. The susceptibility to intergranular SCC of the steel is affected significantly by microstructure as a result of heat treatment. For example, heat-treated microstructures have equiaxial grains, whereas the as-received grains are highly deformed and elongated. The latter deflects the growth of intergranular cracks and hence inhibits crack propagation. Moreover, compared to SCC, the susceptibility of steel to hydrogen embrittlement is less dependent on microstructure.

5.4 MECHANISMS FOR STRESS CORROSION CRACK INITIATION

5.4.1 Electrochemical Corrosion Mechanism of Pipeline Steels in a Thin Layer of Carbonate–Bicarbonate Electrolyte Trapped Under a Disbonded Coating

High-pH SCC of pipelines occurs in aerobic, concentrated carbonate–bicarbonate solutions with a pH around 9.5 trapped under disbonded coating. In particular, the solution chemistry and thickness of electrolyte experience changes with time, which affect significantly the electrochemical corrosion behavior of steel. Moreover, the electrolyte trapped under coating is usually thin, especially in the early stage of coating disbondment. Measurements conducted by immersing a steel electrode in bulk solutions would not be representative of the actual situation.

Figure 5-7 shows polarization curves measured on X70 pipeline steel in carbonate–bicarbonate solutions with various concentrations and solution layer thicknesses. It is seen that the solution layer thickness has a significant influence on the polarization behavior measured. In a low-concentration solution (Fig. 5-7a), the steel could be passivated, with a high active–passive transition current density. With a decrease in solution layer thickness, the transition current density decreases. When the solution layer thickness is reduced to 30 μm, the active–passive transition phenomenon disappears and a stable passivity develops on the steel. Although there is no well-defined dependence of the passive current density, i_p, and pitting potential, E_{pit}, on the solution layer thickness, it is noted that the lowest i_p value and the most negative E_{pit} are measured in solution with a thickness of 30 μm. In an intermediate-concentration carbonate–bicarbonate solution (Fig. 5-7b), the passive current density is approximately independent of solution layer thickness. The E_{pit} value is shifted negatively when the solution layer thickness decreases from that of the bulk solution to 30 μm. With a further increase in the solution concentration, stable passivity is developed on steel in all solutions, as shown in Fig. 5-7c.

The polarization curves shown in Fig. 5-7 demonstrate that a decreasing solution layer thickness would enhance the passivity of steel in a low-concentration carbonate–bicarbonate solution. However, E_{pit} is shifted negatively, indicating that the susceptibility to pitting corrosion of passivated steel increases. This finding is important in terms of the fact that the steel is vulnerable to corrosion, especially pitting corrosion, in the early stage of coating disbondment, where the solution layer developed is thin and the solution is not concentrated sufficiently. There is a quite a high level of sensitivity of steel to pitting corrosion at this stage.

Generally, the electrochemical polarization of pipeline steel (X70 steel in this measurement) in concentrated carbonate–bicarbonate solutions includes several regions.

1. *Active dissolution and active–passive transition regions*. The anodic and cathodic reactions during corrosion of pipeline steels in aerated carbonate–bicarbonate solutions include the steel oxidation and reduction of oxygen, respectively:

$$Fe \rightarrow Fe^{2+} + 2e \tag{5-7}$$

$$O_2 + 2H_2O + 4e \rightarrow 4OH^- \tag{5-8}$$

During anodic polarization, the dissolution current density increases with the potential, and the steel is in an active dissolution state. With further polarization, a layer of undissolvable $FeCO_3$ deposit is formed on the steel surface either electrochemically by oxidation of Fe to Fe^{2+} in form of $FeCO_3$ or due

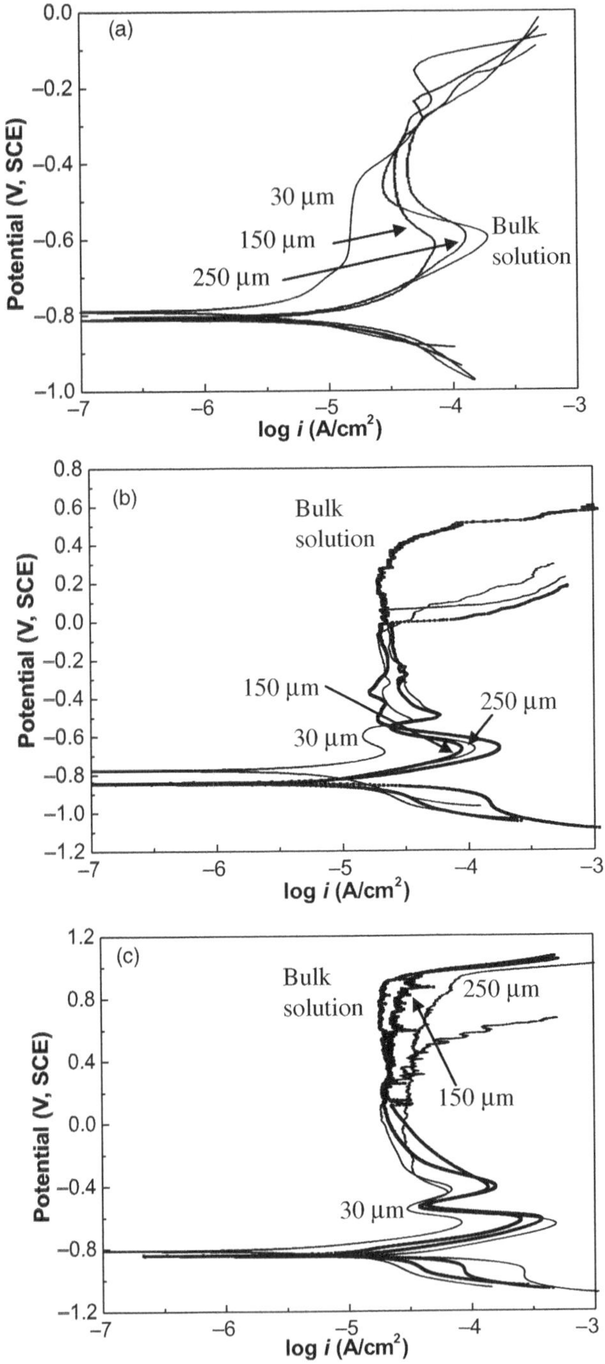

Figure 5-7 *Polarization curves of an X70 steel electrode in low (a), intermediate (b), and high (c) concentrations of carbonate–bicarbonate solutions with various thicknesses. (From Fu and Cheng [2010].)*

chemically to supersaturation of iron carbonate and conversion of $Fe(OH)_2$ to $FeCO_3$ [Davies and Burstein, 1980; Linter and Burstein, 1999]:

$$Fe^{2+} + CO_3^{2-} \rightarrow FeCO_3 \quad (5\text{-}9)$$

$$Fe + HCO_3^- + e \rightarrow FeCO_3 + H \quad (5\text{-}10)$$

$$Fe(OH)_2 + HCO_3^- \rightarrow FeCO_3 + H_2O + OH^- \quad (5\text{-}11)$$

The active–passive transition current peak in Fig. 5-7 is due to the formation of $FeCO_3$ deposit scale electrochemically on the electrode surface, inhibiting further dissolution of the steel.

Generally, the active–passive transition current density decreases when the solution layer thins. It is attributed primarily to a great propensity for formation of the scale because of the low solution volume. Furthermore, the electrochemical polarization behavior of steel in the thin solution layer undercoating is dependent on the carbonate–bicarbonate concentration [Armstrong and Coates, 1974; Davies and Burstein, 1980]. As shown in Fig. 5-7, in intermediate- and high-concentration solutions, there are two distinct active–passive transition current peaks. While the first peak is attributed to formation of the undissolvable $FeCO_3$ and/or $Fe(OH)_2$ deposit, the second peak is due to the further oxidation of ferrous species to Fe_2O_3 and/or Fe_3O_4 [Heuer and Stubbins, 1999]:

$$4FeCO_3 + O_2 + 4H_2O \rightarrow 2Fe_2O_3 + 4HCO_3^- + 4H^+ \quad (5\text{-}12)$$

$$6FeCO_3 + O_2 + 6H_2O \rightarrow 2Fe_3O_4 + 6HCO_3^- + 6H^+ \quad (5\text{-}13)$$

$$4Fe(OH)_2 + O_2 \rightarrow 2Fe_2O_3 + 4H_20 \quad (5\text{-}14)$$

In a low-concentration solution, the second current peak is not apparent, due primarily to the fact that at low CO_3^{2-} and HCO_3^- concentrations, the $FeCO_3$ scale formed is not sufficient to generate an apparent current peak when it is oxidized into ferric species. Consequently, the two current peaks are overlapped to form a widening active–passive transition peak, as shown in Fig. 5-7a.

2. *Passive region*. As the potential increases further, the steel is in stable passivity. The current density measured decreases remarkably and is independent of potential. In a low-concentration carbonate–bicarbonate solution, the passive current density decreases with decreased solution layer thickness, which is associated with the oxidation of Fe(II) to Fe(III). Apparently, enhanced mass transport of dissolved oxygen to a steel surface contributes to the steel oxidation and thus the passivity of steel. The dissolved oxygen is quite easy to diffuse through a thin solution layer to reach the steel surface.

   The effect of solution layer thickness on passive current density is not apparent in intermediate- and high-concentration solutions. It is attributed primarily to the strong passivating ability of a solution at high concentrations. Consequently, the role of oxygen in the passivity of steel becomes less important.

Moreover, the dissolubility of oxygen in a concentrated solution is much less than that in a dilute solution [Raja and Jones, 2006; Revie and Uhlig, 2008].

3. *Transpassive or pitting region.* When a slight positive shift of potential causes a rapid increase in current density, the steel is in the transpassive or pitting corrosion region. It is generally caused by the attack of chloride ions contained in solution on the passive film by displacing oxygen present in the film and generating cation vacancies, resulting in a breakdown of the film and the initiation of corrosion pits [Macdonald, 1992]. Therefore, E_{pit}, which is usually defined as the potential where the anodic current density begins to increase rapidly, depends on the chloride concentration in the solution. Since E_{pit} is shifted negatively with a decrease in solution layer thickness, the film formed in a thin solution layer is more prone to pitting corrosion than that formed in a thick solution layer.

   Furthermore, an increase in carbonate–bicarbonate concentration lessens the effect of solution layer thickness on E_{pit}. In general, the effect of extraneous anions such as HCO_3^- and CO_3^{2-} on the pitting corrosion of steels depends on their ability to compete with Cl^- for sites on the passive surface, making the potential shift to a more noble value in order for Cl^- to displace oxygen. Therefore, the susceptibility of passivated steel to pitting corrosion would decrease with an increase in solution concentration.

   It is worth pointing out that apparently, when the steel is subject to hydrogen entry, its pitting potential is not altered [Xue and Cheng, 2010]. As shown in Fig. 5-8, when cyclic polarization curves of uncharged and charged X80 steel

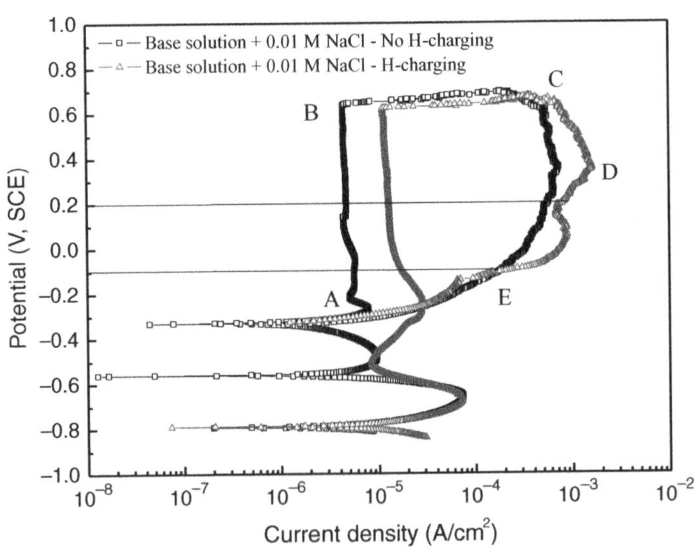

Figure 5-8 *Cyclic polarization curves measured on uncharged and precharged X80 steel electrodes in a carbonate–bicarbonate solution containing 0.01 M NaCl. (From Xue and Cheng [2010].)*

electrodes are present in a chloride-containing carbonate–bicarbonate solution, both curves exhibit a big hysteresis loop, indicating that pitting corrosion is inevitable in testing systems. With an increased passive current density upon hydrogen charging, E_{pit} is not affected. Thus, hydrogen entering the steel does not get involved in pit initiation.

5.4.2 Conceptual Model for Initiation of Stress Corrosion Cracks in a High-pH Carbonate–Bicarbonate Electrolyte Under a Disbonded Coating

A conceptual model has been developed to understand the initiation of SCC on pipelines in a high-pH environment under a disbonded coating [Fu and Cheng, 2010]. According to the model, several steps are critical to the crack initiation, including those noted below.

1. *Development of a high-pH carbonate–bicarbonate electrolyte under a disbonded coating*. Upon application of CP, the CP current would permeate the coating and reach the steel surface, generating OH^- through the reduction of water and thus resulting in an elevated pH for the trapped electrolyte. Simultaneously, CO_2 in the soil generated by the decay of various organic substances permeates the trapped electrolyte and dissolves. At a pH of about 9 to 11, a concentrated carbonate–bicarbonate electrolyte is formed, due to the hydrolysis of H_2CO_3 and HCO_3^- in the alkaline environment [King et al., 2000]. Furthermore, a small amount of Cl^- is contained in the trapped solution from soils [Song, 2008]. In particular, bicarbonate ions play a critical role in the dissolution reactions of pipelines and are a key corrosive agent in anodic and cathodic reactions. Bicarbonate is also proposed to contribute to the formation and complexation of iron carbonate [Linter and Burstein, 1999].
2. *Initiation of corrosion pits in the thin electrolyte developed during an early stage of coating disbondment*. When coating is initially disbonded from the steel, the crevice under the disbonded coating is narrow, and the trapped solution experiences an elevation of the carbonate–bicarbonate concentration, due to the permeation of CO_2. In both low and high concentrations of the solution, there is a considerably lower E_{pit} value in a thin solution layer than that in a thick solution layer. Pit initiation is affected by chloride ions rather than by hydrogen. In the presence of Cl^-, corrosion pits would be initiated at the early stage of coating disbondment, where the trapped solution is usually thin and the solution concentration is relatively low. The role of chloride ions in pitting is versatile, including prevention of stable passivation and increased anodic sensitivity of steels [Jelinek and Neufeld, 1980; Eliyan et al., 2012]. With increased solution layer thickness and solution concentration, the E_{pit} value shifts positively, and the resistance of steel to pitting corrosion increases.

   Furthermore, preapplication of CP on coated steel would increase the anodic current density. Apparently, a CP application history on pipelines would degrade the passive film formed, making it more active. Consequently, it would be easy to be attacked by chloride ions, resulting in the initiation of corrosion pits.

3. *Transition of corrosion pits toward cracks*. While corrosion pits act as stress risers for SCC initiation, the stressing condition that pipelines experience would facilitate the transition from pits to cracks [Van Boven et al., 2007]. As shown in Fig. 5-9, application of tensile stress enhances the anodic current density of steel, resulting in steel being more susceptible to pitting corrosion, as indicated by the negative shift in pitting corrosion. Consequently, there is a strong possibility that the corrosion pits will transit into cracks under stressing conditions. It is thus reasonable to assume that the stresses exerted on pipelines contribute to the transition of corrosion pits toward cracks.

The susceptible potential range for crack initiation in pipeline steel high-pH solutions corresponds to the active-to-passive transition region [Oskuie et al., 2012]. The tendency for SCC to occur at any given potential by the film rupture model can be evaluated on the basis that at least a one-order-of-magnitude difference in current density exists between slow and fast sweeps during potentiodynamic polarization measurements [Holroyd, 1977]. The significance of such a difference between slow and fast scans is that it indicates an intermediate passivation rate that is ideal for a dissolution mechanism [Pilkey et al., 1995]. A high passivation rate causes the crack tip to be protected continually, while a low rate results in excessive dissolution at the crack tip, leading to blunting. Using these criteria, the most probable cracking potential range is found to exist in the region around the active-to-passive transition potential. Moreover, the lower-bound potential of SCC occurs at potentials where passivation

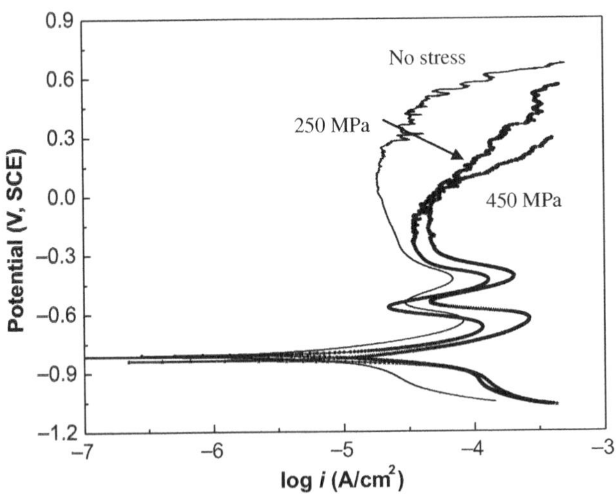

Figure 5-9 *Polarization curves of X70 steel measured in a concentrated carbonate–bicarbonate solution at a thickness of 30 μm as a function of applied stress. (From Fu and Cheng [2010].)*

begins, and this allows selective dissolution [Weng and Li, 2005]. In other words, the sensitive potentials are intermediate values corresponding to the transition from stable passivation to nonselective dissolution of steels. For example, the sensitive potentials of X70 pipeline steels in high-concentration carbonate–bicarbonate solutions at 75°C are determined to be –770 to –810 mV(SCE). This sensitive potential would move in the positive direction as the temperature decreases.

Furthermore, hydrogen, although not involved in the initiation of corrosion pits (does not affect the pitting potential), would accelerate pit growth and the transition toward cracks [Xue and Cheng, 2010], which can be seen from the effect of hydrogen charging on pit propagation rate shown in Fig. 5-10, where the current density profiles of charged and uncharged X80 steels in a potential step cycle in a chloride-containing carbonate–bicarbonate solution are recorded. When the potential is scanned from E_{corr} to 0.2 V(SCE) (i.e., the potential in the passive region), the current density increases initially and then drops gradually to a steady value. Since the steel is in passivity at 0.2 V(SCE), the anodic current density is kept at a low, steady-state value. When the potential is shifted from 0.2 V(SCE) to 0.8 V(SCE), a potential more positive than E_{pit}, which is about 0.6 V(SCE), the current density increases rapidly. When the potential is back to 0.2 V(SCE) in a single step, and held for 10 min, the current density continues to increase and fluctuates around a relatively stable value. After the potential is shifted negatively to −0.1 V(SCE) and held for 10 min, the current density decreases and maintains a low value. When the potential is again increased to 0.2 V(SCE), the current density increases rapidly to a higher value than that recorded previously at the same potential. Moreover, compared to the current density measured on the uncharged electrode, there is a much higher current density obtained on the hydrogen-charged steel electrode.

Apparently, hydrogen is able to accelerate pit growth, as demonstrated by the increased pitting current density when the applied potential is greater than E_{pit} [i.e., up to 0.8 V(SCE)]. Even when the potential is back to a passive potential of 0.2 V(SCE), the current density continues to increase since the pits generated cannot be repassivated at this potential. The increment of current density measured upon hydrogen charging is approximately four times that of uncharged steel during the continuous growth of corrosion pits. Compared to the enhanced anodic dissolution current of the steel, the current increment due to hydrogen oxidation is negligible [Qin et al., 2001; Li and Cheng, 2007]. Therefore, it is reasonable to assume that the current density measured is attributed primarily to dissolution current density rather than oxidation of the charged hydrogen. When the applied potential is farther back, to –0.1 V(SCE), the decrease in current density is due to the partial repassivation of corrosion pits. However, the newly formed passivity demonstrates a higher level of activity on precharged steel than on uncharged steel.

Furthermore, the mode of SCC initiation in high-pH carbonate–bicarbonate solutions can be influenced greatly by the surface condition of the steel [Wang and Atrens, 2003]. For example, for a mechanically polished X65 pipeline steel surface with a deformed layer (5 μm in thickness), where the ferrite grains are elongated parallel to the specimen surface, there are transgranular cracks in the layer. Cracking becomes intergranular when the deformed layer penetrates the steel substrate, and the

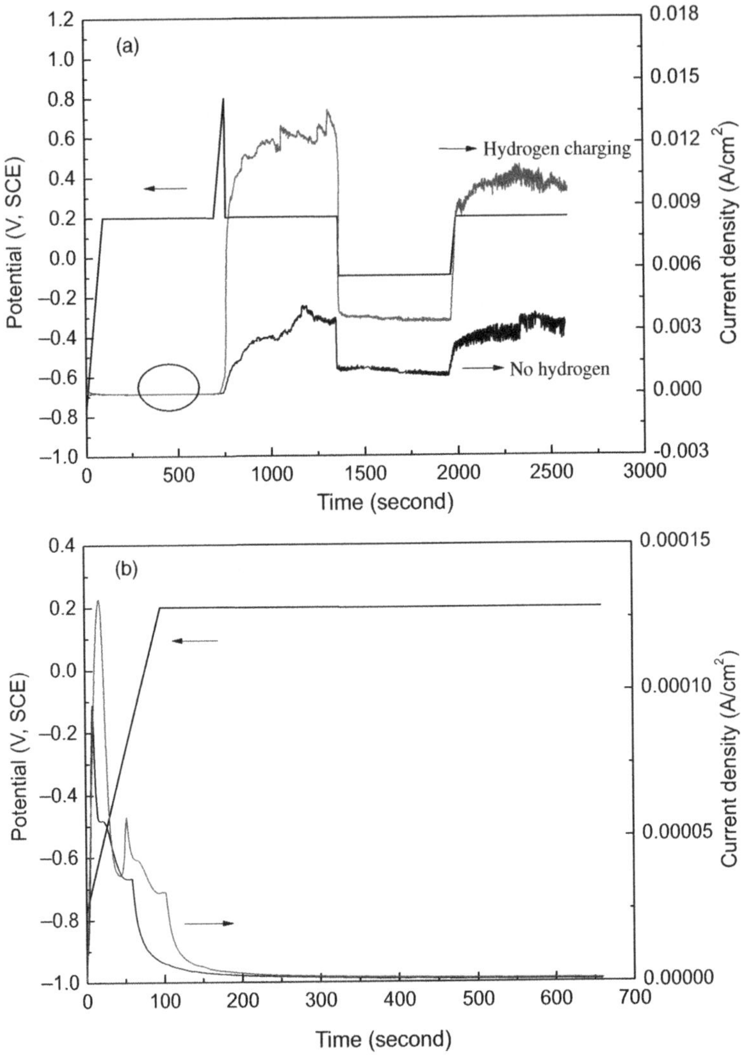

Figure 5-10 *Potential–time and current density–time curves recorded in a potential step cycle in a concentrated carbonate–bicarbonate solution; (b) shows enlarged plots of the passive current densities shown in (a). (From Xue and Cheng [2010].)*

original transgranular cracking stops at ferrite grain boundaries oriented perpendicular to the crack propagation direction. Conversely, when the steel specimen is polished electrochemically, the oxide film is cracked at many locations, but intergranular stress corrosion cracks propagate into the steel substrate only when an oxide crack propagates a grain boundary. Oxide cracking away from grain boundaries would be arrested, and the SCC initiation mechanism observed is not associated with preferential attack of the ferrite grain boundaries. Thus, the influence of the deformed surface layer of steel is important for crack initiation.

5.5 MECHANISMS FOR STRESS CORROSION CRACK PROPAGATION

5.5.1 Enhanced Anodic Dissolution at a Crack Tip

It is well accepted [Armstrong and Coates, 1976; Wang and Atrens, 2003; Sanchez et al., 2007] that high-pH SCC is attributed to anodic dissolution at the grain boundaries and repeated rupture of passive film that forms over the crack tip. Therefore, the properties of passive film are crucial to stress corrosion crack propagation. From the perspective of cracking kinetics, crack propagation requires that the steel be in a certain potential range, which neither allows formation of a permanently stable passive film nor supports constant active dissolution at the crack tip. The propagation rate of cracks actually depends on the competition between the rate of film growth and that of film rupture.

The crack tip works as a stress raiser when tensile stress is applied, whereas the crack wall is free of significant stress concentration. Actually, there is a distinct current density difference in the presence and absence of applied stress, as indicated in Fig. 5-10. The crack tip is subjected to a high dissolution rate, due to the periodic film rupture resulting from applied stress and local stress concentration, while the crack wall may still be in a passive state. As a consequence, the dissolution of steel at the crack tip will be enhanced further. Since the solution electrochemistry and stress state at the crack tip are quite different from those in the region ahead of the crack, the passive film formed at the crack tip would be different from that formed in the surrounding region, affecting the crack propagation.

As shown in Fig. 5-11, systematic electrochemical characterization has been conducted on a precracked X70 pipeline steel specimen to compare corrosion properties

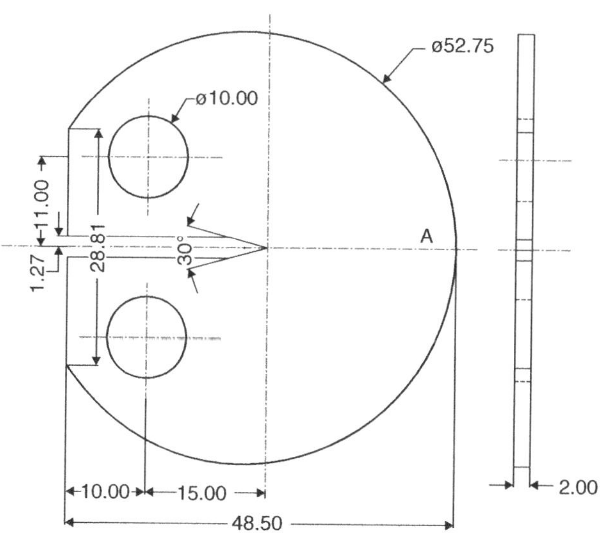

Figure 5-11 *Schematic diagram of a precracked X70 steel compact tension specimen (unit: mm). (From Zhang and Cheng [2010].)*

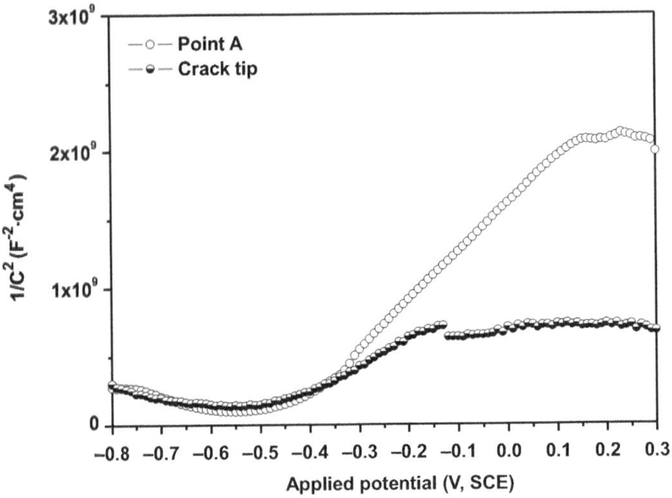

Figure 5-12 *Mott–Schottky plots measured at the crack tip and point A on the specimen in Fig. 5-11 under 1500 N of force in a concentrated carbonate–bicarbonate solution. (From Zhang and Cheng [2010].)*

and the stability of passive film formed at a crack tip and away from the crack tip (i.e., point A in Fig. 5-11) in concentrated carbonate–bicarbonate solutions [Zhang and Cheng, 2010]. Figure 5-12 shows Mott–Schottky plots measured at a crack tip and at point A when a steel specimen is under a 1500-N force in a concentrated carbonate–bicarbonate solution. An approximately linear relationship is observed in both $C^{-2} \sim E_{app}$ (C is the capacitance of the space-charge layer of a passive film, and E_{app} is the applied potential) plots with a positive slope, indicating that the passive film formed on the steel surface is an n-type semiconductor. Moreover, the slope of the line obtained at the crack tip is obviously smaller than that measured at point A. Donor densities in passive films, which are inversely proportional to the Mott–Schottky slope, formed at the crack tip and point A are 5.19×10^{21} and 2.68×10^{21} cm^{-3}, respectively, indicating the highly doped structure of the passive film. The thicknesses of the space-charge layers at the crack tip and point A are 4.16 and 5.34 Å, respectively. From the parameters derived, the passive film formed at the crack tip is more unstable, as indicated by the higher donor density and thinner space-charge layer than those formed at point A.

LEIS measurements at the crack tip and point A of the specimen under various loads are shown in Fig. 5-13. Although there is an identical feature for the impedance plots measured (i.e., one depressed semicircle over the entire frequency range), the size of the semicircle, which represents the charge-transfer resistance, decreases with an increase in the applied load, which is more apparent at the crack tip than at point A. Moreover, the size of the semicircle measured at the crack tip is smaller than that at point A, especially when a load is applied. Furthermore, the LEIS map is also measured on a precracked specimen under various loads, as shown in Fig. 5-14,

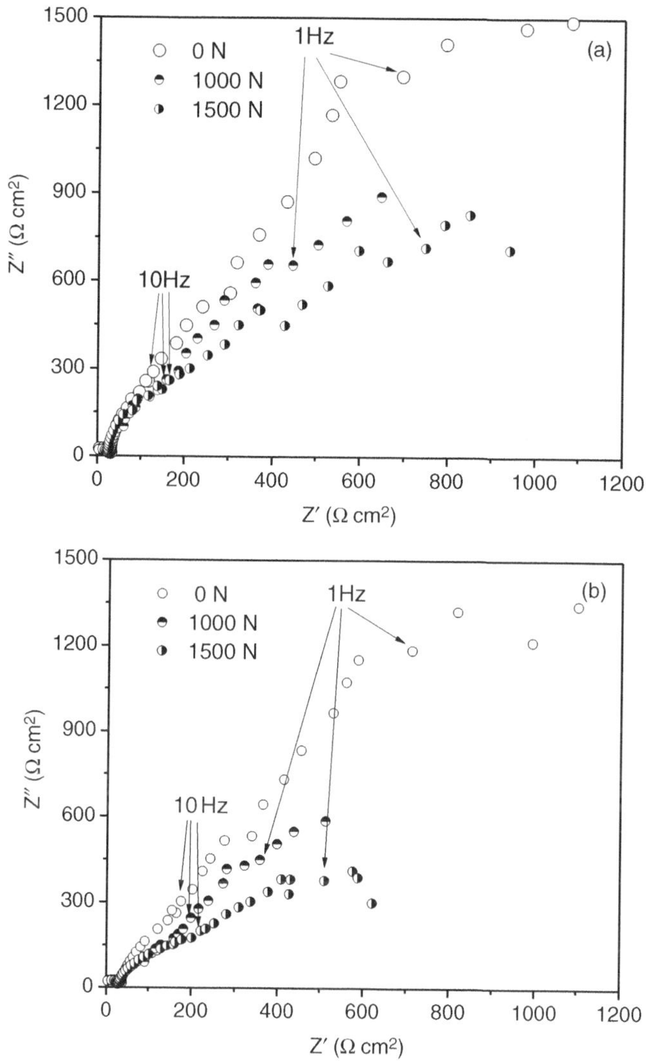

Figure 5-13 *LEIS measured on a precracked X70 steel specimen at (a) point A and (b) a crack tip under various forces in a concentrated carbonate–bicarbonate solution. (From Zhang and Cheng [2010].)*

where there is an obvious impedance valley at the crack tip under an individual load. With an increase in load, the average value of the impedance decreases, especially at the crack tip. All of the results show that there is higher corrosion activity at the crack tip than at any other region in the specimen, such as point A.

Furthermore, to investigate the pitting corrosion behavior of the steel at the crack tip and in the region away from the crack (e.g., point A), the pit propagation rate is measured on the specimen under a 1500-N applied force in the solution. Figure 5-15a

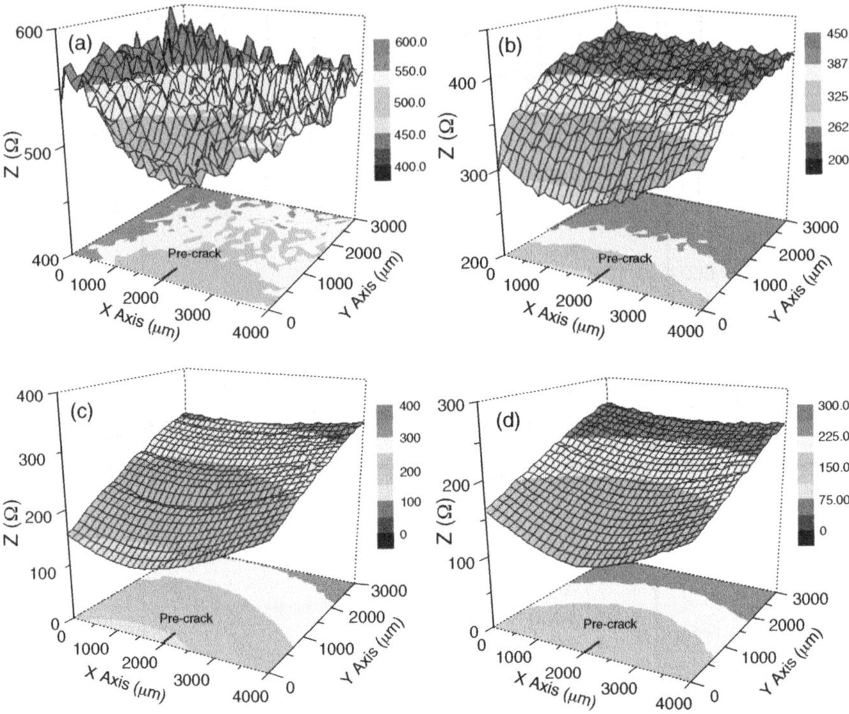

Figure 5-14 *LEIS maps measured on a precracked X70 steel specimen in a concentrated carbonate–bicarbonate solution under various forces: (a) 0 N; (b) 500 N; (c) 1000 N; (d) 1500 N. (From Zhang and Cheng [2010].)*

shows the profile of the applied potential as a function of time, and Fig. 5-15b shows the resulting current density. The potential is first swept to −0.25 V(SCE), which is in the passive range, and is maintained for 10 min to obtain a steady current density. The potential is then shifted to 0.2 V(SCE), which is more positive than E_{pit}. There is a significant increase in current density with the potential shift, due to the initiation of corrosion pits. When the potential is back to −0.25 V(SCE) and held for 10 min, the current density does not recover to the original value but to a much high level, due to the further growth of the pits initiated previously. Since the measured current density (i_{total}) is composed of both passive current density (i_p) and the pitting current density (i_{pit}), the pit propagation rate is determined by

$$i_{\text{pit}} = i_{\text{total}} - i_p \tag{5-15}$$

It is seen from Fig. 5-15 that the pit propagation rates at the crack tip and point A are about 0.49 and 0.36 mA/cm^2, respectively. When the potential is shifted further to −0.3 V(SCE) and maintained for 5 min, the current densities at both the crack tip and point A are not back to the value prior to the pit initiation, indicating that the pits

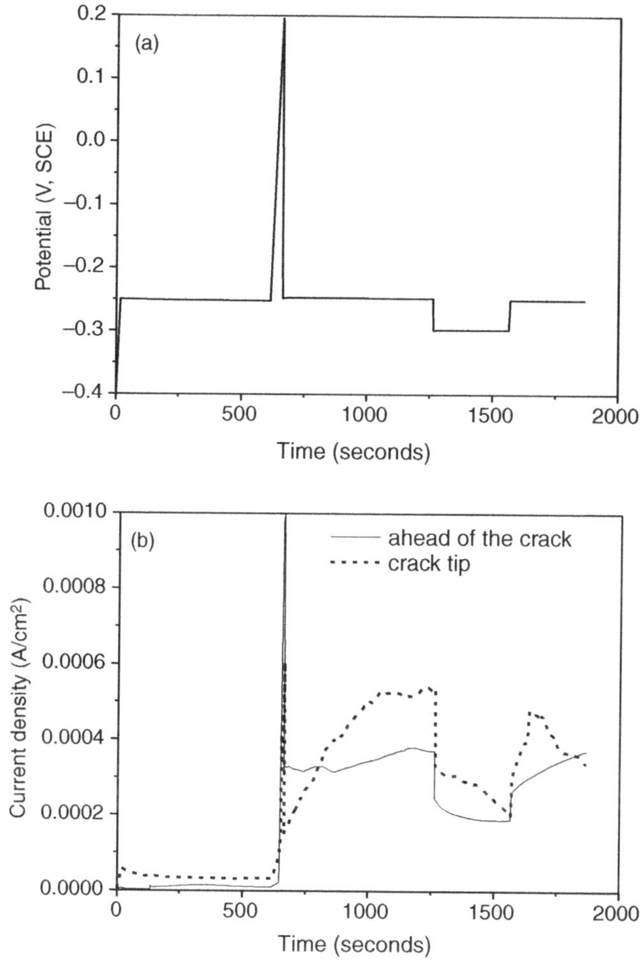

Figure 5-15 *Pit propagation rates measured at the crack tip and point A on the specimen in Fig. 5-11 under 1500 N of force: (a) applied potential profile; (b) recorded current density. (From Zhang and Cheng [2010].)*

cannot be repassivated. The current density measured at the crack tip is much higher than that ahead of the crack (e.g., point A).

Apparently, increasing stress would enhance the anodic current densities at both the crack tip and the region ahead of the crack. Moreover, there is a lower local impedance and higher anodic dissolution rate at the crack tip than in the surrounding region. Therefore, stress concentration at the crack tip could further increase the local dissolution of steel. In terms of the mechanistic aspect of the stress effect on the dissolution of steel, it was proposed [Hirth, 1980] that the stress concentration at a crack tip introduces an additional strain energy to the steel, increasing the internal energy and the lattice strain and resulting in an increasing anodic dissolution rate locally.

The galvanic coupling effect also contributes to enhanced anodic dissolution at the crack tip; that is, the crack tip serves as an anode and the adjacent region as a cathode [Zhang and Cheng, 2010]. The corrosion potential is more negative at the crack tip than at point A, the region ahead of the crack. The galvanic coupling effect causes the crack tip and the adjacent region to be polarized to a galvanic potential E_g. The anodic current density at the crack tip, i_1, and the cathodic current density at the adjacent region from the crack tip, i_2, can be written as

$$i_1 = i_{\text{corr1}}\left[\exp\left(\frac{E_g - E_{\text{corr1}}}{\beta_{a1}}\right) - \exp\left(-\frac{E_g - E_{\text{corr1}}}{\beta_{c1}}\right)\right] \quad (5\text{-}16)$$

$$i_2 = i_{\text{corr2}}\left[\exp\left(-\frac{E_g - E_{\text{corr2}}}{\beta_{c2}}\right) - \exp\left(\frac{E_g - E_{\text{corr2}}}{\beta_{a2}}\right)\right] \quad (5\text{-}17)$$

where E_{corr1}, E_{corr2}, i_{corr1}, and i_{corr2} are corrosion potentials and current densities at the crack tip and the adjacent region, respectively, and β_{a1}, β_{c1}, β_{a2}, and β_{c2} are Tafel slopes of the anodic and cathodic reactions at the crack tip and the adjacent region, respectively. The galvanic current, I_g, between the crack tip and the adjacent region is obtained by

$$I_g = i_1 A_1 = i_2 A_2 \quad (5\text{-}18)$$

where A_1 and A_2 are the areas of the crack tip and the region ahead of the crack, respectively. For the region away from the crack, there is cathodic polarization and, thus, $E_g - E_{\text{corr2}} < 0$. When E_g is quite different from E_{corr1} and E_{corr2}, $e^{-(E_g - E_{\text{corr1}})/\beta_{c1}}$ and $e^{-(E_g - E_{\text{corr2}})/\beta_{a2}}$ are negligible. Combining with Eqs. (5-16) to (5-18), I_g can be determined:

$$\ln(I_g) = \frac{\beta_{c2}}{\beta_{a1} + \beta_{c2}} \ln(A_2 i_{\text{corr2}}) + \frac{\beta_{a1}}{\beta_{a1} + \beta_{c2}} \ln(A_1 i_{\text{corr1}}) + \frac{E_{\text{corr2}} - E_{\text{corr1}}}{\beta_{a1} + \beta_{c2}} \quad (5\text{-}19)$$

Thus, the anodic current at the crack tip is approximately equal to the galvanic current. The anodic current density, i_{a1}, at the crack tip can be expressed as

$$i_{a1} = \frac{I_g}{A_1} \quad (5\text{-}20)$$

Substituting Eq. (5-20) into Eq. (5-19) yields

$$\ln i_{a1} = \frac{\beta_{a1}}{\beta_{a1} + \beta_{c2}} \ln(i_{\text{corr1}}) + \frac{\beta_{c2}}{\beta_{a1} + \beta_{c2}} \ln(i_{\text{corr2}}) + \frac{E_{\text{corr2}} - E_{\text{corr1}}}{\beta_{a1} + \beta_{c2}} + \frac{\beta_{c2}}{\beta_{a1} + \beta_{c2}} \ln \frac{A_2}{A_1} \quad (5\text{-}21)$$

Since the area of the crack tip is much smaller than that of the adjacent region away from the crack, dissolution of the crack tip would be enhanced prominently due to the galvanic coupling effect.

5.5.2 Enhanced Pitting Corrosion at a Crack Tip

Pit propagation rate measurements show that there is different pitting corrosion behavior at the crack tip than at the region ahead of the crack. The SEM morphological observation indicates that the crack tip is more susceptible to pitting corrosion than is the adjacent region, as shown in Fig. 5-16. Generally, the resistance of steels to pitting corrosion is dependent on the property and structure of the passive film formed on the steel surface. The significant stress concentration developed locally at the crack tip is expected to deform the steel plastically, activating dislocations to form slip bands or dislocation pile-ups. As a consequence, the passive film formed would be less compact and less stable than that formed in the region ahead of the crack. This is confirmed by the high passive current density and low impedance measured locally at the crack tip.

Furthermore, Mott–Schottky measurement and analysis also show that the passive film developed at the crack tip is less stable than that formed at other regions, as indicated by a high donor density. The high density of donors (i.e., oxygen vacancies in an n-type semiconductor passive film) results in increasing susceptibility to pitting corrosion. According to the point defect model [Macdonald, 1992], the oxygen vacancies would be replaced by chloride ions in the solution to generate cation vacancies at the film–solution interface. The cation vacancies diffuse toward the steel–film interface to form cation condensate, resulting in local detachment of the film from the steel, and thus the initiation of pits.

When pipelines are subjected to a seasonal wet–dry cycle, CP may reach the steel surface in wet seasons, creating a high-pH carbonate–bicarbonate solution under a disbonded region [Perdomo et al., 2001; Song, 2008]. In the dry seasons, evaporation of water generates concentrated carbonate–bicarbonate solutions, and CP becomes inaccessible to the steel, due to the increased soil resistivity. Consequently, the potential of pipeline steel will shift to a positive direction and may relocate in the potential range in which SCC is susceptible. Moreover, the stability of passive film formed in the absence of CP is much higher than that formed under CP. Therefore, crack growth is more likely to occur in a dry season than in a wet season.

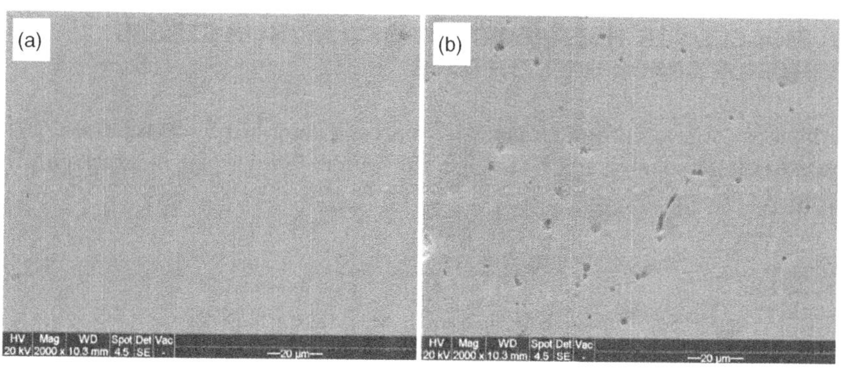

Figure 5-16 *SEM morphology views at point A (a) and the crack tip (b) after 48 h of testing in a concentrated carbonate–bicarbonate solution under a 1500-N force. (From Zhang and Cheng [2010].)*

5.5.3 Relevance to Grain Boundary Structure

The high-pH SCC of pipelines is characterized by intergranular cracking. Undoubtedly, the structure of grain boundaries is essential to facilitating or resisting the cracking process. Generally, random high-angle grain boundaries (HAGBs) have higher energy than that of low-angle grain boundaries (LAGBs) and special coincidence site lattice (CSL) boundaries, and thus provide a relatively easy path for crack propagation [Watanabe, 1984; Lu and Szpunar, 1996]. However, except for the immunity of LAGBs and twin boundaries from cracking, there has been no general consensus as to the specialty of other CSL boundaries in providing resistance to IGSCC, and each material has to be assessed separately [Arafin and Szpunar, 2009].

Studies demonstrate that the character of a grain boundary plays a key role in the intergranular SCC of pipeline steels [Arafin and Szpunar, 2009]. Based on grain boundary character analyses for crack propagation, branching, and deflection, low-angle and special CSL boundaries are crack resistant, while random HAGBs are susceptible to cracking. Crack branching and deflection are controlled primarily by the structure of the grain boundaries at junctions where these deviations from the initial crack propagation path take place. As stated in Section 5.4, the metallurgical structure of the steel and the surface film property are also important to crack propagation. It is thus concluded that the SCC initiation mechanism involves the same factors that are important in SCC propagation [Wang and Atrens, 2003].

Furthermore, the crystallographic texture significantly affects the high-pH, intergranular SCC of pipelines, which is approved by determination of the rotating axes of the boundaries in crack-arrest regions and calculation of the anisotropy of the elastic modulus to assess the relative toughness [Arafin and Szpunar, 2009]. In particular, the boundaries of {110}||rolling plane (RP)– and, to some extent, {111}||RP-textured grains, associated with the <110> and <111> rotation axes, respectively, provide a high level of resistance to IGSCC, while the boundaries of {100}||RP-textured grains are the most susceptible. Therefore, the initiation and subsequent propagation of IGSCC might be avoided either by providing a large fraction of low-angle and special CSL boundaries at the pipe surface or by modifying the surface texture.

5.6 MODELS FOR THE PREDICTION OF A HIGH-PH STRESS CORROSION CRACK GROWTH RATE

The high-pH SCC occurring on pipelines has been described by conventional film rupture or slip-dissolution model [Staehle, 1977; Ford, 1996], where the crack growth velocity, CGR, is expressed by

$$\mathrm{CGR} = \frac{M}{nF\rho} Q_F \frac{\dot{\varepsilon}_{\mathrm{ct}}}{\varepsilon_F} \tag{5-22}$$

where M is the atomic weight, Q_F the electric charge passing between two successive film rupture events, $\dot{\varepsilon}_{\mathrm{ct}}$ the strain rate at the crack tip, and ε_F the rupture ductility of passive film. Crack growth is controlled by the frequency of film rupture events and the repassivation kinetics on a fresh metallic surface.

The repassivation kinetics of steel after film rupture can be written as [Song, 2009]

$$i_a = i_a^* \left(\frac{t}{t_0}\right)^{-z} \qquad t \geq t_0 \tag{5-23}$$

where i_a is the anodic current density, i_a^* the anodic current density immediately after film rupture, z the repassivation exponent and is usually in the range 0.5 to 1, and t_0 is the start time to rupture the film.

A numerical model has been developed for prediction of a high-pH stress corrosion crack growth rate based on the film rupture mechanism [Lu et al., 2010]. According to the model, the crack growth is dominated by repeated film rupture at the crack tip. The crack-tip strain rate due to crack advance changes instantly, and the crack-tip strain rate induced by a cyclic load is dominated by the alternative crack-tip opening displacement. The anodic dissolution rate of a bare metal surface produced by the rupture of passive film can be estimated from the polarization curve of pipeline steel measured in bulk solution with a potential scanning velocity of 1 V/min. One of the limits of this model is the assumption that the dissolution rate of steel in a bulk solution is used to represent the dissolution of bare steel at the crack tip. Actually, the steel has never been in a bare state when immersed in a bulk solution, and will be covered immediately with a layer of oxide film. Apparently, the dissolution rate of steel that is measured is lower than the actual dissolution rate of bare steel at the crack tip. Therefore, this model would lower the estimate of the crack growth rate.

Another relevant model used to predict the growth rate of high-pH stress corrosion cracks incorporates the effects of mechanical stress, the properties of steel, the increase in crack depth with time as a moving boundary condition at the crack tip, mass transport, and the kinetic relations of chemical and electrochemical reactions involved in the cracking process [Song, 2008, 2009]. To express the anodic current density, two methods were used: a potentiodynamic polarization curve and the Butler–Volmer equation. The first method predicts large crack growth rates, and the in-crack chemistry and potential predicted vary significantly with time and space. The second method has the flexibility to predict low crack growth rates, and the in-crack chemistry and potential predicted do not vary with time and space. The mechanism governing the steady-state crack growth that is predicted is due to the balance between the increasing stress intensity factor during cracking (the crack grows and the crack-tip strain rate increases) and the change in the crack-tip condition (for high crack growth rates, a significant negative shift in the crack-tip potential; for low growth rates, an increase in the ferrous ion concentration). While the former tends to accelerate the crack growth, the latter retards the cracking.

REFERENCES

Abedi, SS, Abdolmaleki, A, Adibi, N (2007) Failure analysis of SCC and SRB induced cracking of a transmission oil products pipeline, *Eng. Fail. Anal.* 14, 250–261.

Arafin, MA, Szpunar, JA (2009) A new understanding of intergranular stress corrosion cracking resistance of pipeline steel through grain boundary character and crystallographic texture studies, *Corros. Sci.* 51, 119–128.

Armstrong, RD, Coates, AC (1974) The passivation of iron in carbonate/bicarbonate solutions, Electroanal. *Chem. Interfac. Electrochem.* 50, 303–313.

Armstrong, RD, Coates, AC (1976) A correlation between electrochemical parameters and stress corrosion cracking, *Corros. Sci.* 16, 423–433.

Baker, M, Jr. (2005) *Final Report on Stress Corrosion Cracking Study*, Integrity Management Program Delivery Order DTRS56-02-D-70036, Office of Pipeline Safety, U.S. Department of Transportation, Washington, DC.

Beavers, JA (1992) *Assessment of the Effects of Surface Preparation and Coating on the Susceptibility of Line Pipe to Stress Corrosion Cracking*, Paper L51666, PRCI, Falls Church, VA.

Beavers, JA, Harper, WV (2004) Stress corrosion cracking prediction model, *Corrosion 2004*, Paper 04189, NACE, Houston, TX.

Beavers, JA, Thompson, NG, Coulson, KEW (1993a) Effects of surface preparation and coatings on SCC susceptibility of line pipe: phase 1—laboratory studies, *Corrosion 1993*, Paper 93597, NACE, Houston, TX.

Beavers, JA, Thompson, NG, Coulson, KEW (1993b) Effects of surface preparation and coatings on SCC susceptibility of line pipe: phase 2—field studies, *Proc. 12th International Conference on Offshore Mechanics and Arctic Engineering*, ASME, New York.

Been, J, King, F, Yang, L, Song, F, Sridhar, N (2005) The role of coatings in the generation of high- and near-neutral pH environments that promote environmentally assisted cracking, *Corrosion 2005*, Paper 05167, NACE, Houston, TX.

Danielson, MJ, Jones, RH, Krist, K (2000) Effect of microstructure and microchemistry on the SCC behavior of pipeline steels in a high pH environment, *Corrosion 2000*, Paper 00363, NACE, Houston, TX.

Danielson, MJ, Jones, RH, Dusek, P (2001) Effect of microstructure and microchemistry on the SCC behavior of archival and modern pipeline steels in a high pH environment, *Corrosion 2001*, Paper 01211, NACE, Houston, TX.

Davies, DH, Burstein, GT (1980) The effects of bicarbonate on the corrosion and passivation of iron, *Corrosion* 36, 416–422.

Eliyan, FF, Mahdi, ES, Alfantazi, A (2012) Electrochemical evaluation of the corrosion behaviour of API-X100 pipeline steel in aerated bicarbonate solutions, *Corros. Sci.* 58, 181–191.

Fessler, RR (1969) Stress corrosion cracking, *Proc. 4th Symposium on Line Pipe Research*, Paper L30075, PRCI, Falls Church, VA, pp. F-1 to F-18.

Ford, FP (1996) Quantitative prediction of environmentally assisted cracking, *Corrosion* 52, 375–395.

Fu, AQ, Cheng, YF (2010) Electrochemical polarization behavior of X70 steel in thin carbonate/bicarbonate solution layers trapped under a disbonded coating and its implication on pipeline SCC, *Corros. Sci.* 52, 2511–2518.

Heuer, JK, Stubbins, JF (1999) An XPS characterization of $FeCO_3$ films from CO_2 corrosion, *Corros. Sci.* 41, 1231–1243.

Hirth, JP (1980) Effects of hydrogen on the properties of iron and steel, *Metall. Trans. A* 11, 861–890.

Holroyd, H (1977) Environmental Aspects of the Stress Corrosion Cracking of Ferritic Steels, Ph.D. dissertation, University of Newcastle upon Tyne, UK.

Jack, TR, Erno, B, Krist, K, Fessler, R (2000) Generation of near-neutral pH and high-pH SCC environments on buried pipelines, *Corrosion 2000*, Paper 00362, NACE, Houston, TX.

Jelinek, J, Neufeld, P (1980) Temperature effect on pitting corrosion of mild steel in deaerated sodium bicarbonate–chloride solutions, *Corros. Sci.* 20, 489–496.

King, F, Jack, T, Chen, W, Wilmott, M, Fessler, RR, Krist, K (2000) Mechanistic studies of initiation and early stage crack growth for near-neutral pH SCC on pipelines, *Corrosion 2000*, Paper 361, NACE, Houston, TX.

Li, MC, Cheng, YF (2007) Mechanistic investigation of hydrogen-enhanced anodic dissolution of X70 pipe steel and its implication on near-neutral pH SCC of pipelines, *Electrochim. Acta* 52, 8111–8117.

Li, MC, Cheng, YF (2008) Corrosion of the stressed pipe steel in carbonate–bicarbonate solution studied by scanning localized electrochemical impedance spectroscopy, *Electrochim. Acta* 53, 2831–2836.

Linter, BR, Burstein, GT (1999) Reactions of pipeline steels in carbon dioxide solutions, *Corros. Sci.* 41, 117–139.

Lu, J, Szpunar, JA (1996) Microstructural model of intergranular fracture during tensile tests, *J. Mater. Process. Technol.* 60, 305–311.

Lu, BT, Song, F, Gao, M, Elboujdaini, M (2010) Crack growth model for pipelines exposed to concentrated carbonate–bicarbonate solution with high pH, *Corros. Sci.* 52, 4064–4072.

Macdonald, DD (1992) The point defect model for the passive state, *J. Electrochem. Soc.* 139, 3434–3449.

Mercer, WL (1979) Stress corrosion cracking: control through understanding, *Proc. 6th Symposium on Line Pipe Research*, Paper L30175, PRCI, Falls Church, VA.

Mustapha, A, Charles, EA, Hardie, D (2012) Evaluation of environmentally-assisted cracking susceptibility of a grade X100 pipeline steel, *Corros. Sci.* 54, 5–9.

National Association of Corrosion Engineers (1999a) *White Metal Blast Cleaning*, 1/SSPC-SP 5 (Reaffirmed 1999), NACE, Houston, TX.

National Association of Corrosion Engineers (1999b) *Near-White Metal Blast Cleaning*, 2/SSPC-SP 10 (Reaffirmed 1999), NACE, Houston, TX.

National Energy Board (1996) *Stress Corrosion Cracking on Canadian Oil and Gas Pipelines, Report of the Inquiry*, Report MH-2-95, NEB, Calgary, Alberta, Canada.

Ningshen, S, Mudali, UK, Amarendra, G, Gopalan, P, Dayal, RK, Khatak, HS (2006) Hydrogen effects on the passive film formation and pitting susceptibility of nitrogen containing type 316L stainless steels, *Corros. Sci.* 48, 1106–1121.

Oskuie, AA, Shahrabi, T, Shahriari, A, Saebnoori, E (2012) Electrochemical impedance spectroscopy analysis of X70 pipeline steel stress corrosion cracking in high pH carbonate solution, *Corros. Sci.* 61, 111–122.

Parkins, RN (1974) The controlling parameters in stress corrosion cracking, *Proc. 5th Symposium on Line Pipe Research*, Paper L30174, PRCI, Falls Church, VA.

Parkins, RN (1994) *Overview of Intergranular Stress Corrosion Cracking Research Activities*, Report PRC-232-9401, Line Pipe Research Supervisory Committee of the Pipeline Research Committee of the American Gas Association, AGA, Washington, DC.

Parkins, RN, Zhou, S (1997) The stress corrosion cracking of C–Mn steel in CO_2–HCO_3^--CO_3^{2-} solutions: I. Stress corrosion data, *Corros. Sci.* 39, 159–173.

Perdomo, JJ, Chabica, ME, Song, I (2001) Chemical and electrochemical conditions on steel under disbonded coatings: the effect of previously corrosion surface and wet and dry cycles, *Corros. Sci.* 43, 515–532.

Pilkey, AK, Lambert, SB, Plumtree, A (1995) Stress corrosion cracking of X-60 line pipe steel in a carbonate–bicarbonate solution, *Corrosion* 51, 91–96.

Qin, Z, Norton, PR, Luo, JL (2001) Effects of hydrogen on formation of passive films on AISI 310 stainless steel, *Br. Corros. J.* 36, 33–42.

Raja, KS, Jones, DA (2006) Effects of dissolved oxygen on passive behavior of stainless alloys, *Corros. Sci.* 48, 1623–1638.

Revie, RW, Uhlig, HH (2008) *Corrosion and Corrosion Control*, 4th ed., Wiley-Interscience, Hoboken, NJ.

Sanchez, J, Fullea, J, Andrade, C, Alonso, C (2007) Stress corrosion cracking mechanism of prestressing steels in bicarbonate solutions, *Corros. Sci.* 49, 4069–4080.

Sherar, BWA, Keech, PG, Qin, Z, King, F, Shoesmith, DW (2010) Nominally anaerobic corrosion of carbon steel in near-neutral pH saline environments, *Corrosion* 66, 45001/1–45001/11.

Song, FM (2008) Overall mechanisms of high pH and near-neutral pH SCC, models for forecasting SCC susceptible locations, and simple algorithms for predicting high pH SCC crack growth rates, *Corrosion 2008*, Paper 8129, NACE, Houston, TX.

Song, FM (2009) Predicting the mechanisms and crack growth rates of pipelines undergoing stress corrosion cracking at high pH, *Corros. Sci.* 51, 2657–2674.

Song, FM (2010) Predicting the effect of soil seasonal change on stress corrosion cracking susceptibility of buried pipelines at high pH, *Corrosion* 66, 95004/1–95004/14.

Staehle, RW (1977) Prediction and experimental verification of the slip-dissolution model for stress corrosion cracking of low strength alloys, in *Stress Corrosion Cracking and Hydrogen Embrittlement of Iron-Based Alloys*, R.W. Stahler, J. Hochmann, R.D. McCright, and J.E. Slates, Editors, NACE, Houston, TX, pp. 180–205.

Van Boven, G, Chen, W, Rogge, R (2007) The role of residual stress in neutral pH SCC of pipeline steels: II. Crack dormancy, *Acta Mater.* 55, 29–42.

Wang, JQ, Atrens, A (2003) SCC initiation for X65 pipeline steel in the "high" pH carbonate/bicarbonate solution, *Corros. Sci.* 45, 2199–2217.

Watanabe, T (1984) An approach to grain boundary design for strong and ductile materials, *Res. Mech.* 11, 47–84.

Weng, Y, Li, X (2005) Electrochemical behaviors and SCC sensitive potential of pipeline steels in carbonate–bicarbonate solutions, *Corrosion 2005*, Paper 04201, NACE, Houston, TX.

Wenk, RL (1974) Field investigation of stress corrosion cracking, *Proc. 5th Symposium on Line Pipe Research*, Paper L30174, PRCI, Falls Church, VA.

Xue, HB, Cheng, YF (2010) Passivity and pitting corrosion of X80 pipeline steel in carbonate-bicarbonate solution studied by electrochemical measurements, *J. Mater. Eng. Perf.* 19, 1311–1317.

Yu, JG, Luo, JL, Norton, PR (2001) Effects of hydrogen on the electronic properties and stability of the passive films on iron, *Appl. Surf. Sci.* 177, 129–138.

Zhang, GA, Cheng, YF (2010) Micro-electrochemical characterization of corrosion of pre-cracked X70 pipeline steel in a concentrated carbonate/bicarbonate solution, *Corros. Sci.* 52, 960–968.

6

Stress Corrosion Cracking of Pipelines in Acidic Soil Environments

6.1 INTRODUCTION

In addition to nearly neutral– and high-pH environments, pipeline SCC has been found to occur in acidic environments when the coating is broken or "missing" to expose line pipe steel directly to soils that are acidic in nature [Liu et al., 2008]. For example, in southeastern China, an acidic soil with an average pH of 3.5 to 6.0, called "red soil," is distributed in several provinces. Thousands of kilometers of natural gas pipelines operate in this area. The concern is that construction or other factors may damage the coating, causing the steel to be exposed directly to acidic soils that are supportive of pipeline corrosion and SCC. Moreover, the acidic moisture that originates from the soil could enter the disbonded area by permeating the coating or diffusing from adjacent defects contained in the coating. Consequently, the trapped electrolyte becomes acidic, with quite a low pH. Pipeline corrosion and SCC thus occur in acidic solution environments. Since it is a phenomenon that is distinct from the SCC observed previously on pipelines, it is thus of great significance to understand the potential for pipeline SCC to occur in acidic soil environments.

Although pipeline SCC in nearly neutral– and high-pH electrolytes has been studied worldwide in past decades, the SCC behavior of pipeline steels in acidic soils has not been fully understood. Relevant study is still in its infancy and is conducted in only a few countries in the world.

Stress Corrosion Cracking of Pipelines, First Edition. Y. Frank Cheng.

6.2 PRIMARY CHARACTERISTICS

Acidic soil–extracted solution resulting in pipeline SCC contains vast amounts of chloride ions, with sulfate, bicarbonate, and nitrate ions, and the solution pH is approximately 4.0 to 4.5 [Liu et al., 2009a]. The corrosion potential of pipeline steels, such as X70 steel, in solution is about −750 mV(SCE) [Liu et al., 2009b]. Moreover, the soil solution contains trace amounts of oxygen.

Pipeline SCC in an acidic soil solution is dependent on the applied potential [Liu et al., 2008]. At a relatively reduced negative potential such as −650 mV(SCE), compared to the corrosion potential of steels, the fracture shows a ductile feature, with extensive dimples distributed on the surface, as shown in Fig. 6-1a. When the applied potential is at or slightly more negative [e.g., −850 mV(SCE)] than the corrosion potential, a number of secondary cracks exist on the side faces. Fracture morphology exhibits a brittle characteristic (Fig. 6-1b). Moreover, the fracture faces appears to be transgranular. Therefore, pipeline SCC in an acidic soil solution is basically transgranular cracking. With the further negative shift of applied potential to −1200 mV(SCE), the SCC is dominated completely by a hydrogen-based mechanism, with a riverbed-shaped crack orientation, as shown in Fig. 6-1c.

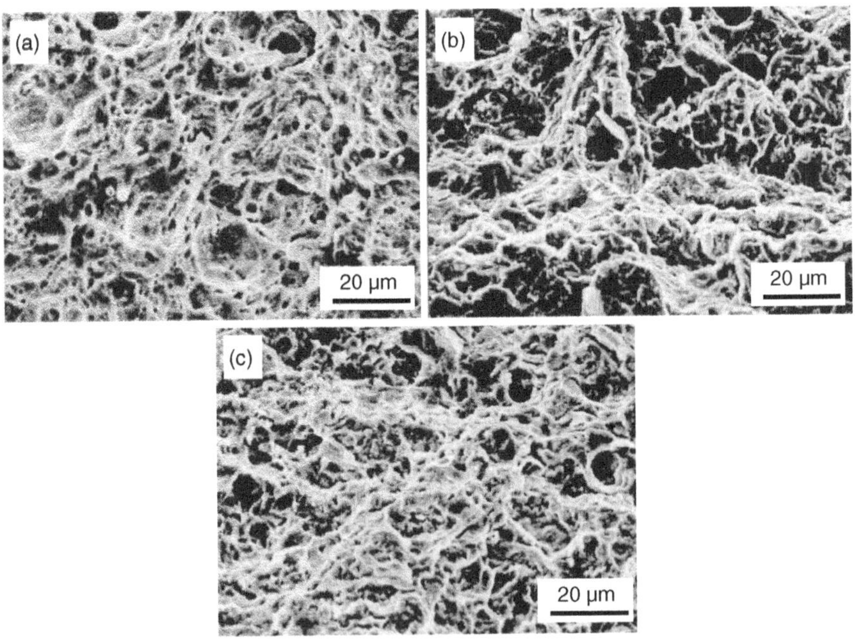

Figure 6-1 *SEM images of the fracture surface of X70 steel at (a) −650 mV(SCE), (b) −850 mV(SCE), and (c) −1200 mV(SCE) in an acidic soil solution. (From Liu et al. [2008].)*

6.3 ELECTROCHEMICAL CORROSION MECHANISM OF PIPELINE STEELS IN ACIDIC SOIL SOLUTIONS

Cathodic and anodic reactions of pipeline steels in acidic soil solution represent primarily the reduction of hydrogen ions and steel oxidation:

$$H^+ + e \rightarrow H \tag{6-1}$$

$$Fe \rightarrow Fe^{2+} + 2e \tag{6-2}$$

With a low solution pH of about 4, the concentration of Fe^{2+} in the solution is expected to increase rapidly. Due to the presence of bicarbonate ions in solution, a layer of $FeCO_3$ corrosion product will form and deposit on the steel surface, due to supersaturation [Qiao et al., 1998; Li and Cheng, 2008]:

$$Fe^{2+} + HCO_3^- \rightarrow FeCO_3 + H^+ \tag{6-3}$$

Since the $FeCO_3$ deposit layer is loose and porous, the "protectiveness" is not as good as that of a real, stable passive film. For example, the passive current density of X70 pipeline steel in a concentrated bicarbonate solution, where the steel is in a stable passivity state, is 10^{-4} to 10^{-3} mA/cm^2 [Li and Cheng, 2007], while that in the acidic soil solution is as high as 10^{-1} mA/cm^2 [Liu et al., 2008].

6.4 MECHANISMS FOR INITIATION AND PROPAGATION OF STRESS CORROSION CRACKS

Measurements of polarization curves at both fast and slow potential sweep rates can be used to study the SCC process of steels mechanistically. During fast sweeping, there is insufficient time for film to form on the steel surface, resulting in generation of a potential range where the steel is in an active dissolution state to simulate crack-tip electrochemistry. During slow sweeping, film formation is favored and the polarization behavior measured can be used to simulate the crack-wall electrochemistry [Parkins, 1980]. The potentiodynamic polarization curves measured on X70 pipeline steel at both fast and slow potential sweepings in an acidic soil–extracted solution are shown in Fig. 6-2. The results show clearly that the steel experiences active dissolution and "passivity" at fast (50 mV/s) and slow (0.5 mV/s) potential sweepings, respectively. Thus, the anodic dissolution reaction plays a significant role in the SCC of X70 steel in the acidic soil solution. Furthermore, the primary cathodic reaction in this system is the hydrogen evolution. It is expected that the hydrogen atoms generated would penetrate steel and participate in the cracking processes. Therefore, SCC pipeline steels in acidic soil solutions are controlled by both anodic dissolution and hydrogen involvement.

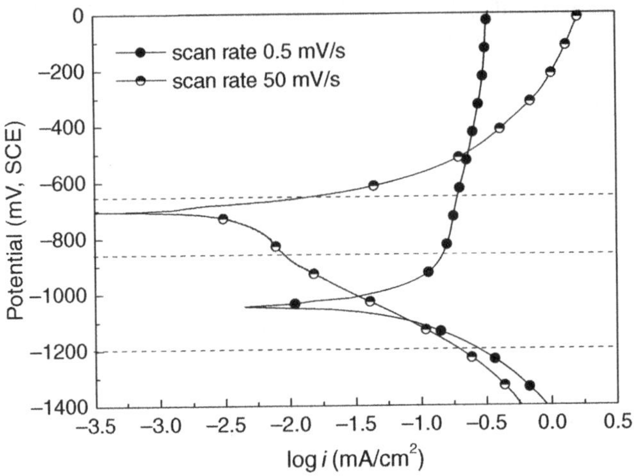

Figure 6-2 *Polarization curves of X70 steel measured at fast (50 mV/s) and slow (0.5 mV/s) sweep rates, respectively, in an acidic soil solution. (From Liu et al. [2008].)*

To further illustrate the SCC mechanism, the potentials applied on the steel in Fig. 6-1 were marked in the polarization curves in Fig. 6-2. It is seen that the steel is in a different polarization status at individual potentials. At –650 mV(SCE), the steel is in anodic dissolution for both slow and fast sweep rates, indicating that both the crack tip and the crack wall are in anodic reaction. When the potential applied is −850 mV(SCE), the steel is in anodic polarization at slow sweep and in cathodic polarization at fast sweep, showing that a combined electrochemical process of anodic reaction on the crack wall and cathodic reaction at the crack tip dominates the SCC. At −1200 mV(SCE), the steel is in cathodic polarization at both fast and slow sweeps. Thus, the cathodic reaction predominates at both the crack tip and the crack wall. Therefore, at relatively lower negative potentials such as –650 mV(SCE), the SCC of pipeline steel follows an anodic dissolution mechanism. The crack tip is in an active dissolution state, while the crack wall is in a less active status. The difference in electrochemical potentials between the crack wall and the crack tip enhances the propagation of crack at the tip. The fracture surface shows primarily the ductile feature. With the negative shift in potential to −850 mV(SCE), for example, the crack wall is still in an active state while the crack tip is dominated by a cathodic reaction, resulting in the generation of hydrogen atoms at the tip. Thus, hydrogen is actively involved in the SCC process, causing a typically observed transgranular cracking feature. With a further negative shift of potential to −1200 mV(SCE), more hydrogen atoms are generated at both the crack tip and the crack wall, contributing to cracking of the steel. The steel SCC follows completely a hydrogen-based mechanism.

Since hydrogen evolution and permeation in steels are a large concern regarding the SCC processes in acidic soils, the interaction of hydrogen with metallurgical defects would play an essential role in crack initiation under these conditions. For example, it has been found [Liu et al., 2009a] that whether a crack initiates at an inclusion depends on its composition, which affects directly interaction of the inclusion with hydrogen. It was established [Zhang, 2006] that inclusions enriched in Al_2O_3 are hard and brittle as well as incoherent to the steel matrix. Quite a sizable lattice deformation occurs adjacent to the inclusions. Consequently, interstices are easy to generate at boundaries between inclusions and the steel substrate. Once entering the steel, hydrogen will be prone to be trapped in these interstices, resulting in crack formation, as shown in Fig. 6-3.

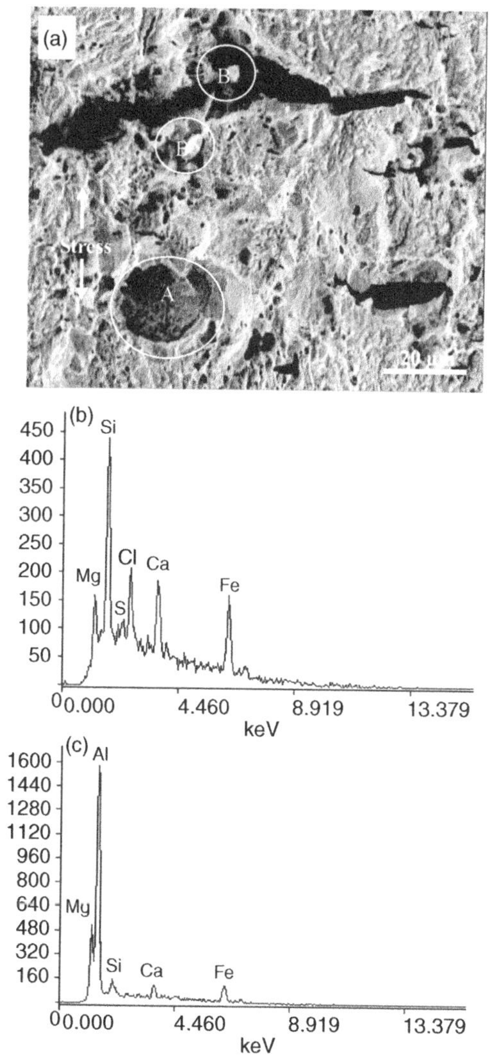

Figure 6-3 *SEM view of crack morphology (a) and EDX spectra on inclusion A (b) and inclusion B (c). (From Liu et al. [2009a].)*

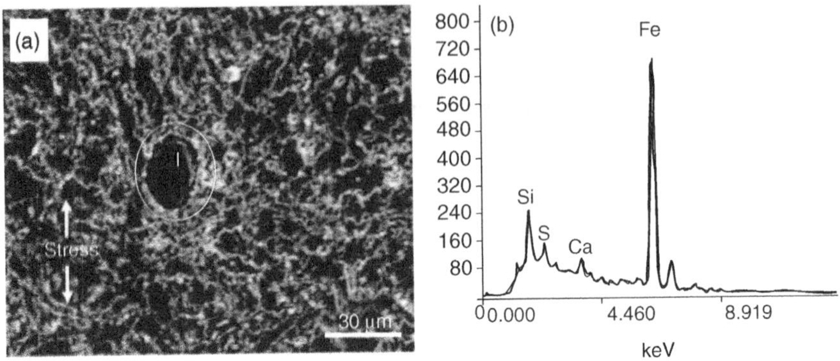

Figure 6-4 SEM view of the morphology of silicon-enriched inclusion I (a), and EDX spectrum of inclusion (b) while the steel is polarized cathodically at −1200 mV(SCE) in an acidic soil solution. (From Liu et al. [2009a].)

Conversely, inclusions enriching silicon are easy to deform, effectively relieving the residual stress [Garet et al., 1998]. Moreover, SiO_2-enriching inclusions are sphere-shaped and are prone to be in a stable state. There is a relatively small local lattice deflection around these types of inclusions. Thus, there is usually no crack associated with inclusions enriching silicon, as shown in Figs. 6-3 and 6-4.

6.5 EFFECT OF STRAIN RATE ON THE SCC OF PIPELINES IN ACIDIC SOILS

Strain rate has been acknowledged to be a critical factor in controlling the SCC process [Rhodes, 2001; Rokuro and Yasuaki, 2004]. For example, for high pH SCC that is based on the slip-rupture model, it is believed [Ford, 1996] that crack propagation is dependent on the competition between the strain rate and the repassivation rate at the crack tip. Figure 6-5 shows the SCC susceptibility of X70 pipeline steel polarized cathodically at −850 mV(SCE) in an acidic soil solution as a function of strain rate. A maximum of SCC susceptibility is observed at the strain rate of 10^{-6} s^{-1}. Apparently, strain rate is critical to the susceptibility of steel to SCC in an acidic soil solution [Liu et al., 2009b].

Moreover, Fig. 6-6 shows potentiodynamic polarization curves measured on X70 steel under stresses of $0.5\sigma_{0.2}$ and various strain rates in an acidic soil solution, where $\sigma_{0.2}$ refers to the applied stress level, resulting in 0.2% of total deformation, and is often used as an approximation of the proof stress of steels. It is seen that the strain rate strongly affects the electrochemical polarization behavior of steel in a soil solution. When stress is applied to a specimen, there is increased cathodic current density at all strain rates. The highest cathodic current density is at an individual potential when the strain rate is 10^{-6} s^{-1}. As the strain rate is increased to 10^{-5} s^{-1}, the cathodic current density decreases. Moreover, a limiting diffusive current density is observed in cathodic polarization curve when the strain rate is 10^{-6} s^{-1}.

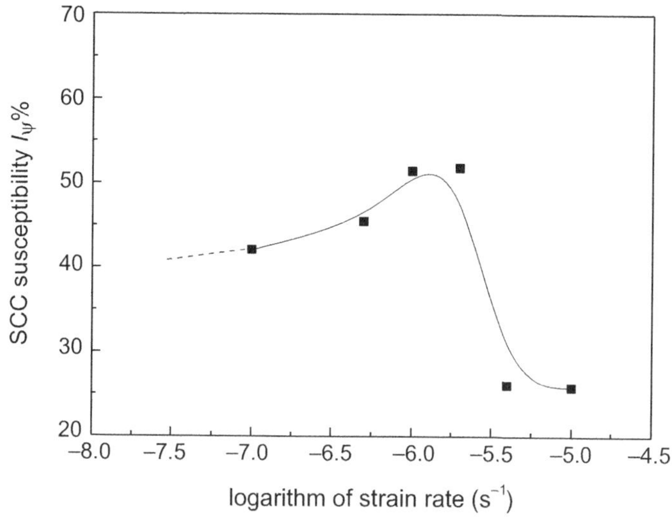

Figure 6-5 *SCC susceptibility of X70 steel polarized cathodically at −850 mV(SCE) in an acidic soil solution as a function of strain rate. (From Liu et al. [2009b].)*

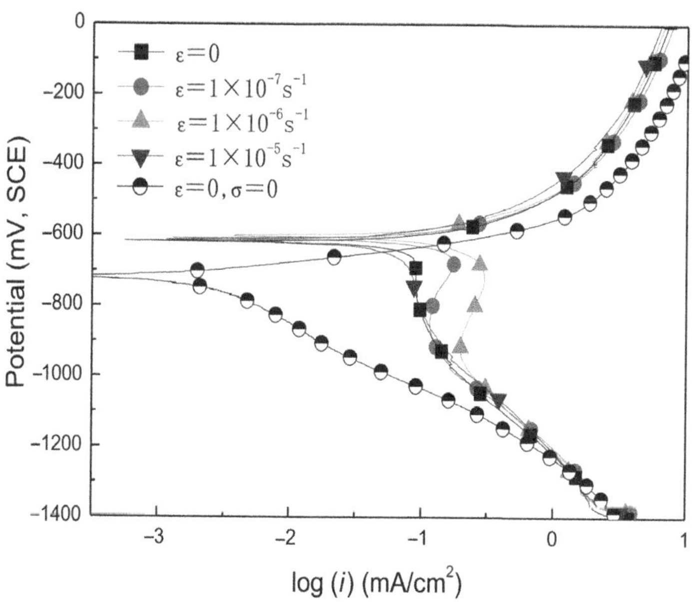

Figure 6-6 *Potentiodynamic polarization curves measured on an X70 steel tensile specimen stressed at* $0.5\sigma_{0.2}$ *in a test solution. (From Liu et al. [2009b].)*

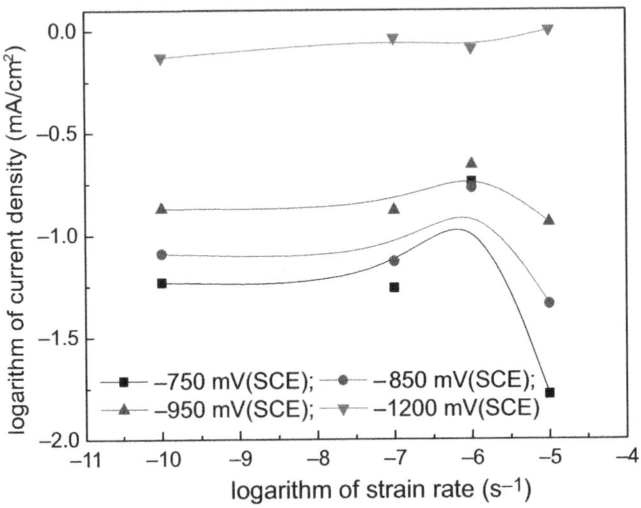

Figure 6-7 *Cathodic current density vs. strain rate for an X70 steel specimen under $0.5\sigma_{0.2}$ in an acidic soil solution. (From Liu et al. [2009b].)*

Details of the dependence of cathodic current density on the strain rate are shown on a cathodic polarization curve as a function of the logarithm of the strain rate at an individual cathodic potential in Fig. 6-7. It is shown that at an individual stress level there is a maximum of cathodic current density at a strain rate of 10^{-6} s^{-1}. However, when the potential is up to −1200 mV(SCE), the cathodic current density is approximately independent of strain rate.

As analyzed, the anodic and cathodic reactions of pipeline steels in an acidic soil solution contain the oxidation of steel and the reduction of H^+, as described in reactions (6-1) and (6-2). An acidic soil solution contains 0.302 g/L (3.595×10^{-3} mol/L) $NaHCO_3$. At pH 4.0, a simple calculation based on the reaction equilibrium constants among H_2CO_3, HCO_3^-, and CO_3^{2-} shows that the equilibrium concentrations of these species are 3.572×10^{-3}, 1.679×10^{-5}, and 0.94×10^{-11} mol/L, respectively. While the contents of HCO_3^- and CO_3^{2-} are negligible, the reduction of H_2CO_3 should be taken into account:

$$H_2CO_3 + e \rightarrow HCO_3^- + H \tag{6-4}$$

At potentials as negative as −1200 mV(SCE), water reduction is also thermodynamically possible:

$$H_2O + e \rightarrow OH^- + H \tag{6-5}$$

Since the strain rate greatly affects the cathodic polarization curves measured in the elastic stress region, it is expected that cathodic reductions of H^+, H_2CO_3, and H_2O at specific potential ranges will be affected by the strain rate.

Even when the steel is in the macroscopic elastic stress region, the local stress concentration may occur at surface microdefects to activate dislocations. In particular, the dislocation emergence points are the dominant sites to generate local additional potential (LAP) [Liu et al., 2009b]. With an increase in strain rate, the number of dislocation emergence points goes up, resulting in a negative shift of LAP. As a result, the cathodic reduction reaction is favorable, and the cathodic current density increases, as shown in Fig. 6-7. When the strain rate goes up to 10^{-6} s^{-1}, a cathodic limiting diffusive current is generated (Fig. 6-6), which is attributed to the fact that the increase in reactive species such as H^+ and H_2CO_3 is not as fast as their consumption, due to the further enhanced LAP. Mass transport thus becomes the rate-limiting step in the cathodic process. However, when the strain rate is sufficiently high, such as at 10^{-5} s^{-1}, the mobility of dislocation emergence points is so high that the reactive species do not have much chance to adsorb at these active sites for reductive reaction. Thus, the cathodic current density decreases at high strain rates. Furthermore, the dependence of cathodic current density on the strain rate becomes less important when the potential applied is shifted more negatively. For example, when the polarization potential is −1200 mV(SCE), the proposed cathodic reactions (6-1), (6-4), and (6-5) are all thermodynamically possible. The strain-induced LAP effect is negligible with respect to a sufficiently negative potential of the steel.

The SCC of pipeline steels in an acidic soil solution is a hydrogen-based mechanism. With an increase in strain rate, it has been estimated that the cathodic current density increases. Since the cathodic reactions proposed involve hydrogen evolution, it is expected that more hydrogen atoms are generated and penetrated a steel upon an increase in strain rate, resulting in an elevated SCC susceptibility. However, when the strain rate is sufficiently high, such as 10^{-5} s^{-1}, the cathodic current density decreases and, consequently, hydrogen evolution is inhibited. Correspondingly, the SCC susceptibility decreases, as shown in Fig. 6-5.

REFERENCES

Ford, FP (1996) Quantitative prediction of environmentally assisted cracking, *Corrosion* 52, 375–395.

Garet, M, Brass, AM, Haut, C, Guttierez-Solana, F (1998) Hydrogen trapping on nonmetallic inclusions in Cr–Mo low alloy steels, *Corros. Sci.* 40, 1073–1086.

Li, MC, Cheng, YF (2007) Mechanistic investigation of hydrogen-enhanced anodic dissolution of X70 pipe steel and its implication on near-neutral pH SCC of pipelines, *Electrochim. Acta* 52, 8111–8117.

Li, MC, Cheng, YF (2008) Corrosion of the stressed pipe steel in carbonate–bicarbonate solution studied by scanning localized electrochemical impedance spectroscopy, *Electrochim. Acta* 53, 2831–2836.

Liu, ZY, Li, XG, Du, CW, Zhai, GL, Cheng, YF (2008) Stress corrosion cracking behavior of X70 pipe steel in an acidic soil environment, *Corros. Sci.* 50, 2251–2257.

Liu, ZY, Li, XG, Du, CW, Lu, L, Zhang, YR, Cheng, YF (2009a) Effect of inclusions on initiation of stress corrosion cracks in X70 pipeline steel in an acidic soil environment, *Corros. Sci.* 51, 895–900.

Liu, ZY, Li, XG, Du, CW, Cheng, YF (2009b) Local additional potential model for effect of strain rate on SCC of pipeline steel in an acidic soil solution, *Corros. Sci.* 51, 2863–2871.

Parkins, RN (1980) Predictive approaches to stress corrosion cracking failure, *Corros. Sci.* 20, 147–166.

Qiao, LJ, Luo, JL, Mao, X (1998) Hydrogen evolution and enrichment around stress corrosion crack tips of pipeline steels in aqueous acid-solution, *Corrosion* 54, 115–120.

Rhodes, PR (2001) Environment assisted cracking of corrosion resistant alloys in oil and gas production environments: a review, *Corrosion* 57, 923–965.

Rokuro, N, Yasuaki, M (2004) SCC evaluation of type 304 and 316 austenitic of type 304 and 316 austenitic stainless steels in acidic chloride solutions, *Corros. Sci.* 46, 769–785.

Zhang, LF (2006) Inclusion and bubble in steel: a review, *J. Iron Steel Res. Int.* 13, 1–8.

7

Stress Corrosion Cracking at Pipeline Welds

7.1 INTRODUCTION

SCC failures occur frequently in pipelines at welds, even when the proper weld metal has been selected, industry codes and standards followed, and welds deposited that possess full weld penetration and proper shape and contour. It is not unusual to find that although a metal or alloy is resistant to SCC in a particular environment, the welded counterpart is not. However, there have also been many instances in which the weld exhibits SCC resistance that is superior to that of the base metal [Davis, 2006]. Apparently, SCC at welds is quite a complex phenomenon.

Field observations have found that pipeline SCC occurs frequently at welds and in adjacent areas [National Energy Board, 1996; Parkins, 2000; Baker, 2005]. It has been acknowledged [Lu et al., 2010; Zhang and Cheng, 2010] that welding would increase the corrosion activity and SCC susceptibility of pipeline steels, due to the metallurgical changes and residual stresses introduced as well as a series of the phase transformations taking place in both the weld zone and the heat-affected zone (HAZ). Therefore, a complete understanding of SCC occurring at pipeline welds requires knowledge of steel metallurgy, corrosion, and the mechanical stress with which pipelines are associated.

Stress Corrosion Cracking of Pipelines, First Edition. Y. Frank Cheng.

7.2 FUNDAMENTALS OF WELDING METALLURGY

An understanding of welding metallurgy is important for corrosion, cracking, and failure analysis at welds. The field of welding metallurgy is vast, and in this section we summarize only those relevant to SCC.

7.2.1 Welding Processes

The welding process involves melting and solidification of metals in a fashion similar to that occurring during metal manufacturing or heat treatment, but at a much faster rate. A welded joint consists of weld metal, the HAZ, and the unaffected base metal. A few types of welding processes are described below [Rampaul, 2003].

1. *Gas welding*: a process that joins metals by heating them with a flame produced by reaction between a fuel gas and oxygen. Oxyacetylene welding is the most common gas welding process. The flux melts, solidifies, and forms a slag skin on weld metal. Depending on requirements, the flames used could be neutral, reducing, or oxidizing in nature.
2. *Arc welding*: a process involving heat produced by an arc established between an electrode and a metal. The electrode is connected to one terminal of an electrical power supply, and the workpiece is connected to the other terminal. The core of the electrode provides filler metal for a joint. The heat of the arc causes the electrode tip to melt, forming a molten metal pool, and solidifies into the weld metal.
3. *Electron beam welding* (EBW): a process that joins metals by melting with heat from an electron beam. When heated, the cathode of the electron beam gun emits electrons. These electrons are accelerated by an electric field between the negatively charged bias electrode and the anode, which produces enough heat for a weld joint.

7.2.2 Welding Solidification and Microstructure

It is inevitable that welds possess compositional and microstructure heterogeneities. On a large scale, a weld consists of a transition from wrought base metal through a HAZ and into the solidified weld metal, including five microstructurally distinct regions: the fusion zone, an unmixed region, a partially melted region, the HAZ, and the unaffected base metal [Savage, 1969]. The unmixed region is actually part of the fusion zone, and the partially melted region is part of the HAZ. Generally, all five zones are not present in any given weld.

The most important change in microstructure occurs in the fusion zone, where the metal experiences complete melting and resolidification. Parameters such as temperature gradient, cooling rate, and alloy composition determine the development of microstructures at welds [Lancaster, 1999].

Weld solidification was derived from a theory regarding the freezing of castings, ingots, and single crystals at low thermal gradients and slow growth rates. Nowadays, rapid solidification theories have been used to explain weld solidification at very fast cooling rates [Kou, 2003]. The microstructure developed in the fusion zone is very complicated because of a variety of physical processes that occur locally, including melting, heat and fluid flow, solid-state transformation, vaporization, stresses, solidification, and distortion. The interactions of these processes affect the weld pool solidification and resulting microstructure significantly. To date, modeling of welding processes has provided deep insight into an understanding of both the welding process and the welded materials, and is capable of explaining various processes that occur during welding [Evans and Bailey, 1997].

An important area for understanding weld solidification is the dynamics of weld pool development and its geometry. The shape of a weld is associated directly with the grain structure and the dendrite growth process. The temperature gradient and grain growth vary significantly throughout the weld pool. The growth rate is lowest along the fusion line, and it is fastest at the centerline of the weld. As a result, the microstructure differs significantly from the edges of the weld to its center.

The shape of the solid–liquid interface controls the development of microstructural features, while growth of the solid occurs inside the weld pool. The nature of the solid–liquid interface is determined by the thermal and constitutional conditions that exist surrounding the interface. The interface grows by planar, cellular, or dendritic growth. Computational thermodynamic models are now available that can help to describe the phase evolution of multicomponent systems during weld solidification [David et al., 2003]. The solidification phases form as a result of solute partitioning during solidification and the stability of these phases. Moreover, the chemical composition of metals affects the phases formed during welding.

The HAZ is the portion of the weld joint that has experienced temperatures high enough to produce solid-state microstructural changes, but the temperature is low and cannot cause melting. Every position in the HAZ relative to the fusion line experiences a unique thermal experience during welding, in terms of both maximum temperature and cooling rate, and each element experiences a typical time–temperature cycle. Thus, each position has its own microstructural features. For example, compositional gradients on the scale of a few micrometers (i.e., microsegregation) exist within individual weld zones, due to segregation of major and trace elements during solidification [Buchheit et al., 1990].

The partially melted region is usually one or two grains into the HAZ relative to the fusion line. It is characterized by grain boundary liquation, which may result in liquation cracking. These cracks, which are found in the grain boundaries one or two grains below the fusion line, have been identified as potential initiation sites for HIC in high-strength steel [Wahid et al., 1993].

The part that has not undergone any metallurgical change is the unaffected base metal. Although metallurgically unchanged, both the unaffected base metal and the entire weld joint are likely to be in a state of high residual stress, either transversely or longitudinally, depending on the degree of restraint imposed on the weld.

7.2.3 Parameters Affecting the Welding Process

A number of parameters affect the welding process.

1. *Weld shape*. The shape of a weld is determined by the heat input and welding speed. As both heat intensity and welding speed increase or decrease, the weld becomes elongated or spreads, respectively.
2. *Cooling rate*. An increasing heat input would decrease the cooling rate. Preheating is an essential practice in reducing the risk of HAZ cracking.
3. *Power density distribution*. When the heat input and welding speed are kept constant, the weld penetration is directly proportional to the power density of the heat source. The penetration depth increases with increasing power density.
4. *Heat-sink effect of the workpiece*. The cooling rate for a weld increases when the workpiece is thick. The higher thickness acts as a larger and better heat sink to cool down the weld.

7.2.4 Defects at the Weld

Defects existing at welds are usually due to incomplete joining, internal defects such as inclusions, and lack of fusion in the middle area of the weld [Otegui et al., 2002]. One typical form of defect at the weld is a hook crack, in which an inclined crack is formed several millimeters away from the weld centerline, following the lamination planes [Kiefner, 1997]. During the welding process, the heated material is ejected from the center of the thickness toward the surfaces of the pipe. Lamination defects, originally parallel to the surface, are reoriented through the thickness. This decreases the resistance of the weld to circumferential stresses.

Another typical form of weld defect is due to elimination of the ejected material mechanically, which is widely called *shaving*. This process may leave scratches or grooves on surfaces. Subsurface plastic deformation due to shaving also creates a local state of residual stresses in the through-thickness direction.

Moreover, there exist extensive metallurgical defects at the welding area, including nonmetallic inclusions, voids, and impurity segregation. During welding, the grains at the weld grow rapidly, generating a coarse, large-grained structure, especially at the HAZ. Furthermore, the remarkable metallurgical changes at the weld also include the generation of a wide variety of microscopic phases in this area. For example, the base X70 pipeline steel contains ferrite and/or pearlite as the microphase, while the HAZ may generate bainite microstructure, resulting in metallurgical structure heterogeneity.

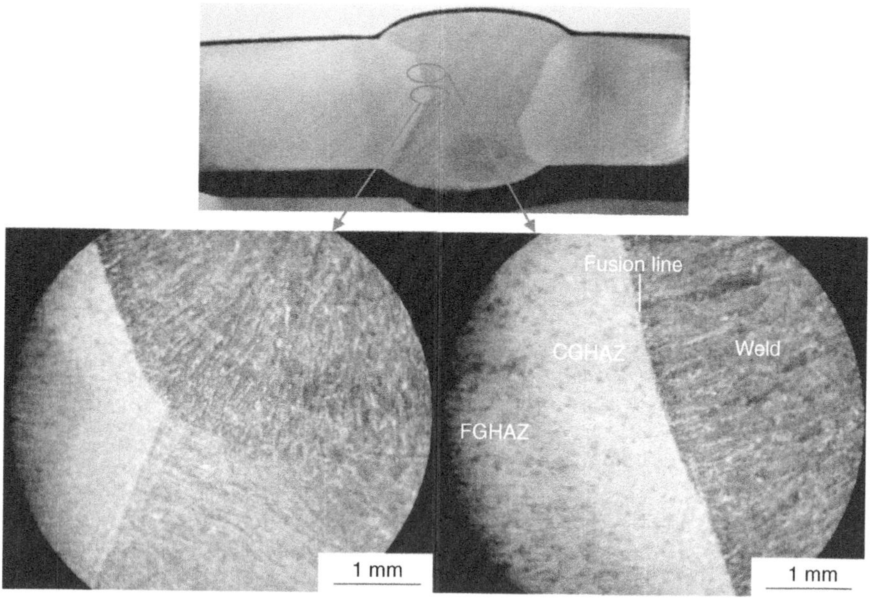

Figure 7-1 Welded X70 pipeline steel specimen observed by optical microscopy. (From Zhang and Cheng [2009a].)

7.3 PIPELINE WELDING: METALLURGICAL ASPECTS

7.3.1 X70 Steel Weld

The welded X70 pipeline steel specimen shown in Fig. 7-1 is used for metallographic characterization, and an optical view of the microstructure of various zones (i.e., weld metal, HAZ, and base steel) is shown in Fig. 7-2. It is seen that the X70 base steel typically contains ferrite and pearlite (Fig. 7-2a). The microstructure of a HAZ is a mixture of acicular ferrite (AF) and bainitic ferrite (BF), as shown in Fig. 7-2b. The weld metal consists of AF and grain boundary ferrite (GBF), where the AF nucleates within the austenite grain or grain boundary and grows in any direction, while GBF nucleates at the austenite grain boundary and is elongated along the boundary and grows into the grain to form equiaxial or nearly equiaxial ferrite grains.

7.3.2 X80 Steel Weld

Optical and electronic views of the microstructure of various zones in an X80 steel weld are shown in Figs. 7-3 and 7-4. The HAZ contains a mixture of bainite (B) and ferrite (F) with white cementite (C), as shown in Fig. 7-4b. The weld metal is featured with AF and GBF (Fig. 7-4c). The microstructure of an X80 steel substrate typically contains polygonal ferrite and bainite bundles (Fig. 7-4d).

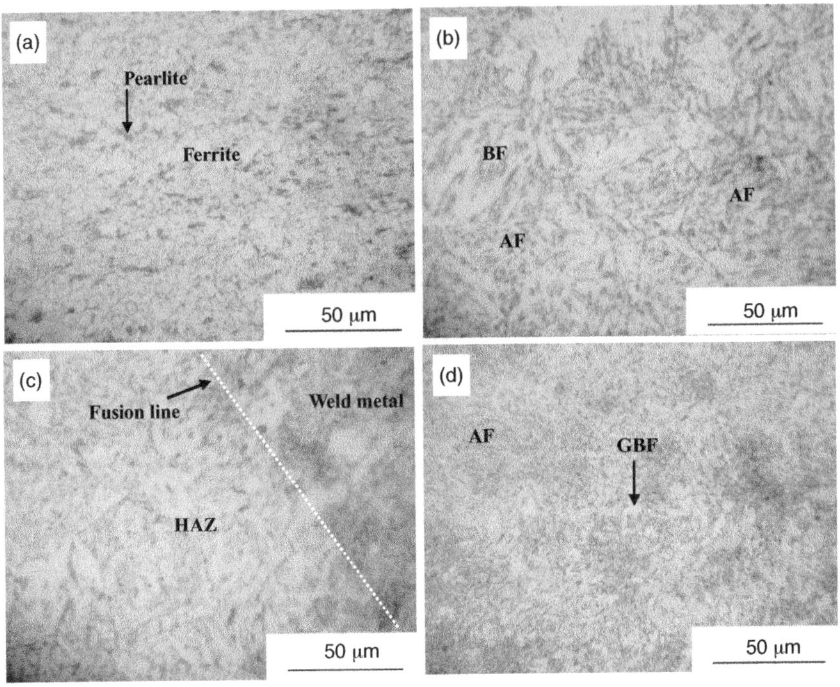

Figure 7-2 Optical view of the microstructure of various zones of a welded X70 steel specimen: (a) base steel; (b) HAZ; (c) HAZ/weld metal; (d) weld metal. (From Zhang and Cheng [2009b].)

7.3.3 X100 Steel Weld

Figure 7-5 shows optical views of the microstructure of weld metal, the HAZ, and base X100 steel, respectively, on the specimen shown in Fig. 7-6, where points A, B, and C on base steel indicate various distances from the welding zone, as marked in the figure. The weld metal is typically featured with allotriomorphic ferrite and cementite. The HAZ shows a microphase of polygonal ferrite with an average grain size of about 10 to 15 μm mixed with cementite. The microstructure of X100 base steel is dependent on the distance from the HAZ. At point A it consists of polygonal ferrite with an average grain size of 5 to 10 μm and bainite. While the microphase at point B is featured with AF and granular bainite, point C consists primarily of bainite. Apparently, the welding operation results in generation of the ferrite phase on the steel substrate adjacent to the weld.

7.4 PIPELINE WELDING: MECHANICAL ASPECTS

7.4.1 Residual Stress

Several factors can contribute to the generation of residual stresses in a weld, including the welding process, heat input, thickness of the base metal, restraint of the weld,

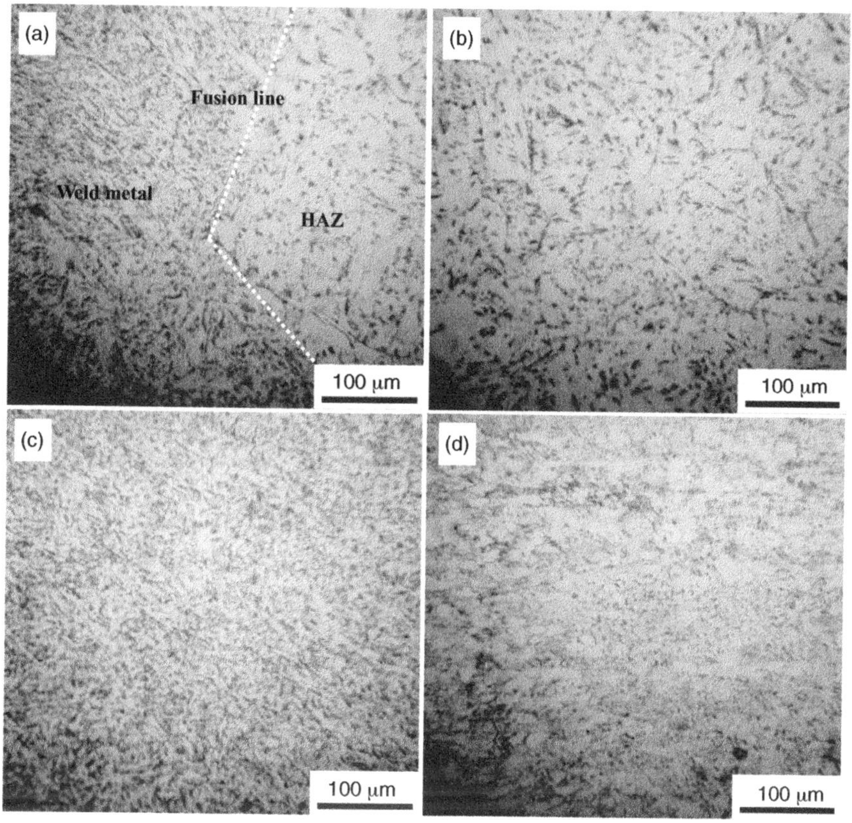

Figure 7-3 *Optical views of the microstructure of the welded X80 pipeline steel (a) HAZ/weld metal, (b) HAZ, (c) weld metal, and (d) base steel. (From Xue and Cheng [2012].)*

and cooling rate. During the welding process, heating and cooling cause expansion and contraction of the weld area. The adjacent base metal acts as a restraint for any expansion or contraction that the weld area undergoes. This leads to residual stress buildup when it cools down to room temperature.

Moreover, the shaving process may result in plastic deformation of the subsurface and generating residual stress locally. If the shaving tool used to remove the weld flash is not aligned or shaped properly in accordance with the profile of the structural surface, longitudinal defects will be generated. The longitudinal steps on the outer surface of structural components produced during manufacturing through improper shaving of a weld flash could act as an effective stress raiser [Hasan et al., 2007].

Generally, there are three principal ways to reduce welding residual stresses.

1. *Peening operation.* A peening hammer is often used after each weld layer to distort the hot weld metal to reduce residual stress. Care has to be taken during this process because it can lead to cracking in weld metal if operated incorrectly.

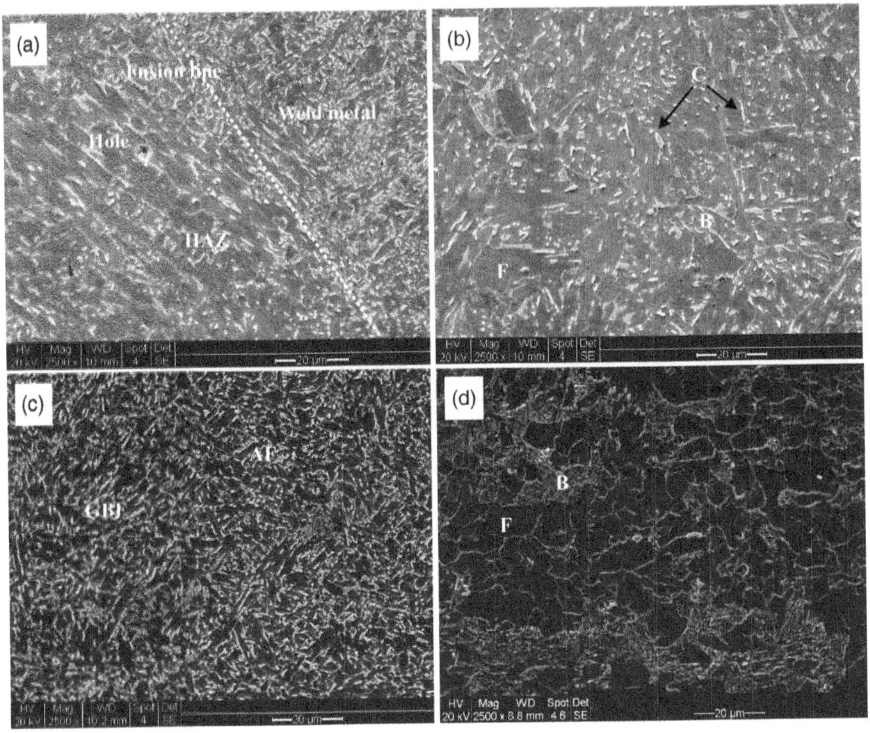

Figure 7-4 *SEM observation of the microstructure of welded X80 steel (a) HAZ/weld metal, (b) HAZ, (c) weld metal, and (d) base steel. (From Xue and Cheng [2012].)*

2. *Postweld heat treatment* (PWHT). The finished weld is reheated to an appropriate temperature to reduce the residual stress. In carbon and low-alloy steels, the welding residual stresses can be reduced from levels of 30,000 psi to around 5000 psi after this process [Wallace, 1979].
3. *Vibratory stress relief* (VSR). Based on the weight of the welded piece, the VSR method introduces high-amplitude, low-frequency vibrations for a given period of time. This relieves residual stress without distortion or alteration of tensile strength, yield strength, or resistance to fatigue. The most efficient vibrations are the resonant vibrations, because in resonance frequency vibrations, stress is better distributed.

7.4.2 Hardness of the Weld

Hardness testing of welds provides an indication of two parameters significant to the determination of a successful weld joint: the strength and microstructure of a known material. For example, steels that have been tempered to achieve the desired properties can undergo HAZ softening as a result of welding [Wang, 2006]. This leads to localized loss of the material strength, which can have a devastating effect

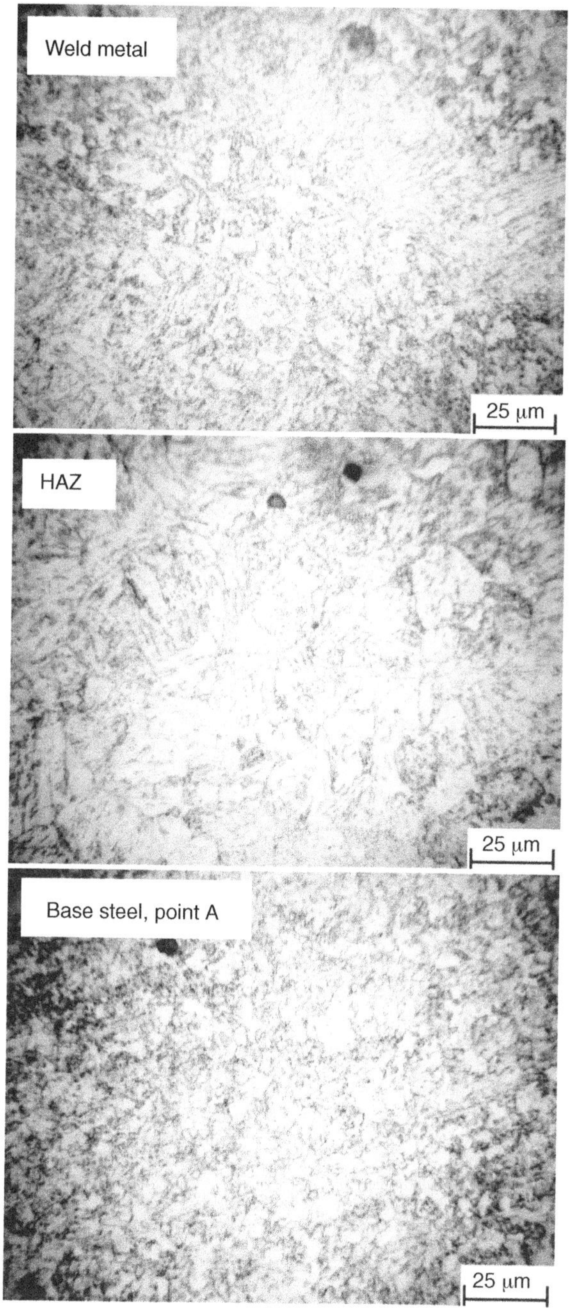

Figure 7-5 *Optical views of the microstructures of a welded X100 steel containing weld metal, the HAZ, and various points in the base steel as shown in Fig. 7-6. (From Zhang and Cheng [2010].)*

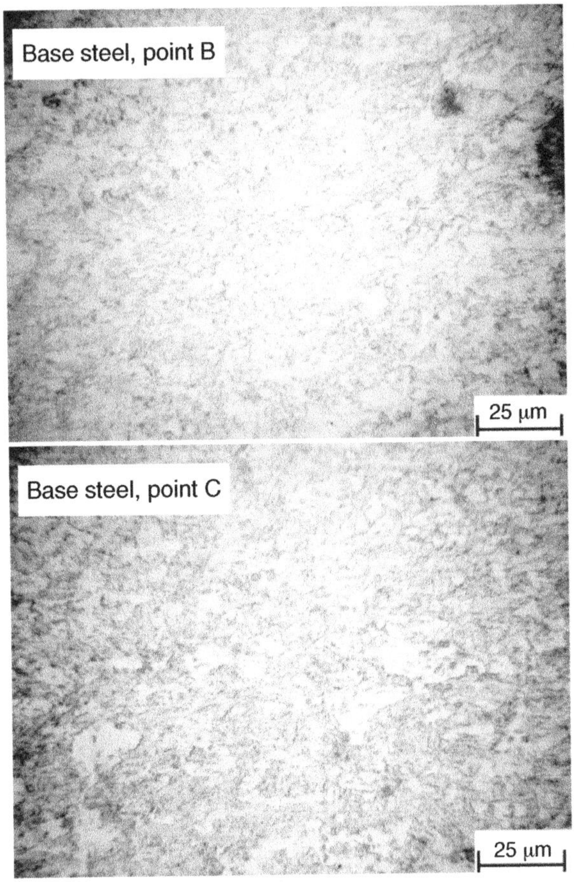

Figure 7-5 (Continued)

on structural performance. The hardness of the HAZ provides an estimation of the level of softening that may have occurred as a result of welding.

Vickers hardness is the predominant measurement test method for the welding zones, where a range of loads from 1 to 100 kg of diamond indenters can be utilized. The higher the load, the better the accuracy of the hardness reading, since the diamond makes a larger impression on the surface of the steel. Because it is important to determine the hardness of the HAZ, which is generally 1 or 2 mm thick, it is necessary to use lighter loads (e.g., 1, 2, or 5 kg).

The aim of weld hardness testing is to identify [Kou, 2003]:

- The hardness of the parent metal, to make an approximate determination of the material tensile strength to assure that the correct material is being welded.
- The hardness of the weld, to ensure that the weld metal meets or exceeds the strength requirements of the parent metal.

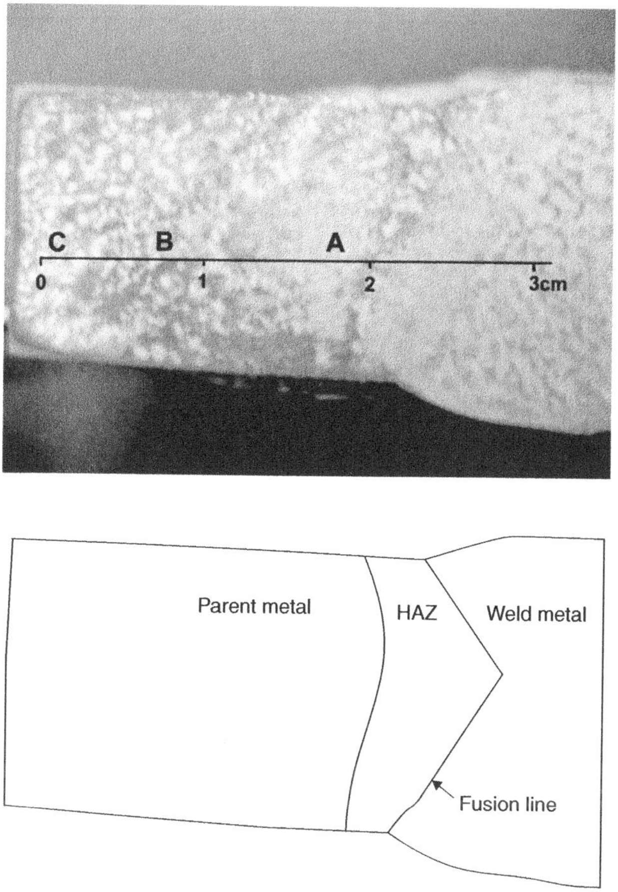

Figure 7-6 *Schematic diagram of a welded X100 pipeline steel specimen. (From Zhang and Cheng [2010].)*

- The hardness of the HAZ, to ensure that the welding heat input, preheat and interpass temperature have been controlled sufficiently to produce a HAZ with the appropriate strength and toughness.

When the hardness of substructures within the HAZ zone (e.g., coarse grains, fine grains, etc.) is required, it is necessary to conduct microhardness tests with loads in the range 0.5 to 2 kg. With these techniques, the area of interest is identified under a microscope and the hardness is measured directly in the area of interest. Such techniques are applied to critical welding applications such as temper bead welding, where a primary basis of acceptance of the welding technique is the hardness test [Sperko, 2005].

Figure 7-7 shows the Vickers microhardness distribution of various zones around an X70 steel weld. It is seen that there is the maximum microhardness at weld metal

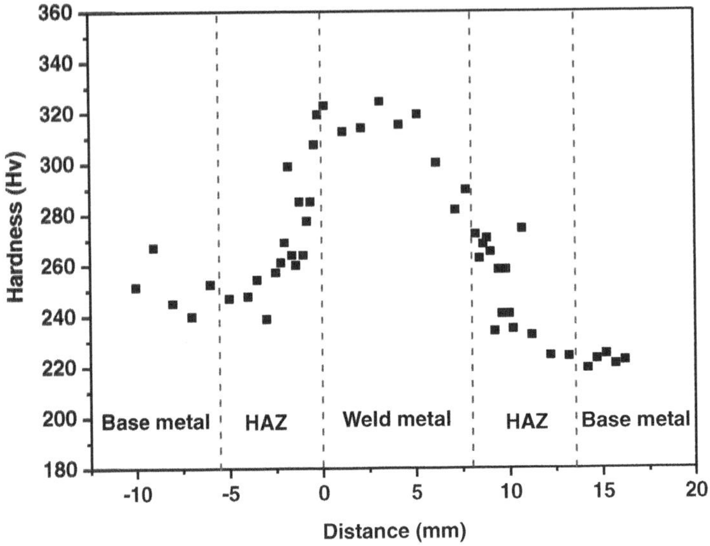

Figure 7-7 Microhardness distribution of a welded X70 steel. (From Zhang and Cheng [2009a].)

of about 310 to 320 Hv (Vickers hardness scale). The microhardness of the HAZ is higher than that of X70 base steel, which is approximately 220 to 240 Hv. The microstructural characterization shows that there are low-temperature transformation products such as bainitic ferrite at the HAZ. These products enhance the hardness of the HAZ compared to the ferrite–pearlite microstructure of the base steel.

The hardness measured on an X100 steel welded specimen (described schematically in Fig. 7-6) is shown in Fig. 7-8. There is lower hardness at the weld metal and HAZ than at the base steel. The maximum hardness is found near point B in base steel. The HAZ softening phenomenon occurs on welded X100 steel. This is due to the improved microstructure and carbon distribution, which usually occurs during postwelding treatment. Moreover, there is a high hardness level in the base steel which is associated with the presence of the hardening phase of bainite. In particular, the highest microhardness is at point B, due to the enhanced hardness by acicular ferrite.

7.5 PIPELINE WELDING: ENVIRONMENTAL ASPECTS

Environmental conditions prevalent around the welding area affect the welding processes as well as the weld properties.

7.5.1 Introduction of Hydrogen into Welds

Hydrogen could be introduced into a weld during the welding process. The mechanism of hydrogen introduction is that, during welding, a molten weld pool has a high

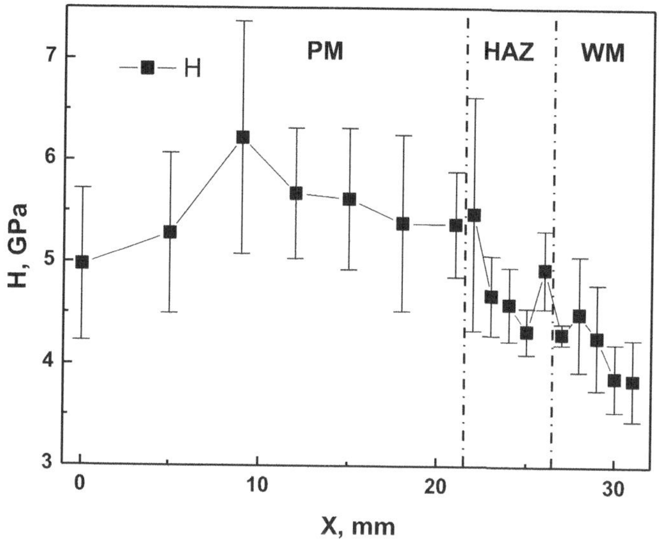

Figure 7-8 *Microhardness distribution on the welded X100 steel specimen shown in Fig. 7-6. (From Zhang and Cheng [2010].)*

solubility value for atomic hydrogen formed by the dissociation of water vapor or by a hydrocarbon in the welding arc. The rate of diffusion of hydrogen in steel at or near its melting temperature is quite high. Therefore, a molten weld pool can rapidly pick up atomic hydrogen from the hot gas in the arc. Hydrogen atoms can diffuse rapidly from the weld metal into the HAZ of the base metal, where a significant number of traps exist to accumulate hydrogen.

The sources of hydrogen introduced into welds include:

- Moisture retained in electrode flux as well as moisture introduced directly into the weld metal from the atmosphere.
- Hydrogen present in the weld metal.
- Oil or dirt present around the welding area.

In particular, the primary source of moisture is the ambient air. Generally, the intense energy of the arc causes the moisture in ambient air, which is ionized in the arc into atomic hydrogen, to be absorbed into the weld metal and, subsequently, leads to cracking.

Different types of electrodes are used in welding, including cellulosic, rutile, and basic electrodes. Each type of electrode has a different hydrogen content. For example, cellulosic electrodes exhibit the highest hydrogen content, due to the high proportion of hydroxyl groups in the electrode covering. One of the main objectives of the basic electrode is to achieve the lowest possible hydrogen content, ensuring crack-free seams even when welding high-strength steels. Furthermore, each electrode has different hydrogen-diffusible contents, which is basically a measure of moisture absorption of the electrode when moving it from storage into humid air. It is a common practice when not welding, not to expose the electrode to the atmosphere for more than 10 to 15 min. If an electrode is exposed for more than the time specified, it has to be re-baked to release the absorbed hydrogen following the electrode manufacturer's regulations and recommendations.

The introduction of hydrogen into welds could lead to many consequences that could cause failure of the welding area, such as HE, which degrades the strength of the joint and steel around the weld. Moreover, hydrogen in the weld usually leads to cold cracking in the HAZ. Reduction of hydrogen input in the weld metal can be achieved through a number of practices, including:

- Optimization of electrode coverings by modifying the coating composition, which will reduce the basic hydrogen content present in the electrode.
- Use of low-hydrogen-diffusible electrodes.
- Proper cleaning of the weld groove prior to the start of welding.
- Proper preheat maintained throughout welding.
- Proper welding variables followed.

If welding is expected to be interrupted, it is necessary to ensure that the weld is slowly cooled under insulation. This ensures that if hydrogen is trapped in the weld, it might have a chance of escaping. If the welding is for piping that has been in operation, hydrogen must be baked out prior to the start of fabrication. This will ensure that any hydrogen present in the parent metal is removed. This also reduces the probability of cracking during welding. A number of guides and reports of practical experiences can be helpful to avoid hydrogen entry and hydrogen cracking of welded steels [Bailey et al., 1993].

7.5.2 Corrosion at Welds

When welding similar or dissimilar metal structural components, the weld metal usually differs considerably from the parent metal in composition, microstrucuture, and properties. Moreover, there is usually a high level of welding residual stress at the weld. Due to the heterogeneous properties and stress distribution, and, more important, the fact that the weld metal is electrochemically anodic to the parent metal, the weld frequently suffers from preferential corrosion. In particular, the fusion zone is generated by melting, which fuses the base metal and filler metal to produce a zone whose composition is often different from that of the base metal. This compositional difference produces a galvanic couple, which can influence the corrosion process at the vicinity of the weld. This dissimilar-metal couple can produce macroscopic galvanic corrosion. Moreover, the fusion zone itself offers a microscopic galvanic effect, due to microstructural segregations, resulting from solidification [Buchheit et al., 1990].

One of the primary contributors to the accelerated corrosion of a weld is the high residual stress in the weld. Therefore, to reduce the stresses that are introduced during welding and to improve the microstructure and corrosion resistance, stress relief should be carried out. Moreover, the weld seam corrosion is affected by other factors that will influence the corrosion potential. Even a small amount of chemical composition variation within the welded region may increase or decrease corrosion. For example, an increase in sulfur concentration has been associated with the increased corrosion of weld seams in EBW pipes, while the addition of copper would reduce seam corrosion [Davis, 2006]. It has also been noted that addition of minor constitutions such as chromium and nickel may reduce the likelihood of weld seam corrosion by forming a passive surface oxide.

7.5.3 Electrochemistry of Localized Corrosion at Pipeline Welds

Field observations found [National Energy Board, 1996; Baker, 2005] that pipeline corrosion and SCC occur frequently at welds and the adjacent area. There have been extensive studies on the problem and explanations provided to understand this phenomenon. Postweld heat treatment was believed to affect corrosion resistance and hydrogen embrittlement of the HAZ [Hemmingsen et al., 2002]. Moreover, welding would increase the corrosion activity and SCC susceptibility of pipelines due to the metallurgical microstructural changes, residual stress generation, and a series of phase transformations taking place in the welding zone, especially the HAZ [Zhang et al., 2002].

To characterize the electrochemical corrosion behavior of various zones at the weld, the anodic dissolution current density is mapped on a welded X70 steel specimen in a nearly neutral–pH solution using the SVET technique, and the result is shown in Fig. 7-9, where the SVET scanning region includes the weld metal (0 to 2 mm in the figure), HAZ (2 to 8 mm), and base steel (8 to 10 mm). It is seen that the corrosion current of the weld metal is about 5 μA. The current increases gradually and reaches a maximum of approximately 25 μA in the HAZ (i.e., approximately at the HAZ side of the interface between the HAZ and the base steel). The smallest current, about 2 μA, is measured in the base steel. Apparently, the highest corrosion activity is at the HAZ in the solution, which is associated with the metallurgical phases present in this zone. Microstructurally, low-temperature transformation products such as bainitic ferrite exist in the HAZ. Generally, corrosion is appreciably more severe when the steel contains hardened structures [Voruganti et al., 1991; Davis, 2006]. Moreover, the weld metal and the HAZ usually contain a high density of lattice defects, which also contribute to the high corrosion activity in the weld metal and the HAZ compared to base steel.

For a welded X100 steel specimen immersed in a nearly neutral–pH solution, the SVET current shows a quite different map. Figure 7-10 shows the measured SVET map starting at point C (0 mm) along the black line in Fig. 7-6. It is seen that there is a maximum of current density up to 120 $\mu A/cm^2$ in the base steel centered at 1 cm (point B in Fig. 7-6), and that the smallest current density, about 45 $\mu A/cm^2$, occurs in the HAZ. It is believed that the improved microstructure and carbon distribution,

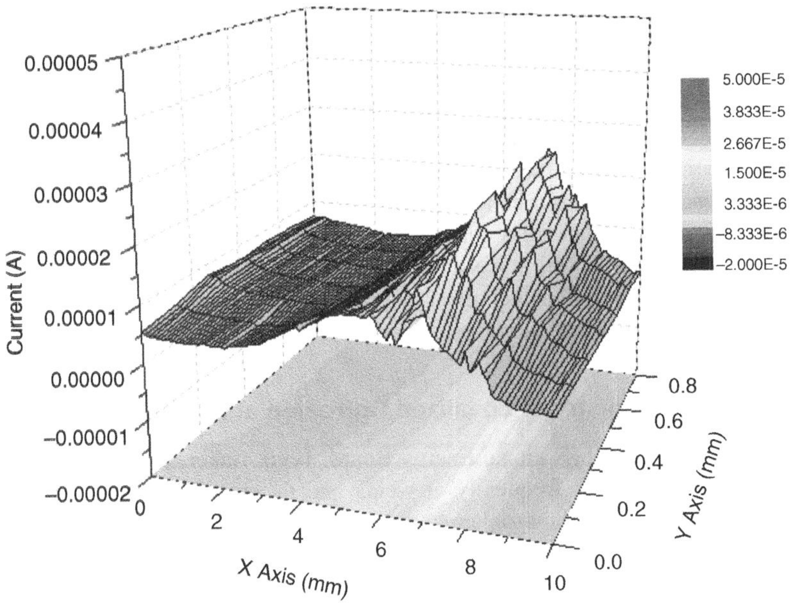

Figure 7-9 *SVET mapping of the corrosion current on a welded X70 steel specimen containing weld metal (0 to 2 mm), HAZ (2 to 8 mm), and base steel (8 to 10 mm) zones in a nearly neutral–pH solution. (From Zhang and Cheng [2009a].)*

which usually occurs during postwelding treatment for high-strength steels, lowers the corrosion activity of the HAZ. This is a very important finding regarding the corrosion of welded high-strength pipeline steel, because previous work conducted on low-grade pipeline steel welds found the highest corrosion current at the HAZ [Du et al., 2009; Zhang and Cheng, 2009a]. Furthermore, the highest current density is found at point B in the base steel, which is due primarily to the presence of a significant number of inclusions in this area [Zhang and Cheng, 2010]. The presence of inclusions provides active spots for anodic dissolution. It is thus assumed that postwelding treatment on X100 steel results in segregation of inclusions in base steel adjacent to the weld and provides potential sites for localized corrosion occurrence.

It has been demonstrated that hydrogen would enhance the corrosion of steels. After hydrogen enters welded steel, it prefers to concentrate at sites with high stress, such as welds in which a significant number of hydrogen traps, such as inclusions and voids, exist. Figure 7-11 shows the SVET maps measured on welded X70 steel in nearly neutral–pH solution after 2 h of hydrogen charging at various current densities. Generally, the current measured on all zones increases with increasing hydrogen-charging current density, indicating the enhanced dissolution activity of the steel upon hydrogen charging. Moreover, the highest currents are measured at the HAZ, indicating the accumulation of hydrogen in this zone due to the coarse grain structure acting as a hydrogen trap. Therefore, for low-grade pipeline steels, such as

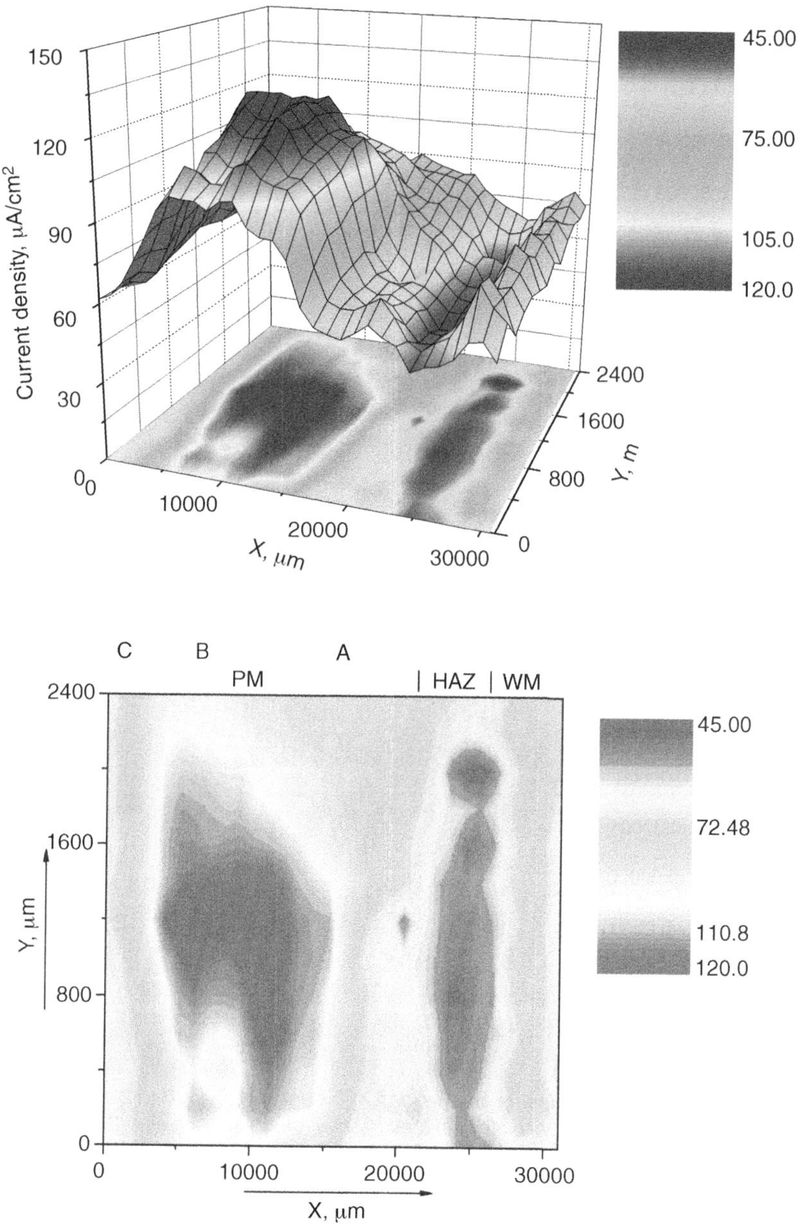

Figure 7-10 *SVET current density map measured on a welded X100 steel specimen in a nearly neutral–pH solution. (From Zhang and Cheng [2010].)*

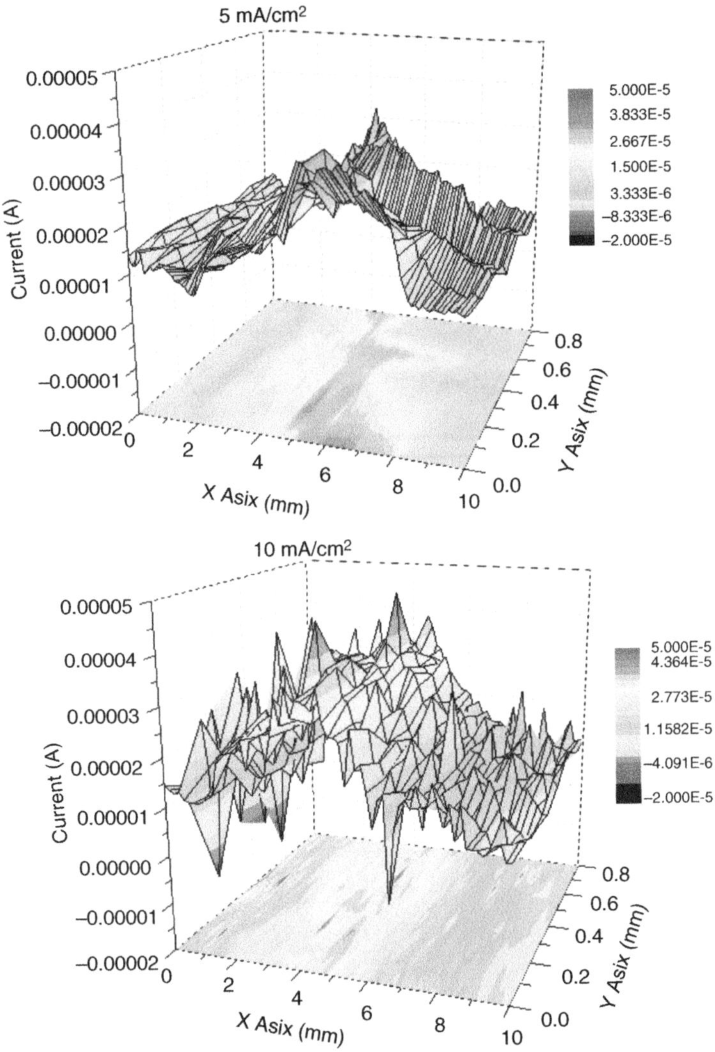

Figure 7-11 SVET measurements on a welded X70 steel specimen (0 to 2 mm: weld metal; 2 to 8 mm: HAZ; 8 to 10 mm: base steel) in a nearly neutral–pH solution after 2 h of charging at 5 and 10 mA/cm^2 of charging current densities. (From Zhang and Cheng [2009a].)

X70 steel, the HAZ shows the highest corrosion activity in the nearly neutral–pH, diluted bicarbonate solution before and after hydrogen charging.

The hydrogen-enhanced corrosion behavior of welded X100 steel is also characterized by SVET mapping, as shown in Fig. 7-12. The highest current density is observed at the HAZ rather than at the base steel as measured in the absence of hydrogen charging in Fig. 7-10. It is very interesting that hydrogen charging changes

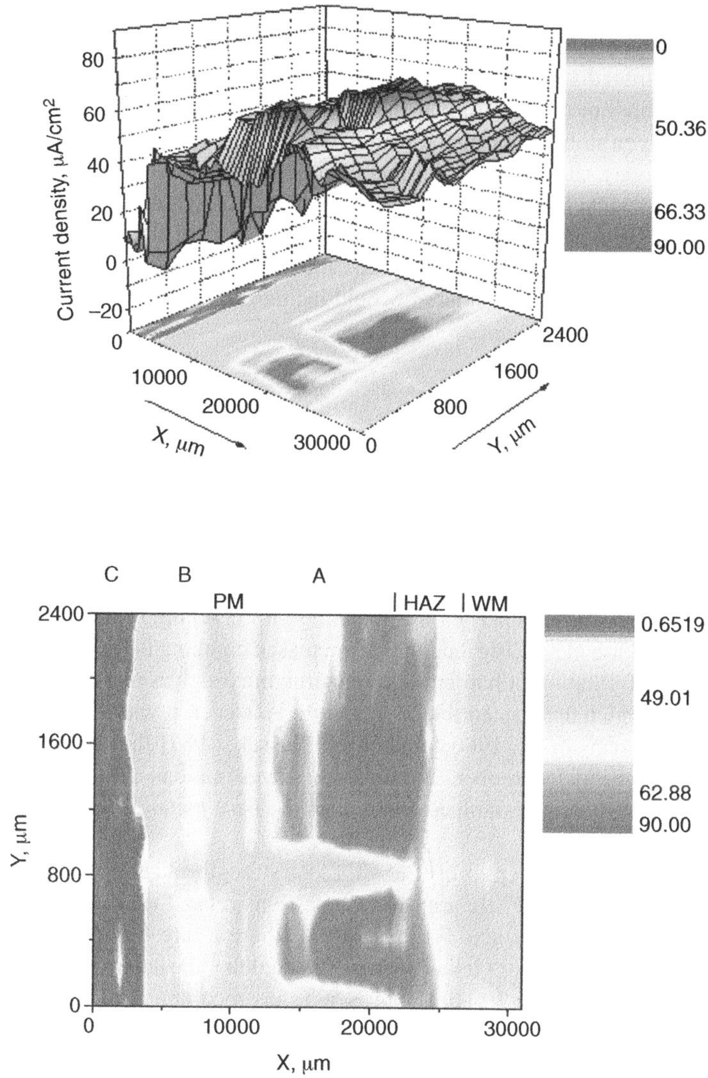

***Figure 7-12** SVET current density maps measured on a welded X100 steel specimen after 2 h of hydrogen charging at −2 V(SCE) in a nearly neutral–pH solution. (From Zhang and Cheng [2010].)*

the corrosion activity of welded steel, as indicated by the dissolution current density distribution measured on the welded steel specimen. The high current density at a HAZ is attributed to the accumulation of hydrogen locally. The microstructural characterization does not find a hardening phase in a HAZ, as shown in Fig. 7-5, and furthermore, the microhardness of a HAZ is even lower than that of base steel. Thus, the accumulation of hydrogen at a HAZ is not attributed to the metallurgical microstructural factor, but probably to the presence of welding residual stress. Indeed, the maximal longitudinal residual stress exists nearby the weld edge after calculation of the residual stress distribution [Zhang et al., 2009]. Even after stress-relieving treatment, a significant amount of residual stress still exists at the HAZ. Upon hydrogen charging, hydrogen atoms entering the steel would accumulate at sites with high stress concentration zones, such as welds [Oriani et al., 1985]. The synergism of hydrogen and local stress results in high corrosion activity at a HAZ. Therefore, different from low-grade pipeline steels, a high-strength steel weld shows a different corrosion activity distribution after hydrogen charging (i.e., the HAZ, which shows low corrosion activity before hydrogen is considered, would increase its local corrosion due to the local accumulation of hydrogen).

7.6 SCC AT PIPELINE WELDS

7.6.1 Effects of Material Properties and Microstructure

Numerous field experiences and lab studies indicate that SCC occurs primarily in the HAZ of pipelines and/or pressure vessel steel welds [Merrick, 1988; Kimura et al., 1989; Mahmoud et al., 1991]. The threshold stress for cracking is much lower in the HAZ than in the base metal. The tendency to increased cracking in this zone is usually attributed to the formation of hardening microstructures such as martensite/austenite (M/A) islands, which are susceptible to cracking. However, it was also found [Ume et al., 1985; Tromans et al., 1986; Pircher and Sussek, 1987] that substantial SCC occurs in weld metal. Moreover, a number of investigations [Taira et al., 1984; Onsoien et al., 1990] have established that even the base metal itself is not immune to SCC.

The influence on the susceptibility of SCC of the strength and hardness of materials has long been recognized. There are several questions related to the use of hardness as a measure of SCC resistance. For example, is there a threshold hardness below which cracking will not occur? It has been established that the hardness limit to avoid cracking ranges from 20 to 30 HRC, but NACE has recommended 22 HRC as the cutoff hardness, above which the weld metal can be susceptible to SCC [National Association of Corrosion Engineers, 2003]. It is worth pointing out that the hardness limit above which steels are susceptible to SCC depends on the environment and other variables. Moreover, reduction of the hardness at the HAZ does not always result in an improvement in the SCC resistance of a weld [Pircher and Sussek, 1987].

Furthermore, the SCC susceptibility of materials increases with certain microstructures. For example, tempered bainite has a higher susceptibility to cracking than one with a mixed microstructure. Acicular ferrite with a fine and interlocking microstructure usually results in good mechanical properties in terms of strength and toughness, while grain boundary ferrite represents a brittle phase with low resistance to crack propagation [Glover et al., 1977; Abson and Pargeter, 1986; Groug and Matlock, 1986]. Moreover, an increase in the pearlite content in microstructure has a detrimental effect on the SCC resistance of pipeline steels with a ferrite–pearlite structure [Lu and Luo, 2006].

Other variables, such as degree of alloying segregation and the type, size, shape, and distribution of inclusions, also influence resistance to SCC. For example, in investigations of SCC in X52 and X70 steels with longitudinal welds in an H_2S-saturated solution, it was found that the cracks were associated with MnS inclusions that were globular in morphology and of various sizes (1 to 5 μm) [Contreras et al., 2004].

7.6.2 Effects of the Welding Process

The welding process has an important effect on the SCC susceptibility of a weld. The welding processes commonly used for joining pipelines are electrical resistance welding and submerged arc welding (SAW), where wide HAZs are created. As a result, a high probability of cracking results, due to the heterogeneous zone generated. Therefore, a welding process with a narrower HAZ, called the indirect electric arc welding (IEAW) process, is currently under development [Garcia et al., 2002, 2003]. A comparative study of the SCC resistance of X65 steel welds by the IEAW, SAW, and metal inert gas (MIG) processes found that the IEA process produces a homogeneous columnar microstructure with a fine grain and small HAZ at the weld [Natividad et al., 2007]. This weld shows better resistance to SCC than do the MIG and SAW processes. Furthermore, it was found that the MIG and metal active gas (MAG) processes could create different microstructures at the welds of X70 and X80 pipeline steels, affecting the strength of the steels and their SCC resistance [Omweg et al., 2003].

The heat input during a welding process could control the cooling rate and affect the susceptibility of welds to SCC. An increasing welding heat input would cause a decrease in the impact toughness of the grain-coarsened HAZ of high-strength steel welds. For example, the lowest toughness of the grain-coarsened HAZ of an X80 steel weld was obtained at a high welding heat input [Smith et al., 1989]. In investigations of the cracking characteristics of A516 pressure vessel steel welds prepared by SAW under various welding heat inputs (i.e., 15, 30, and 45 kJ/cm) in a H_2S saturated solution, it was found that environmentally assisted cracking in the HAZ occurred only in a weld welded at a high heat input of 45 kJ/cm [Huang et al., 1994]. However, it was proposed that the resistance of a grain-coarsened HAZ to SCC could be improved by a decrease in the cooling rate by increasing the welding heat input, which would facilitate the formation of ferrite instead of martensite [Onsoien et al., 1990]. Postweld heat treatment can reduce the SCC of a weld by redistributing the localized stress and reducing the residual tensile stress available to induce cracking [Krishnan and Rao, 1990].

7.6.3 Hydrogen Sulfide SCC of Pipeline Welds

Hydrogen sulfide stress corrosion cracking (HSSCC) is a widely known phenomenon that often leads to failure of oil and gas pipelines and other oilfield facilities in petroleum production, where hydrogen sulfide is present in varying amounts. In general, this type of degradation has been attributed to hydrogen evolution as H_2S reacts with steel to form a FeS film and hydrogen. The hydrogen atoms produced tend to be absorbed and to penetrate the steel in regions of high internal stress of a tensile nature. Thus, HSSCC is a type of HIC.

It has been shown that both SCC and HIC could occur at a carbon steel weld. In the base steel, HIC was found to propagate not only along the ferrite–pearlite interface, but also through the ferrite and pearlite. The susceptibility of the weld to cracking in H_2S-containing environments increases with an increasing amount of grain boundary ferrite in the weld metal or acicular ferrite in the HAZ through the increase in welding heat input. At a welding heat input of 45 kJ/cm, the weld metal, HAZ, and base steel were all prone to fracture in H_2S environments [Huang et al., 1994]. The amount of hydrogen pickup in weld metal depends on the grain size and nature of the grain boundaries of the constituents in the weld metal. The small grain size and high-angle grain boundaries of acicular ferrite contain more hydrogen-trapping sites than does grain boundary ferrite. It is thus expected that the critical concentration of hydrogen for crack formation in acicular ferrite is higher than that in grain boundary ferrite. An increase in the volume fraction ratio of grain boundary ferrite versus acicular ferrite would occur by increasing the welding heat input. Thus, hydrogen-induced SCC occurring in weld metal at high welding heat inputs might be attributed to the high volume fraction of grain boundary ferrite in the weld metal.

Furthermore, there are apparent effects of microstructure on the HIC of the weld. Finely solidified structures with finely dispersed spherical oxide inclusions in the weld metal, resulting from rapid cooling and solidification during the welding process, possess excellent resistance to HIC [Taira et al., 1981]. The microstructures of the weld metal and the HAZ are, however, dependent on the welding process and can vary considerably. The HIC characteristics associated with complex microstructures of the weld metal and the HAZ require detailed investigations before welded steels are used in sour environments.

REFERENCES

Abson, DJ, Pargeter, RJ (1986) Factors influencing as-deposited strength, microstructure, and toughness of manual metal arc welds suitable for C–Mn steel fabrications, *Int. Met. Rev.* 31, 141–194.

Bailey, N, Coe, FR, Gooch, TG, Hart, PHM, Jenkins, N, Pargeter, RJ (1993) *Welding Steels Without Hydrogen Cracking*, 2nd ed., Woodhead Publishing, Cambridge, UK.

Baker, M (2005) *Final Report on Stress Corrosion Cracking Study*, Integrity Management Program Delivery Order DTRS56-02-D-70036, Office of Pipeline Safety, U.S. Department of Transportation, Washington, DC.

Buchheit, RG, Moran, JP, Stoner, GE (1990) Localized corrosion behavior of alloy 2090: the role of microstructural heterogeneity, *Corrosion* 46, 610–617.

Contreras, A, Albiter, A, Angeles-Chavez, C, Pérez, R (2004) Mechanical and microstructural effects on the stress corrosion cracking of weld beads of X-52 and X-70 pipeline steels, *Rev. Mex. Fis.* 50S1, 49–53.

David, SA, Babu, SS, Vitek, JM (2003) Welding: solidification and microstructure, *JOM*, June.

Davis, JR (2006) *Corrosion of Weldments*, ASM International, Materials Park, OH.

Du, CW, Li, XG, Liang, P, Liu, ZY, Jia, GF, Cheng, YF (2009) Effects of microstructure on corrosion of X70 pipe steel in an alkaline soil, *J. Mater. Eng. Perf.* 18, 216–220.

Evans, GM, Bailey, N (1997) *Metallurgy of Basic Weld Metal*, Woodhead Publishing, Cambridge, UK.

Garcia, R, Lopez, VH, Bedoll, E, Manzano, A (2002) MIG welding process with indirect electric arc, *J. Mater. Sci. Lett.* 21, 1965–1967.

Garcia, R, Lopez, VH, Bedoll, E, Manzano, A (2003) A comparative study of the MIG welding of Al/TiC composites using direct and indirect electric arc processes, *J. Mater. Sci.* 38, 2771–2779.

Glover, AG, McGarth, JT, Tinkler, MJ, Weatherly, GC (1977) The influence of cooling rate and composition on weld metal microstrutures in C–Mn and HSLA steel, *Weld. J.* 56, 267–273.

Groug, O, Matlock, DR (1986) Microstructural development in mild and low-alloy steel weld metals, *Int. Met. Rev.* 31, 27–48.

Hasan, F, Iqbal, J, Ahmed, F (2007) Stress corrosion failure of high-pressure gas pipeline, *Eng. Fail. Anal.* 14, 801–809.

Hemmingsen, T, Hovdan, H, Sanni, P, Aagotnes, NO (2002) The influence of electrolyte reduction potential on weld corrosion, *Electrochim. Acta* 47, 3949–3955.

Huang, HH, Tsai, WT, Lee, JT (1994) Cracking characteristics of A516 steel weldment in H_2S containing environments, *Mater. Sci. Eng. A* 188, 219–227.

Kiefner, F (1997) Failure analysis of pipelines, in *ASM Handbook*, ASM, Metals Park, OH.

Kimura, M, Totsuka, N, Kurisu, T, Amano, Matsuyama, J, Nakai, Y (1989) Sulfide stress-corrosion cracking of line pipe, *Corrosion* 45, 340–346.

Kou, S (2003) *Welding Metallurgy*, 2nd ed., Wiley, Hoboken, NJ.

Krishnan, KN, Rao, KP (1990) Room-temperature stress corrosion cracking resistance of post-weld heat-treated austenitic weld metals, *Corrosion* 46, 734–742.

Lancaster, JF (1999) *Metallurgy of Welding*, 6th ed., William Andrew Publishing, Cambridge, UK.

Lu, BT, Luo, JL (2006) Relationship between yield strength and near-neutral pH stress corrosion cacking resistance of pipeline steels: an effect of microstructure, *Corrosion* 62, 129–140.

Lu, B, Luo, J, Ivey, D (2010) Near-neutral pH stress corrosion cracking susceptibility of plastically prestrained X70 steel weldment, *Metall. Mater. Trans. A* 41, 2538–2547.

Mahmoud, SE, Petersen, CW, Franco, RJ (1991) Overview of hydrogen induced cracking of pressure in upstream operations, *Corrosion 1991*, Paper 10, NACE, Houston, TX.

Merrick, RD (1988) Refinery experience with cracking in wet H_2S environments, *Mater. Perf.* 27, 30–36.

National Association of Corrosion Engineers (2003) *Materials Resistant to Sulfide Stress Cracking in Corrosive Petroleum Refining Environments*, Standard MR0103-2003, NACE, Houston, TX.

National Energy Board (1996) *Stress Corrosion Cracking on Canadian Oil and Gas Pipelines, Report of the Inquiry*, Report MH-2-95, NEB, Calgary, Alberta, Canada.

Natividad, C, Salazar, M, Espinosa-Medina, MA, Pérez, R (2007) A comparative study of the SSC resistance of a novel welding process IEA with SAW and MIG, *Mater. Character.* 58, 786–793.

Omweg, GM, Frankel, GS, Bruce, WA, Ramirez, JE, Koch, G (2003) The performance of welded high-strength low-alloy steels in sour environments, *Corrosion* 59, 640–653.

Onsoien, MI, Akselsen, OM, Grong, O, Kvaale, PE (1990) Prediction of cracking resistance in steel weldments, *Weld. J.* 69, 45–51.

Oriani, RA, Hirth, JP, Smialowski, M (1985) *Hydrogen Degradation of Ferrous Alloys*, Noyes Publications, Park Ridge, NJ.

Otegui, JL, Chapetti, MD, Motylicki, J (2002) Fatigue assessment of an electrical resistance welded oil pipeline, *Fatigue* 24, 21–28.

Parkins, RN (2000) A review of stress corrosion cracking of high-pressure gas pipelines, *Corrosion 2000*, Paper 363, NACE, Houston, TX.

Pircher, H, Sussek, G (1987) Testing the resistance of welding in low-alloy steels to hydrogen induced stress cracking, *Corros. Sci.* 27, 1183–1196.

Rampaul, H (2003) *Pipe Welding Procedures*, 2nd ed., Industrial Press, New York.

Savage, WF (1969) New insight into weld cracking and a new way of looking at welds, *Weld. Des. Eng.*, Dec.

Smith, NJ, McGrath, JT, Gianetto, JA, Orr, RF (1989) Microstructure/mechanical property relationships of submerged arc welds in HSLA 80 steel, *Weld. J.* 68, 112–120.

Sperko, WJ (2005) Exploring temper bead welding, *Weld. J.*, Aug.

Taira, T, Tsukada, K, Kobayashi, Y, Inagaki, H, Watanabe, T (1981) Sulfide corrosion cracking of linepipe for sour gas service, *Corrosion* 37, 5–16.

Taira, T, Kobayashi, Y, Matsumoto, K, Tsukada, K (1984) Resistance of line pipe steels to wet sour gas, *Corrosion* 40, 478–486.

Tromans, D, Ramakrishna, S, Hawbolt, EB (1986) Stress corrosion cracking of ASTM A516 steel in hot caustic sulfide solutions: potential and welding effects, *Corrosion* 42, 63–70.

Ume, K, Taira, T, Hyodo, T, Kobayashi, Y (1985) Initiation and propagation morphology of sulfide stress corrosion cracking, *Corrosion 1985*, Paper 240, NACE, Houston, TX.

Voruganti, VS, Luft, HB, De Geer, D, Bradford, SA (1991) Scanning reference electrode technique for the investigation of preferential corrosion of weldments in offshore applications, *Corrosion* 47, 343–351.

Wahid, A, Olson, DL, Matlock, DK (1993) Corrosion of weldments, in *Welding, Brazing and Soldering*, ASM Handbook, Vol. 6, D.L. Olson, T.A. Siewert, S. Liu, and G.R. Edwards, Editors, ASM, Metals Park, OH, p. 1065.

Wallace, JF (1979) *A Review of Welding Cast Steels and Its Effects on Fatigue and Toughness Properties*, Steel Founders' Society of America, Rocky River, OH.

Wang, YY (2006) Optimizing weld integrity for high-strength steels, *Advanced Welding and Joining Technical Workshop*, Boulder, Co.

Xue, HB, Cheng, YF (2012) Hydrogen permeation and electrochemical corrosion behavior of the X80 pipeline steel weld, *J. Mater. Eng. Perf. DOI*: 10.1007/s11665-012-0216-1.

Zhang, C, Cheng, YF (2010) Corrosion of welded X100 pipeline steel in a near-neutral pH solution, *J. Mater. Eng. Perf.* 19, 834–840.

Zhang, GA, Cheng, YF (2009a) Micro-electrochemical characterization of corrosion of welded X70 pipeline steel in near-neutral pH solution, *Corros. Sci.* 51, 1714–1724.

Zhang, GA, Cheng, YF (2009b) Micro-electrochemical characterization and Mott–Schottky analysis of corrosion of welded X70 pipeline steel in carbonate/bicarbonate solution, *Electrochim. Acta* 55, 316–324.

Zhang, HJ, Zhang, GJ, Cai, CB, Gao, HM, Wu, L (2009) Numerical simulation of three-dimension stress field in double-sided double arc multipass welding process, *Mater. Sci. Eng. A* 499, 309–314.

Zhang, W, Elmer, JW, DebRoy, T (2002) Modeling and real time mapping of phases during welding of 1005 steel, *Mater. Sci. Eng. A* 333, 321–330.

8

Stress Corrosion Cracking of High-Strength Pipeline Steels

8.1 INTRODUCTION

Nowadays, with new gas and conventional oil discoveries, as well as the development of technologies making the commercialization of oil sands and heavy oil reserves feasible economically, continuous development of pipeline projects has been envisioned throughout the world. Pipelines have expanded significantly, and new routes are designed and constructed every year. In North America, Russia, and northern Europe, there has been increasing interest in oil and gas exploration and production in the sub-Arctic and Arctic regions [Prolog Canada Inc., 2003].

To meet technical challenges and to reduce the cost related to the transmission of large volumes of oil and natural gas, assessments of advanced high-strength pipeline steels have been carried on API X80, X100, and X120 steels, which will potentially be used in future pipelines, such as the Mackenzie Delta pipeline in northern Canada and the Alaska pipeline starting at the North Slope in Alaska. In Canada, TransCanada Pipelines and its partners have been involved in a series of technology programs on high-strength steels (i.e., X80 and X100 steels) which will enable the development of high-pressure long-distance transmission pipelines. For example, the proposed Mackenzie Gas Project will carry 1.0 billion cubic feet (bcf) per day of natural gas from the Mackenzie Delta of the Northwest Territory, Canada to northern Alberta via a 798-mile pipeline with an operating pressure of 2050 psi [Prolog Canada Inc., 2003]. X80 steel is proposed for use with a pipeline diameter of 762 mm (30 in.) and a wall thickness of 15.8 mm (0.625 in.). The pipeline has the potential to be expanded up

Stress Corrosion Cracking of Pipelines, First Edition. Y. Frank Cheng.

to 1.5 bcf per day and is expected to be in service in 2015. Furthermore, a 1400-mile portion of the proposed 2140-mile Alaska gas pipeline will pass through Canada, possibly running through the Yukon, then into British Columbia, and linking into an existing pipeline network in Alberta [Palmer, 2004]. For the Alaska gas pipeline project, TransCanada Pipelines has selected a pipe platform 48 in. in diameter with a 2500-psi operating pressure to transport an initial volume of 4.5 bcf per day with a potential expansion up to approximately 5.6 bcf per day. Although TransCanada has years of experience with X80 grade steel, this project will use X100 steel, a higher grade of steel.

8.2 DEVELOPMENT OF HIGH-STRENGTH STEEL PIPELINE TECHNOLOGY

8.2.1 Evolution of Pipeline Steels

In recent years, with the continuing growth in energy consumption, extensive attention has been paid to supplying oil and natural gas in a more economical and safer way. Development of high-strength steel pipelines has enabled the energy industry to realize significant savings in the total cost of long-distance oil and gas transmission in view of pipeline wall thickness and operating pressure [Corbett et al., 2004]. The highest-strength grades of pipelines that have so far been brought to commercial application or to trials by pipeline operators are X80 and X100 [Kalwa et al., 2002].

Over the past 60 years the trend toward increased transportation efficiency has been achieved largely by the development and processing of materials for pipelines. The primary effort put in by pipeline and metallurgical industries has been to develop high-strength pipeline steels, such as grade X80, X100, and X120 steels.

The evolution in the development of commercial pipeline steel technologies has taken place over 70 years [Asahi, 2004]. The materials used initially for pipeline networks were X52 steel or lower grades in the 1950s. The increasing grades of steel have been used with the improvement in steel processing technologies, such as the X60 and X65 steels used in the 1970s, the X70 steel that began to be used at the end of the 1970s, and the X70 steel that dominates nowadays. While the highest-strength grade of line pipes in the worldwide range so far been brought to commercial application is X80 steel by the API standard, the evolution of steel grades for pipeline applications is closely linked to developments in steelmaking processing technology.

To produce adequate formability and weldability, high-strength steels usually have a low carbon content (e.g., 0.05 to 0.25 wt% C). Small amounts of alloy elements such as chromium, nickel, molybdenum, copper, vanadium, niobium, and titanium are added to facilitate the formation of required microstructures, and thus to improve the properties of steels and to meet the demand for high strength and low-temperature toughness [Kalwa et al., 2002].

One important advance in the development of pipeline steels occurred in the 1970s with the introduction of thermomechanical controlled processing (TMCP) to substitute for traditional heat treatment production routes. Subsequent introduction of accelerated cooling after controlled rolling results in the possibility of producing steels with high strength levels [De S. Bott et al., 2005]. For example, the accelerated cooling process enables steels up to X70 to be produced from steels that have reduced carbon content. An improved processing method involving thermomechanical (TM) rolling and subsequent accelerated cooling emerged in the 1980s. By this method, a higher-strength steel such as X80 steel was produced, having an even lower carbon content and, thereby, excellent weldability. After an initial trial in the early 1990s, X80 steel was implemented in the eastern Alberta system in 1994. Since then, more than 460 km of X80 steel has been installed in Canada, including projects dealing with discontinuous permafrost in northwestern Alberta.

As the development of X80 steel is finished, this grade is the state of the art for high-pressure gas pipelines. The addition of molybdenum, copper, and nickel would enable the strength level of X80 steel to be raised to that of grade of X100 when the steel is processed by TM rolling plus accelerated cooling. Grade X100 steel has been in full-scale testing. The use of materials of even higher strength, such as grade X120 steel, is the challenging next step [Hillenbrand et al., 2004]. X120 pipeline steel requires far-higher production technology than that for X80 and X100 steels, in all the technical fields, such as material design, steelmaking, casting, plate production, and pipe production. This steel is still undergoing laboratory testing.

8.2.2 High-Strength Steels in a Global Pipeline Application

In the early 1970s, grade X70 steel was introduced in Germany in the construction of gas transmission pipelines [Hillenbrand et al., online source]. Since then, X70 steel has proven to be a very reliable material in numerous pipeline projects.

Following satisfactory experiences with X70 steel applications, grade X80 steel line pipe came into use for the first time in 1985 in a 3.2-km section of pipeline on a trial basis in Germany [Hillenbrand et al., online source]. In 1992 and 1993, X80 steel was first employed in a 250-km pipeline project constructed by Ruhrgas AG in Germany [Hillenbrand et al., online source]. In 2001 and 2002, X80 steel pipes were manufactured in the UK, and several parts of a gas pipeline network were ordered to a total length of 158 km. Until now, grade X80 steel has been used in a number of pipeline projects, as shown in Table 8-1. Both X70 and X80 steels offer wall thickness advantages for typical design and operational pressure. In Canada, the TransCanada pipeline system currently has more than 6300 km of large-diameter X70 steel pipes and almost 500 km of large-diameter X80 steel line pipes.

With the continuously increasing operating pressure, a potential for even higher pressure is forecast, and higher-strength pipeline steels have been considered for this purpose. Following an initial research and development project in 1995, a detailed program was begun in 1997 to study the benefits and applications of X100 steel in new, higher-pressure pipeline systems [Glover et al., 1999]. The program was accelerated in 1999 when interests in the Alaskan and Mackenzie Delta pipeline projects emerged. As a series of pipeline projects, X100 steel was implemented successfully in the West

TABLE 8-1 Projects Executed with Line Pipes Made of X80 Steel

Project	Dimensions	Quantity	Realization
Megal II, Germany	1118 × 13.6 mm	3.2 km	1985
Fourth Transit gas pipeline, Czechoslovakia	1420 × 15.5 mm	1.5 km	1985
Werne–Schlüchtern pipeline, Ruhrgas, Germany	48 in. × 18.4 and 19.3 mm	250 km	1992–1993
Nova pipeline; Matzhiwn project, Alberta, Canada	48 in. × 12.1 mm	54 km	1994
TransCanada pipeline,	48 in. × 12.0 and 16.0 mm	118 km	1997
Transco, UK	48 in. × 15.1 and 21.8 mm	42 km	2001
Canadian Natural Resources, Canada	24 in. × 25.4 mm	18 km	2001

Source: Graf et al. [2003].

Path project in the fall of 2002 and was utilized on a winter construction project at Godin Lake, Alberta, with 2 km of X100 pipe included in the loop [Mohitpour et al., 2001; Glover et al., 2004]. In 2006, IPSCO in Regina successfully produced and supplied helical-welded X100 steel pipes of 42 in. × 0.5 in. wall thickness over 2 km in length. IPSCO also supplied X100 steel pipes of 30 in. × 0.375 in. wall thickness for the Fort McKay project [Horsley, 2007].

X100 steel is a significant improvement on its predecessors. It has a minimum yield strength of 690 MPa or 100,000 psi, from which it takes the name X100, making it 43% and 25% stronger than X70 and X80 steels, respectively. If X100 steel is used, the pipeline wall thickness required for gas under the very high pressures envisaged for the Alaskan pipeline would be reduced significantly, typically 20.6 mm compared with 25.8 mm for X80 steel pipe. Thus, stronger steels allow thinner walls or a smaller pipe diameter, with all the advantages of manufacturing and construction, plus the economic benefit. Other future possibilities for X100 steel include a proposed pipeline from eastern Siberia to China of a size similar to that of the Alaskan gas project.

Since 1996, Nippon Steel has carried out a joint project with ExxonMobil for the development of X120 ultrahigh-strength steel pipe, designed to achieve a substantial cost reduction in a pipeline project. In 2004, for a demonstration line (1 mile) in Canada, for the first time globally, Nippon Steel made a delivery of this jointly developed X120 steel pipe, which proved successful in laying and operation [Nippon Steel Corporation, 2006].

Evolution of grades of high-strength steels has made a significant contribution to cost reduction through increased operating pressure and decreased pipe wall thickness. For a pipeline project of 250 km, use of X100 pipeline steels would result in material savings of approximately 19,000 and 39,000 tons compared to the use of X80 and X70 steels, respectively [Graf et al., 2003]. Obviously, with increasing grades of high-strength pipeline steel, the saving on materials is significant. This preliminary economic evaluation highlights the fact that X100 steel high-pressure pipelines could provide investment cost savings of about 7% with respect to grade X80. Other studies claim cost savings of up to 30% when X70 and X100 steel are compared.

8.3 METALLURGY OF HIGH-STRENGTH PIPELINE STEELS

8.3.1 Thermomechanical Controlled Processing

In modern oil and gas pipeline steel technology, a suitable microstructure is an important factor for improvement in the strength and toughness of steel. Thermomechanically controlled rolling processes were developed in the 1970s, changing the metallographic microstructure of steels from ferrite–pearlite to acicular ferrite and bainite.

Thermomechanically controlled processing is controlled rolling to produce desirable, fine-grained microstructure. Meanwhile, with accelerated cooling it can provide a high strength–toughness combination. The microstructure of steels obtained depends strongly on the processing parameters, such as the reheating temperature, cooling rate, and cooling temperature [Shanmugam et al., 2008]. The cooling system can be put into operation twice during rolling. The first cooling operation enhances the grain refinement of ferrite, and the second operation prevents the formation of pearlite during cooling, thereby improving the homogeneity of the final microstructure. Important variables in cooling operations are the cooling rate and the cooling stop temperature. A few pearlite islands may be present in the midwall region of thermomechanically rolled steel plates. As a result of two-stage cooling, not only is the ferrite grain size refined further but the pearlite is also replaced with bainite. Based on the alloying design and TMCP in the steelmaking, different combinations of microstructures could be obtained. In particular, a bainitic ferrite microstructure with martensite–austenite (M–A) islands as the second phase is designed as the primary microstructure of grade X100 steel.

The predominant microstructure in X80 pipeline steel is bainite with fine ferrite, which is obtained by a combination of TMCP and accelerated cooling [Stalheim and Muralidharan, 2006; Johnson et al., 2008; Kim and Bae, 2008]. In the accelerated cooling process, the cooling rate is controlled at 15 to 20°C/s until a temperature of around 550°C is reached. Air cooling is then used to change the microstructure from a fine ferrite base to a dominant bainite content. For X100 high-strength steel, the microstructure of the steel includes lower-temperature transformation products such as lower bainite, with a small amount of M–A as the second phase, to achieve both higher strength and appropriate toughness [Ishikawa et al., 2008]. For the production of X100 and X120 steel plates, more advanced technologies are required for use in material design and steel making, casting, and plate production. With a low-level carbon content and carbon equivalents, combined with the two-stage TMCP process in the austenite single region and the austenite + ferrite two-phase region, and the consequent optimized accelerated cooling to temperatures below 300°C, the best mechanical properties and appropriate weldability of the steels can be obtained.

8.3.2 Alloying Treatment

In high-strength pipe steels, the basic chemical composition includes C, Mn, Si, Ni, Cr, Mo, Nb, and Ti. Increasing the carbon content or carbon equivalent would increase the material strength and hardness but decrease the ductility and weldability of the steel. The *carbon equivalent* is an empirical formula in weight percent that relates the combined effects of different alloying elements used in the making of steels to an equivalent amount of carbon [Yurioka, 2001]. In terms of welding, it is an excellent tool to use to estimate the weldability of steels in quantity because it governs the hardenability of the parent metal or weld metal. The concentration of each solute is scaled by a coefficient that expresses its ability, relative to carbon, to retard transformation from the austenite phase to the ferrite phase. Steels with a carbon equivalent in excess of about 0.4 wt% cannot be welded easily because of their increased tendency to form martensite [Wu, 2006].

The properties of pipeline steels can be improved the by addition of alloying elements that affect the phase transformation and thus the mechanical properties of steels [American Petroleum Institute, 2004]. Alloying additions such as Mn, Nb, V, Ti, Mo, Ni, Cr, and Cu are commonly used to obtain the desired microstructure and mechanical properties [Lee et al., 2000; Sun et al., 2002]. For example, Mn is an important alloying element for solid strengthening, but an Mn content above 0.3% in steels could cause hydrogen-induced blister cracking when subjected to a sour service environment. Steels containing a low Mn content with an additional strength requirement can be achieved by adding Cu. Furthermore, to meet the demand is steels for high strength and high toughness, a combination of low carbon content and low sulfur is necessary. For X80 and X100 steels, Mo is usually added to control the microstructure and to obtain excellent properties. It was found that Mo- and Nb-containing steels exhibit superior strength and toughness combinations to those with Nb–V-added steels. Adding small amounts of grain refining elements such as Nb, V, Ti, and Al can produce fine ferrite grain sizes in the steel, increasing the strength and reasonable ductility and decreasing the brittle transition temperature. The role of alloying elements in acicular ferrite transformation is generally dependent on their effects on hardenability enhancement. Some elements, such as Mn, Mo, Cr, and Ni, favor the phase transformation, whereas others, such as S, P, and Cu, inhibit it.

For X120 pipeline steel, the carbon content should be kept at a lower level, with addition of the element boron (B) to obtain high toughness. The addition of boron in steels suppresses the ferrite formation and lowers the bainitic transformation temperature. The steel contains Cu, Ni, Cr, Nb, Ti, and V, in addition to boron. By interruptured direct quenching (IDQ) or heavy accelerated cooling (> 20°C/s) to temperatures below 300°C, the formation of lower bainite dominates the microstructure, with M–A constituents as the second phase [Bai et al., 2008; Pourkia and Abedini, 2008].

8.3.3 Microstructure of High-Strength Steels

Fine-grained structures are achieved by TMCP plus modified accelerated cooling for X80 and X100 steels, and TMCP plus heavy accelerated cooling for X120 steel

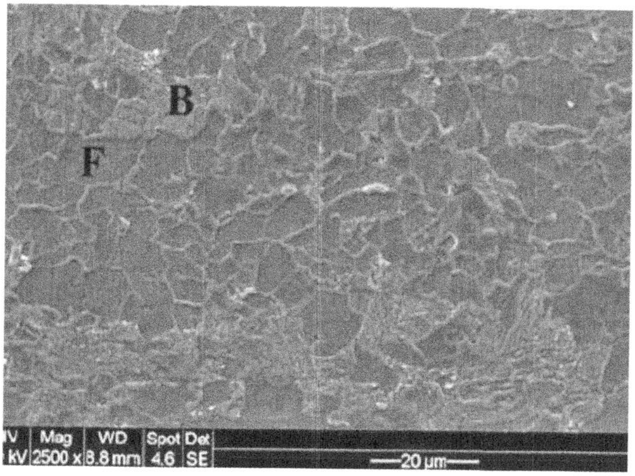

Figure 8-1 SEM view of the microstructure of X80 pipeline steel F and B refer to ferrite and bainite, respectively. (From Xue and Cheng [2012].)

[Hillenbrand et al., 2004]. As a comparison, Fig. 8-1 shows the metallographic view of the microstructure of X80 steel, where the polygonal ferrite and bainite structure is observed. Figures 8-2 and 8-3 show the microstructure of X100 and X120 steels, respectively. Whereas X100 steel contains a lathlike bainite matrix, X120 steel features a needle-type bainite.

The chemistry variation among the three steels is one factor affecting the differences in their microstructure. Generally, higher-grade steels such as X100 and X120

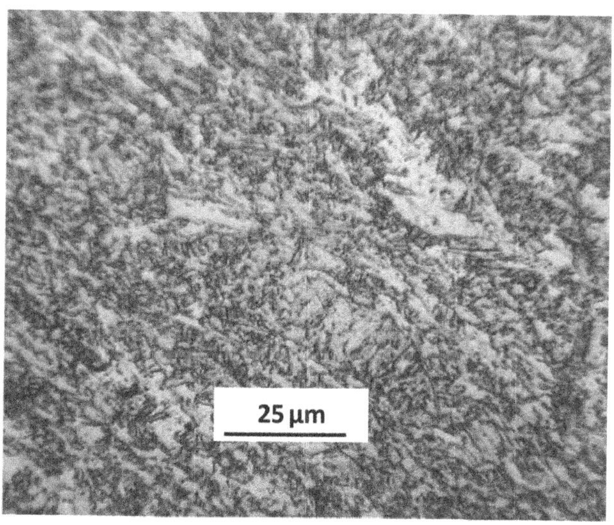

Figure 8-2 Optical view of the microstructure of X100 pipeline steel. (From Jin et al. [2010].)

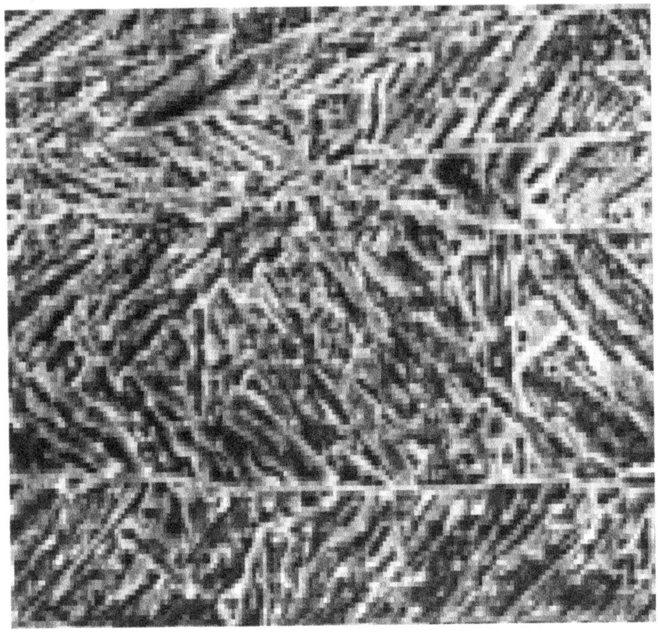

Figure 8-3 *Optical view of the microstructure of X120 pipeline steel. (From Pourkia and Abedini [2008].)*

possess higher Cu, Mn, Mo, and Ni contents. These alloying elements tend to increase the metal hardenability and shift the continuous cooling transformation (CCT) diagram line to the right, increasing the formation of low-temperature transformation microstructures such as bainite and martensite under the same cooling rate.

8.3.4 Metallurgical Defects

During the steel production process, metallurgical defects such as nonmetallic inclusions are inevitably introduced. The inclusions contained in steels are formed from metallic elements such as iron, manganese, magnesium, silicon, aluminum, and calcium, and nonmetallic elements such as nitrogen, sulfur, and oxygen [Anderson et al., 2003; Payandeh and Soltanieh, 2007; Nayak et al., 2008]. The inclusions contain primarily sulfide and various oxides, such as the elongated MnS inclusion commonly distributed in low-grade steels. The oxide inclusions vary in size and composition, including aluminum oxide and silicon oxide. The size, quantity, and distribution of inclusions are controlled by the steel purity. By utilizing an advanced metallurgy technique it is possible to reduce the content of some elements, such as oxygen and sulfur, less than 10 ppm. The size of inclusions decreases drastically, even below 1 μm in diameter. However, if the content of these elements is decreased to a very low level, the cost of steel manufacturing will be very high.

8.4 SUSCEPTIBILITY OF HIGH-STRENGTH STEELS TO HYDROGEN DAMAGE

Generally, the susceptibility of steels to hydrogen damage, including hydrogen embrittlement, hydrogen blistering, and HIC, increases with their strength or grade [Marsh and Gerberich, 1992]. For high-strength steels, hydrogen-induced failure at lower stresses or stress-intensity levels is characterized by reduced plasticity and plastic zone size, and the corresponding fractography is that of quasi-cleavage or brittle cleavage [Hirth, 1980].

Hydrogen could be introduced into pipeline steels from various sources. The steels encounter hydrogen either during the transport of sour crude oil and other petroleum products or from the soil environment. Hydrogen permeation through steels has been studied extensively to gauge the mechanism of hydrogen entry, the amount of hydrogen coverage, the hydrogen diffusivity through steels, and hydrogen trapping and detrapping [Casanova and Crousier, 1996]. Cathodic hydrogen charging is commonly used as the source of hydrogen input, but it is not truly representative of the hydrogen entry conditions except when a CP system is in place. Hydrogen entry in most situations occurs under freely corroding conditions.

8.4.1 Hydrogen Blistering and HIC of High-Strength Pipeline Steels

High-strength pipeline steels exposed to hydrogen sulfide environments tend to develop fissures and cracks [Tresseder, 1977]. Most of the failures can be related to the susceptibility of steel to HE by either HIC or sulfide stress corrosion cracking (SSCC). The development of cracks in high-strength steels under aggressive H_2S media and stressed conditions is influenced strongly by the actual microstructure and hardness of the steel. Hence, corrosion-resistant alloys currently used for H_2S service have been limited to yield strengths of 690 MPa or hardness values of 22 HRC [Tresseder, 1977]. While there have been attempts at developing improved grades of pipeline steels resistant to SSCC in sour environments, it was found that API grades X60 to X80 and X100 of pipeline steels are prone to HIC in the absence of external stresses [Margot-Marette et al., 1987; Jin et al., 2010]. Moreover, the inherent SSCC resistance is strongly affected by metallurgical factors that include inclusions and second-phase precipitates, grain boundary segregation, and hard bands produced during controlled rolling.

Hydrogen could induce blistering and cracking of high-strength steels, even in the absence of external stress. As shown in Fig. 8-4, hydrogen-induced blisters can be observed on the surface of X100 steel that is under hydrogen charging at 50 and 100 mA/cm^2 for 20 h [Jin et al., 2010]. The number of blisters increases with the cathodic charging current density. Moreover, some blisters are ruptured, leaving apparent signs on the steel surface, as shown in Fig. 8-4c and d.

Hydrogen-induced cracks would also be induced in high-strength steels in the absence of external stress. Figure 8-5 shows the optical (a) and SEM (b) views of cracks generated on the hydrogen charging at 100 mA/cm^2 on the surface of an X100 steel specimen. Moreover, the hydrogen-induced crack is associated with a

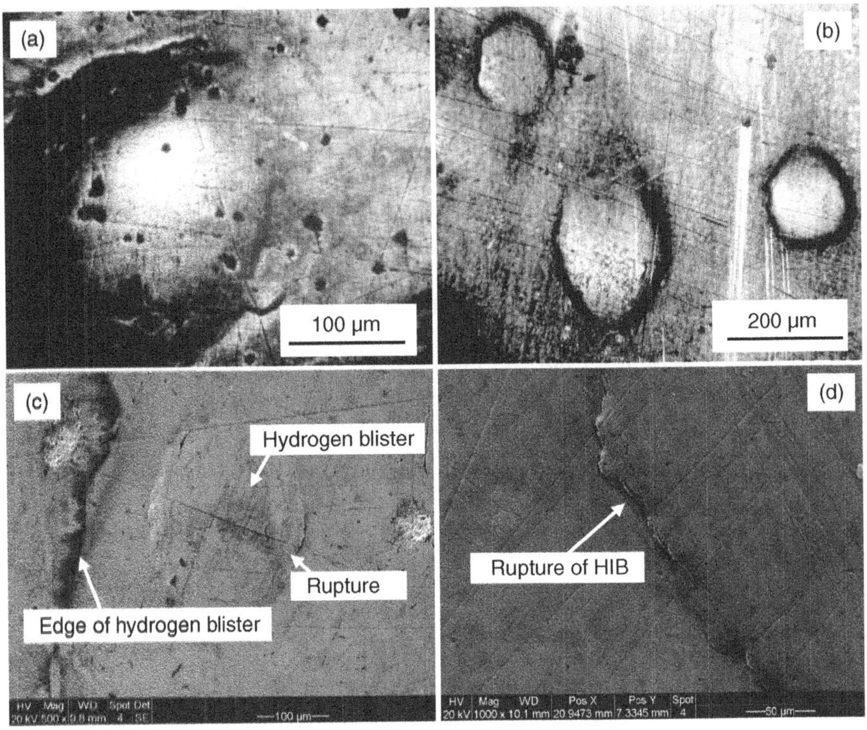

Figure 8-4 *Surface morphology of hydrogen blisters on X100 steel after 20 h of charging at (a) 50 mA/cm² and (b) 100 mA/cm² as well as the ruptured blisters (c) and (d). (From Jin et al. [2010].)*

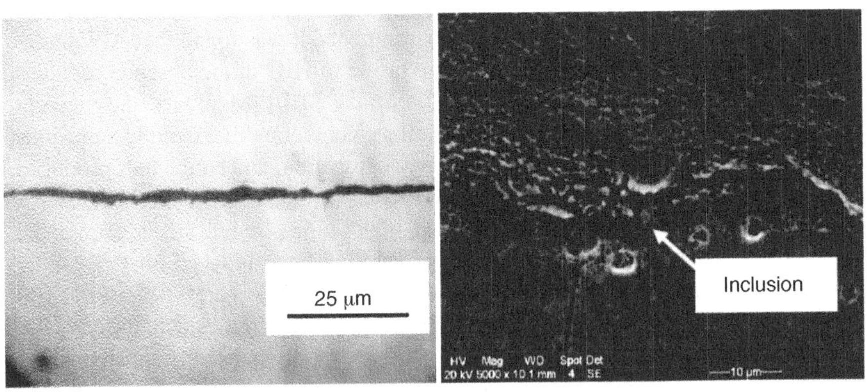

Figure 8-5 *Optical (a) and SEM (b) views of the cracks on an X100 steel specimen, associated with an inclusion, after 20 h of hydrogen charging at 100 mA/cm². (From Jin et al. [2010].)*

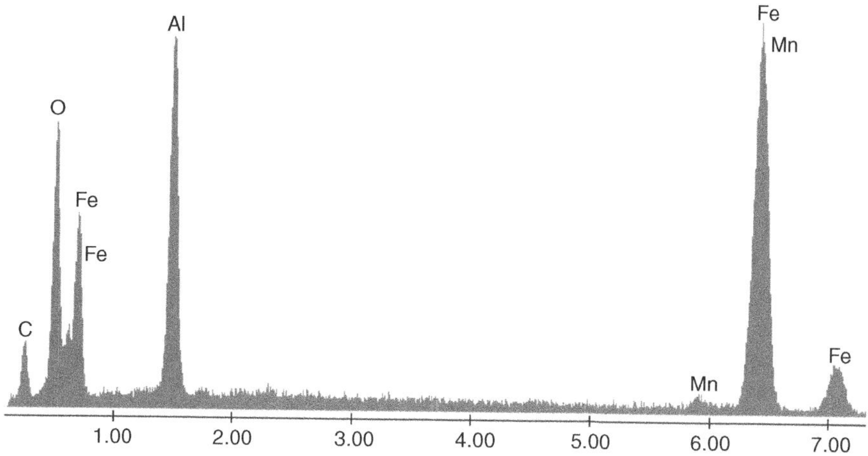

Figure 8-6 *EDX spectrum obtained on the inclusion marked in Fig. 8-8b. (From Jin et al. [2010].)*

spherical inclusion, as marked in Fig. 8-5b. Figure 8-6 shows the energy-dispersive x-ray (EDX) spectrum obtained at the inclusion, where an Al-enriched inclusion is identified [Jin et al., 2010].

The hydrogen-induced cracks were also found to be associated with titanium oxide inclusions in X100 steel. As shown in Fig. 8-7, upon hydrogen charging a crack is initiated at titanium oxide and aluminum oxide inclusions [Dong et al., 2009].

Generally, the HIC resistance of steels decreases as the inclusion level increases for an identical microstructure. Even when the inclusion level is sufficiently low that HIC does not occur at inclusions, clustering of M–A constituents could induce HIC

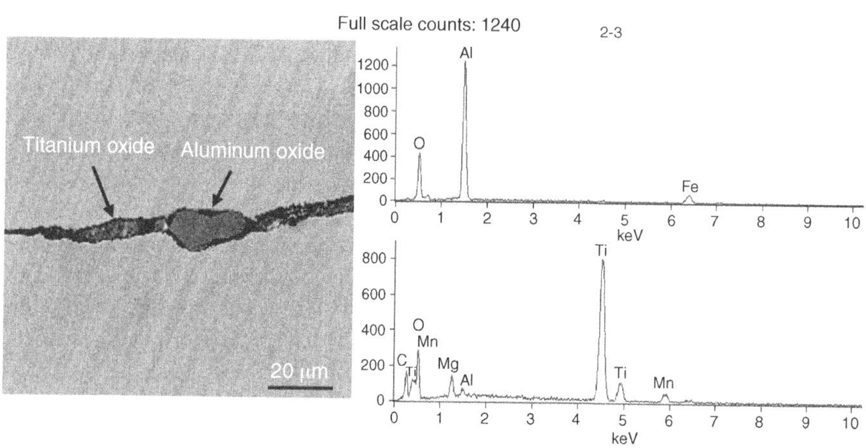

Figure 8-7 *Crack initiated at titanium oxide and aluminum oxide inclusions in X100 steel. (From Dong et al. [2009].)*

if they are agglomerated to a certain degree (e.g., 100 μm in length and 20% of area along the transverse direction to the rolling plane) [Koh et al., 2008].

Local irregularities such as cavities, dislocations, grain boundaries, and inclusions are primarily the hydrogen traps in the steel, which are of great importance for HIC susceptibility. Various types of inclusions have been identified in high-strength pipeline steels (e.g., X80 and X100 steels), including aluminum oxide, titanium oxide, silicon oxide, and a trace amount of calcium and magnesium oxides and magnesium sulfide [Jin et al., 2010; Xue and Cheng, 2011]. Upon hydrogen charging, the charged hydrogen atoms tend to accumulate at these irregularities, where a high stress concentration is usually developed [Oriani et al., 1985]. Hydrogen atoms would result in the initiation of microcracks locally either by decreasing the cohesion of Fe–Fe bonds or by increasing the local brittlement of the steel. As shown in Figs. 8-6 and 8-7, cracks are always found at or across the inclusions contained in the steel. Once the microcracks are initiated, the accumulated hydrogen atoms also contribute to the crack propagation by increasing the local dissolution rate of steel at the crack tip [Jin and Cheng, 2011] or by decreasing the surface energy of the newly formed planes to lower the work of fracture [Oriani et al., 1985]. Therefore, hydrogen-charged steels have high susceptibility to SCC.

The HIC resistance depends strongly on the microstructure of steels. Of the typical microstructures found in high-strength steels, the acicular ferrite is most resistant to HIC. Thus, steels with a ferrite and AF microstructure are not severely embrittled even when the diffusible hydrogen content in the steel is high. The ferrite + bainite microstructure is susceptible to embrittlement by diffusible hydrogen [Koh et al., 2008]. Therefore, with the enhanced strength level of high-strength steels with an increasing percentage of bainite in their microstructures, the HIC susceptibility is actually increased.

8.4.2 Hydrogen Permeation Behavior of High-Strength Pipeline Steels

According to the ISO method for the measurement of hydrogen permeation and determination of hydrogen uptake and transport in metals [International Standard Organization, 2004], an electrochemical hydrogen-permeation curve provides important information about hydrogen diffusion and trapping, such as the hydrogen-permeation rate ($J_H L_H$), hydrogen diffusivity (D_{eff}), apparent hydrogen solubility (C_{app}), subsurface hydrogen concentration, and hydrogen-trapping density in the steel (N_T). Figure 8-8 shows a hydrogen-permeation current curve measured on an X80 pipeline steel 0.5 mm thick conducted in a Devanathan–Stachurski cell [Devanathan and Stachurski, 1962, 1964; Devanathan et al., 1963], with the charging side filled with 0.5 M H_2SO_4 solution and the detection side with 0.1 M NaOH solution under an anodic potential of 200 mV(SCE) [Xue and Cheng, 2011].

The hydrogen flux, J_H, through the steel specimen (mol H/m^2·s) is measured by the steady-state hydrogen-permeation current density, i_p^∞, by [Park et al., 2008]

$$J_H = \frac{i_p^\infty}{nF} \qquad (8\text{-}1)$$

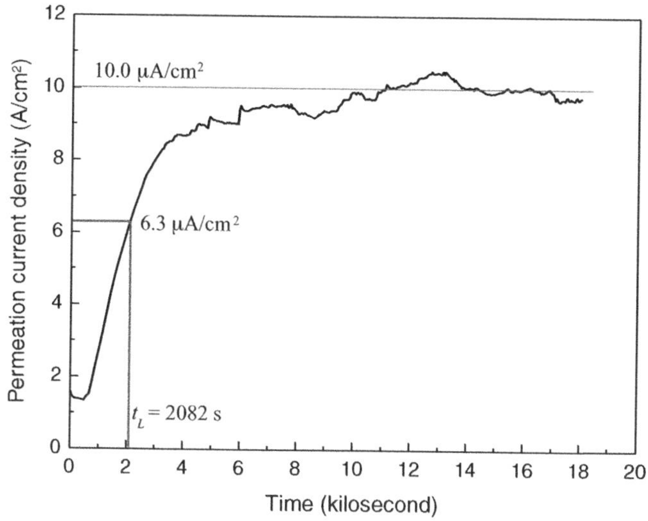

Figure 8-8 *Transient hydrogen-permeation current density curve measured on X80 steel (0.5 mm in thickness). (From Xue and Cheng [2011].)*

where n is the number of electrons transferred and F is Faraday's constant. The hydrogen-permeation rate (mol H/m·s) is defined by

$$J_{\mathrm{H}}L = \frac{i_p^{\infty} L_{\mathrm{H}}}{nF} \tag{8-2}$$

where L_{H} is the specimen thickness. The effective hydrogen diffusivity, D_{eff}, can be calculated by [Cheng, 2007a; Banerjee and Chatterjee, 2001]

$$D_{\mathrm{eff}} = \frac{L^2}{6t_L} \tag{8-3}$$

where t_L is the time lag, corresponding to the time point on the hydrogen-permeation current curve at which $i_t = 0.63 i_p^{\infty}$. If the surface hydrogen is in thermodynamic equilibrium with the subsurface hydrogen, the apparent hydrogen solubility, C_{app} (mol H/m^3), is defined by

$$C_{\mathrm{app}} = \frac{J_{\mathrm{H}}L}{D_{\mathrm{eff}}} \tag{8-4}$$

The hydrogen-permeation rate, diffusivity, and apparent hydrogen solubility in X80 steel are then determined as

$$J_{\mathrm{H}}L = \frac{i_p^{\infty}}{nF} L = 5.2 \times 10^{-10} (\mathrm{mol\ H/m{\cdot}s})$$

$$D_{\mathrm{eff}} = \frac{L^2}{6t_L} = 2.0 \times 10^{-11} (\mathrm{m^2/s})$$

$$C_{\mathrm{app}} = \frac{J_{\mathrm{H}}L}{D_{\mathrm{eff}}} = 26\ (\mathrm{mol\ H/m^3})$$

The hydrogen-trapping density in X80 steel can be estimated by [Yen and Huang, 2003]

$$N_T = \frac{C_{\text{app}}}{3}\left(\frac{D_l}{D_{\text{eff}}} - 1\right) \tag{8-5}$$

where N_T is the number of hydrogen-trapping sites per unit volume (m^{-3}), and D_l is the lattice diffusion coefficient of hydrogen (m^2/s), with a value of 1.28×10^{-8} m^2/s in α-Fe [Dong et al., 2009], due to the unavailability of this parameter in X80 steel. N_T is then calculated by

$$N_T = \frac{C_{\text{app}}}{3}\left(\frac{D_l}{D_{\text{eff}}} - 1\right) = 3.33 \times 10^{27}(m^{-3}) \tag{8-6}$$

D_{eff} represents the apparent lattice diffusivity of dissolved and reversibly trapped hydrogen, and the apparent solubility, C_{app}, corresponds to the hydrogen in lattice and reversible traps. It is generally believed [Banerjee and Chatterjee, 2001; Park et al., 2008] that an increase in C_{app} and a decrease in D_{eff} and $J_H L$ are indicative of more hydrogen trapping in the steel. The diffusivity of hydrogen in α-Fe is on the order of 10^{-8} m^2/s. However, the D_{eff} calculated in this work is 2×10^{-11} m^2/s, three orders of magnitude lower than the lattice diffusivity of hydrogen. Moreover, the hydrogen-trapping density is calculated as up to 3.33×10^{27} m^{-3} in X80 steel. Apparently, local irregularities such as dislocations, grain boundaries, inclusions, and precipitates would serve as the sites for hydrogen trapping [Venegas et al., 2009; Nanninga et al., 2010], and affect the hydrogen diffusivity in the steel. Moreover, X80 steel is structured with the polygonal ferrite and bainitic ferrite containing M–A islands. The presence of M–A islands provides sites for the hydrogen trapping and results in low effective diffusivity [Luu and Wu, 1996].

In addition to reversible hydrogen-trapping sites, the irreversible trapping of hydrogen is associated with various nonmetallic inclusions in the steel and plays an important role in decreasing the hydrogen diffusivity. It has been established that crack initiation is usually associated with the inclusions. Therefore, the effect of irreversible hydrogen trapping at inclusions raises an important concern: hydrogen trapped at irreversible traps, such as inclusions, in the steel is primarily responsible for hydrogen-induced cracking.

Furthermore, the resistance to hydrogen absorption decreases with decreasing steel strength (yield stress) [Capelle et al., 2010]. By investigating the hydrogen-permeation behavior of X52, X70, and X100 steels in NS4 solution, it was found that there is no practical difference between the values of subsurface hydrogen concentration for unstressed metals, although this value is slightly higher for X52 steel than for X70 and X100 steels. Under loading of specimens by tensile stress to simulate the operating conditions of pipelines, the X70 and X100 steels possess approximately equal resistance to hydrogen absorption, and this resistance decreases significantly for X52 steel. Therefore, although high-strength steels are sensitive to HIC, the hydrogen absorption is not high compared to that of low-strength steels under identical hydrogen-generating conditions.

8.5 METALLURGICAL MICROELECTROCHEMISTRY OF HIGH-STRENGTH PIPELINE STEELS

8.5.1 Microelectrochemical Activity at Metallurgical Defects

The presence of various metallurgical defects, such as inclusions and grain boundaries, introduces not only structural instability but also increasing local corrosion activity. It has been confirmed that there exists a difference in the electrochemical corrosion activity between inclusions and the adjacent steel substrate [Vignal et al., 2007; Muto et al., 2009]. In particular, nonmetallic inclusions serve as cathodic sites and the surrounding steel matrix as anodic sites. Consequently, pits are detected in the steel matrix, whereas cathodic reactions take place at inclusions and the matrix far from these defects.

For high-strength pipeline steels such as X100 steel, at least four types of inclusion are contained in the steel [Jin et al., 2010]. Figure 8-9a shows the localized electrochemical impedance (LEIS) scanning on an X100 steel electrode surface, where an inclusion is identified and is present in the scan direction. The EDX analysis shows that the inclusion is aluminum-enriched, as shown in Fig. 8-9b. The LEIS linear scanning result is shown in Fig. 8-10, where the impedance fluctuations occur at the inclusion. The impedance is approximately constant on the steel matrix, drops at the boundary between the steel matrix and inclusion, and then increases at the inclusion itself. The impedance drops again at another interface between inclusion and the steel and then returns to an obviously constant value.

Apparently, an aluminum oxide–enriched inclusion has a higher local impedance than that obtained on a steel substrate. Thus, compared to the steel, the inclusion is more stable. In the galvanic couple formed between an inclusion and the adjacent steel matrix, the inclusion serves as a cathode and the adjacent steel as an anode. Thus, corrosion, once begun, would occur preferentially on the steel matrix, as indicated by the drop in local impedance at the interfaces between inclusion and steel. The local dissolution of the steel causes a drop-off of the inclusion, and consequently, generates a microvoid. Further dissolution of the microvoid results in a corrosion pit. The importance of this research is to measure directly the electrochemical impedance values, which indicate the relative corrosion activities of inclusion and steel in a test solution at a microscopic level. The use of metallurgical microelectrochemistry is critical to an understanding of the fundamentals of localized corrosion and crack initiation in high-strength pipeline steels.

A similar LEIS scan is conducted along another direction, where a silicon-enriched inclusion is identified, as shown in Fig. 8-11. It is apparent that there is a lower impedance on the silicon-enriched inclusion than on the steel matrix. Thus, the electrochemical behavior of inclusions differs, depending on their composition and the associated roles in corrosion pit initiation. The low local impedance associated with

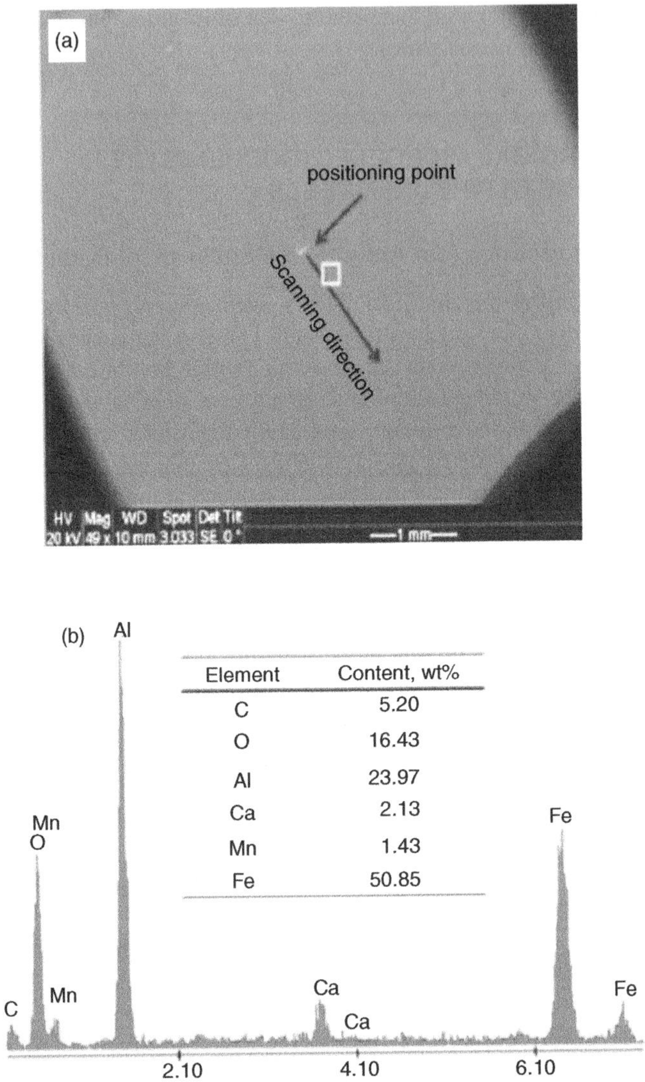

Figure 8-9 *LEIS scan line on an X100 steel electrode surface (a) and the EDX spectrum (b) obtained on the defect as marked by a square in (a). (From Jin and Cheng [2011].)*

a silicon-enriched inclusion indicates that there is a high degree of electrochemical corrosion activity at the inclusion in the test solution. Therefore, a galvanic couple forms between the inclusion and the adjacent steel matrix, where the former serves as an anode and the latter as a cathode. Preferential dissolution of the silicon-enriched inclusion generates a local microvoid, which continues to dissolve due to its high electrochemical activity. A pit is thus formed and could be the site to initiate stress corrosion crack in the presence of applied stress.

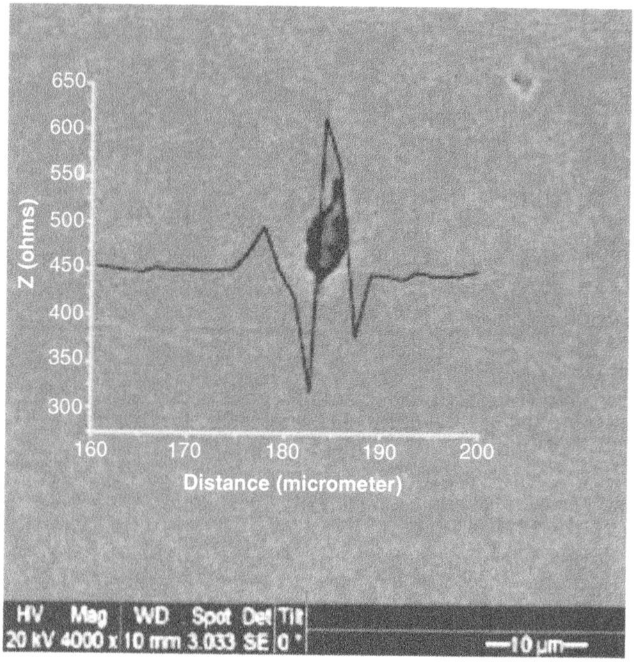

Figure 8-10 *LEIS linear scanning map measured on an X100 steel substrate and around the inclusion in a nearly neutral–pH bicarbonate solution. (From Jin and Cheng [2011].)*

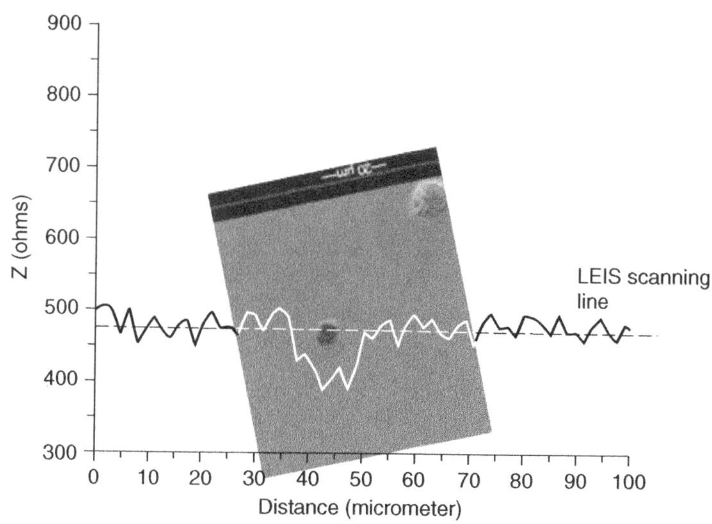

Figure 8-11 *LEIS linear scan map measured around a silicon-enriched inclusion in a nearly neutral–pH solution. (From Jin and Cheng [2011].)*

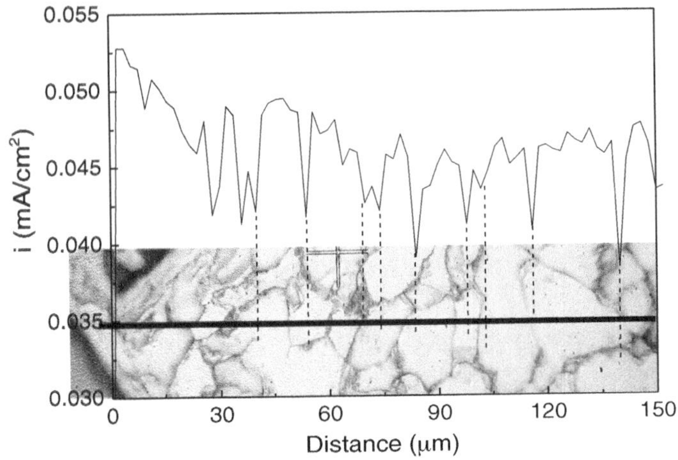

Figure 8-12 SVET map measured along the black line on an X70 steel electrode in a nearly neutral–pH solution. (From Liu et al. [2010].)

Furthermore, grain boundaries present metallurgical nonuniformity in steels. Of particular importance to pipeline SCC are the interganular and transgranular SCC that occur in high- and nearly neutral–pH environmental conditions, respectively [National Energy Board, 1996]. Figure 8-12 shows anodic current density maps measured by a scanning vibrating electrode technique (SVET) and the corresponding metallographic views of X70 pipeline steel in a nearly neutral–pH bicarbonate solution [Liu et al., 2010]. It is seen that the steel contains a polygonal ferrite matrix, and the current density at the grain boundary is always lower than that measured on a grain. The surface morphologies of the steel electrode after SVET measurement are shown in Fig. 8-13a, where a corrosion pit is formed on the grain surface rather than

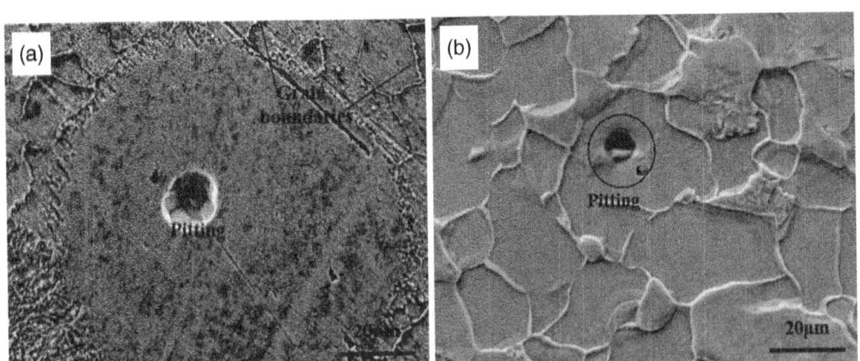

Figure 8-13 SEM views of the surface morphology of an X70 steel electrode after SVET measurement in Fig. 8-12 (a), and after etching in a 4 wt% HNO_3–ethanol solution (b). (From Liu et al. [2010].)

at the grain boundary. Moreover, the presence of the pit immunizes the adjacent zone from corrosion within the grain. Figure 8-13b shows the surface morphology of the electrode etched in 4 wt% HNO_3–ethanol solution to present a clearer image of the position of the corrosion pit. The pit is initiated on the grain surface, and there is no pit at the grain boundary.

The SVET measurement result and the SEM observation show a difference in corrosion activity between the grain and the grain boundary. Since the SVET current density is proportional to the local potential, it is thus believed that the grain is anodic and the grain boundary is cathodic. The initiation of corrosion pits on the grain rather than at the grain boundary demonstrates that there is a high level of anodic dissolution activity on the grain. It is speculated that the enhanced stability at a grain boundary of X70 pipeline steel is due to the manufacturing of "clean" steel, with quite a few impurity elements segregating locally. Once localized corrosion (e.g., pitting corrosion) occurs on the grain, a galvanic coupling effect forms between the corrosion pit and the adjacent grain zone. Consequently, corrosion would occur preferentially at the pit, while the adjacent zone is protected.

It has been accepted [Qiao et al., 1998; Jack et al., 2000; Cheng, 2007a] that hydrogen-facilitated anodic dissolution at a crack tip dominates the nearly neutral–pH SCC of pipelines. Cheng [2007b] proposed that the growth rate of nearly neutral–pH stress corrosion cracks in pipeline steels is dependent on the effects of stress, hydrogen, and their synergism on the anodic dissolution current density of steel at the crack tip. For a stress corrosion crack initiating from the corrosion pit, when it grows within a grain, the anodic current density at the crack tip would be higher than that at the grain boundary, as demonstrated by SVET measurements. The electrochemical driving force for the crack growth within the grain is thus higher than that at the grain boundary. The SCC growth would follow a transgranular rather than an intergranular mode under nearly neutral–pH conditions. This microelectrochemical measurement is of great significance, since it may explain the nature of the transgranular SCC of pipeline steels under nearly neutral–pH solutions, where the grains corrode preferentially. This would be direct evidence, at a microscopic level, demonstrating that corrosion of steel occurs preferentially at grains rather than at grain boundaries.

8.5.2 Preferential Dissolution and Pitting Corrosion Around Inclusions

Corrosion pits generate at or around inclusions, due to preferential dissolution occurring locally. It has been found [Reformatskaya and Freiman, 2001] that the size of nonmetallic inclusions in X80 and X100 steels are associated directly with local pitting corrosion. The average inclusion diameter in X100 steel is 8.0 μm, which is higher than that in X80 steel, 3.5 μm. The effect of inclusions on localized corrosion is associated with preferential dissolution of the surrounding steel matrix. Corrosion of ferrite at inclusion–matrix interfaces develops more rapidly than that of ferrite in the matrix. Due to the high dissolution rate in that area, a narrow corroded area forms around the inclusion. The interface between inclusion and the steel matrix dissolves, forming a local pit at the inclusion site. Because of the larger inclusion diameter in

X100 steel, the inclusion–matrix interface is larger, which leads to a higher corrosion rate than for X80. This is accomplished by providing a larger cathodic area (inclusion) relative to any single nucleation site in the surrounding metal. It was found that for both steels, the surface consists of randomly distributed pits. However, the corrosion pits in X100 steel are bigger than those in X80 steel.

Figures 8-14, 8-15, and 8-16 are SEM images of X100 pipeline steel after anodic polarization at −500 mV(SCE) for 30 s in a nearly neutral–pH solution. It is evident that corrosion pits always seem to be associated with inclusions contained in the steel, although in very rare cases a corrosion pit was initiated at a site where there was inclusion (Fig. 8-16). Furthermore, there are quite different features among corrosion pits formed at various inclusions. As characterized by EDX, pit 1 shown in Fig. 8-14 is initiated at an MnS inclusion, and the pitting attack occurs preferentially on the steel matrix at the inclusion–steel interface. The inclusion remains inside the pit after it forms. Another corrosion pit, marked as pit 2, is initiated at a MnS–Al_2O_3 inclusion, where a partial dissolution of the inclusion is observed.

Apparently, the corrosion pit initiated at an MnS inclusion is featured with a preferential dissolution of the steel matrix at the inclusion–steel interface. The anodic dissolution reactions of MnS and Fe are [Wranglen, 1974]

$$\mathrm{MnS} \rightarrow \mathrm{S} + \mathrm{Mn}^{2+} + 2\mathrm{e}^-, \qquad E^\circ = -0.345\ \mathrm{V(SCE)} \tag{8-7}$$

$$\mathrm{Fe} \rightarrow \mathrm{Fe}^{2+} + 2\mathrm{e}^-, \qquad E^\circ = -0.485\ \mathrm{V(SCE)} \tag{8-8}$$

where E° is the standard electrode potential. Apparently, the dissolution of an MnS inclusion has a standard electrode potential less negative than that of Fe dissolution. Therefore, the steel matrix is more anodic than MnS inclusion and dissolves preferentially in the solution. While the steel forms anodic sites (area of attack) during the corrosion process, a cathodic reaction occurs on the MnS inclusion. Thus, a galvanic effect exists between MnS inclusion and the adjacent steel, where pits form due to the preferential dissolution of the steel at the inclusion–steel interface.

Partial dissolution of an inclusion occurs when MnS and Al_2O_3 are combined, as shown in Fig. 8-14. This duplex inclusion is structured with the attachment of MnS to Al_2O_3 [Lim et al., 2001]. As analyzed, the anodic reaction of steel generates corrosion pits at the inclusion–steel interface, and aggressive solution chemistry is developed inside the pit, resulting in dissolution of the Al_2O_3 constituent in the inclusion since local acidification occurring in the pit could cause the solution pH value to drop to 3 or lower. Al_2O_3 is stable until the pH is below 3.2, when Al^{3+} dominates [Baker and Castle, 1992]. During pit growth, the pH value of the pit solution decreases, and the Al_2O_3 constituent of the duplex inclusion dissolves. The Al^{3+} generated diffuses out of the pit, as detected by the EDX spectrum shown in Fig. 8-15.

Corrosion pits are also found to initiate at an Al_2O_3 inclusion, as shown in Fig. 8-15, where a pit is formed due to preferential dissolution of the steel matrix at the interface between the inclusion and the adjacent steel, and the inclusion remains inside the pit. Furthermore, it is seen from Fig. 8-16 that the pit is formed by direct dissolution of the inclusion. Thus, the inclusion is not observable after the pit is

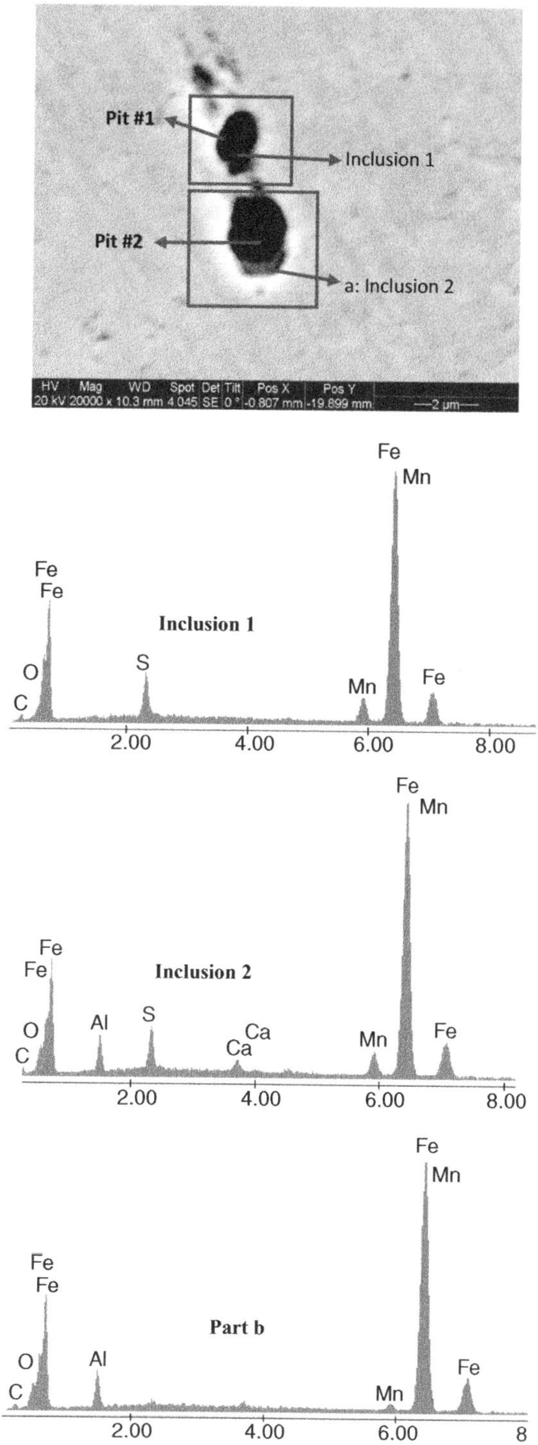

Figure 8-14 *SEM image of corrosion pits initiated at an MnS inclusion (pit 1) and an MnS/Al_2O_3 inclusion (pit 2) as well as relevant EDX spectra. (From Jin [2011].)*

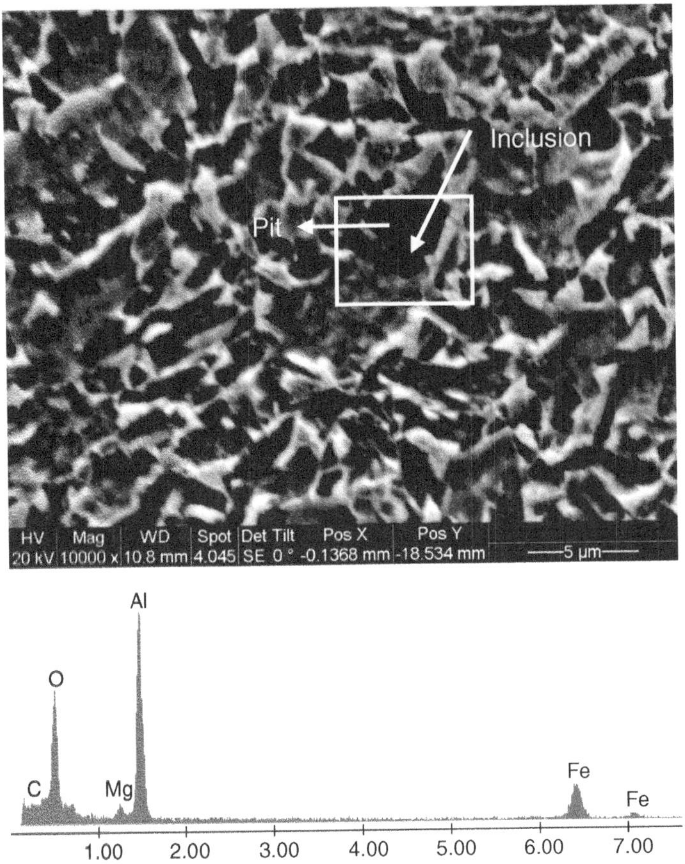

Figure 8-15 *SEM image of a corrosion pit initiating at an Al_2O_3 inclusion as determined by an EDX spectrum. (From Jin [2011].)*

generated. However, elemental analysis by EDX finds that this is a silicon-enriched inclusion.

The variety of mechanisms for pit initiation at inclusions with different compositions are attributed to the metallurgical microelectrochemistry of the steel. Oxide inclusions show lower electrochemical corrosion activity than that at the adjacent steel matrix in a nearly neutral–pH solution [Jin and Cheng, 2011]. When corrosion occurs, the steel matrix is attacked preferentially, acting as the anodic site, and a cathodic reaction occurs on the inclusion. Generally, oxide inclusions are inert, showing a high level of stability during corrosion. Corrosion pits form due to the preferential dissolution of steel at the oxide inclusion–steel interface, with the inclusion remaining in the pits. However, when corrosion pits are initiated at the silicon-enriched inclusion, the inclusion is dissolved, leaving an open mouth. It has been demonstrated [Jin and Cheng, 2011] that the electrochemical activity is higher in a silicon-enriched inclusion than in the adjacent steel matrix. Thus, the silicon-enriched inclusion serves as

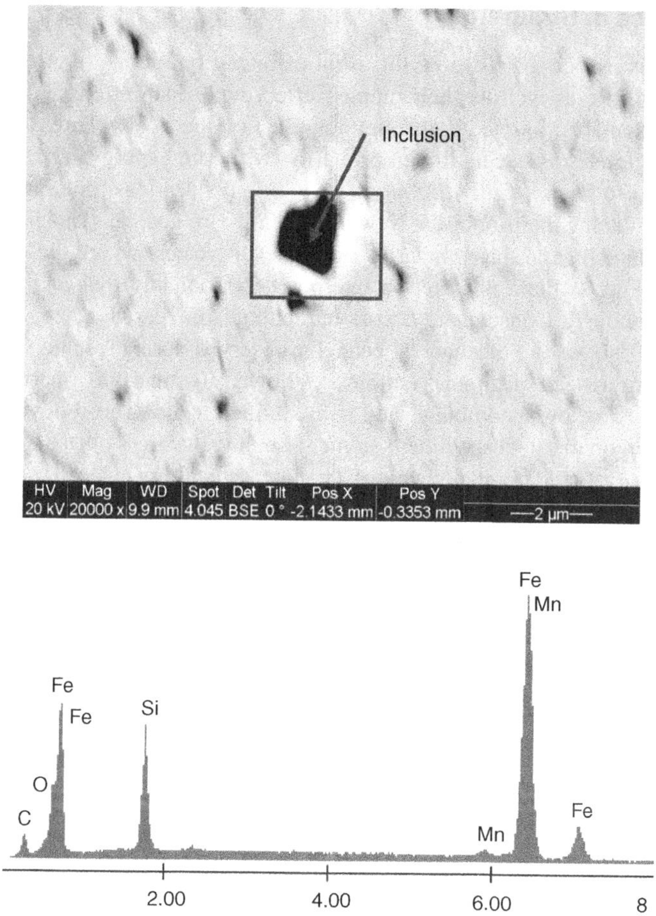

Figure 8-16 SEM view of a corrosion pit associated with a silicon-enriched inclusion as determined by an EDX spectrum. (from Jin [2011].)

an anodic site and the steel matrix as a cathode. A corrosion pit is formed due to the preferential dissolution of the silicon inclusion itself.

8.6 STRAIN AGING OF HIGH-STRENGTH STEELS AND ITS IMPLICATION ON PIPELINE SCC

Strain aging is a phenomenon in which a metal hardens as a result of aging after plastic deformation. Strain aging shows an additional increase in flow stress, occurring after or during straining. If strain aging occurs after deformation, it is called *static strain aging*; if it occurs during deformation, it is called *dynamic strain aging*. The increase in flow stress after static strain aging is known as the *strain aging index* [Reed-Hill, 1973].

8.6.1 Basics of Strain Aging

Strain aging is associated with the diffusion of interstitial atoms (e.g., carbon and nitrogen) in ferrite as well as their pinning effect with dislocations. Formation of fresh dislocations is promoted by prestrain applied on the steel, whereas movement of the interstitial atoms occurs driven primarily by thermal energy. The interstitials are segregated to form a *Cottrell atmosphere* along dislocations, pinning the dislocations and reducing their mobility. New dislocations are generated at sites of stress concentration such as grain boundaries or solute interfaces, where the dislocation density is greater if the aging step is interposed periodically than if the strain is applied continuously. Consequently, work hardening is increased.

Figure 8-17 shows the mechanical behavior observed during a static strain aging process, where a steel specimen is strained, allowed to stand for a period of time at temperatures as low as ambient, and then strained again. The following effects usually happen on the steel specimen: an increase in yield stress (Y_3), a return of the Lüders strain (i.e., yield point elongation), an increase in ultimate tensile stress (ΔU), and a decrease in total elongation (Δe).

During dynamic strain aging, solute atoms are able to make a diffusive jump while an alloy is being deformed, and therefore this process is sensitive to temperature and strain rate [Hosford, 2005]. Thus, dynamic strain aging usually occurs at high temperatures during plastic deformation, whereas static strain aging occurs at ambient temperatures after a long period of prestrain. Dynamic aging has some distinctive characteristics, such as a high rate of work hardening and a serrated stress–strain curve that is attributed to the locking and unlocking of dislocations by solute atoms during deformation.

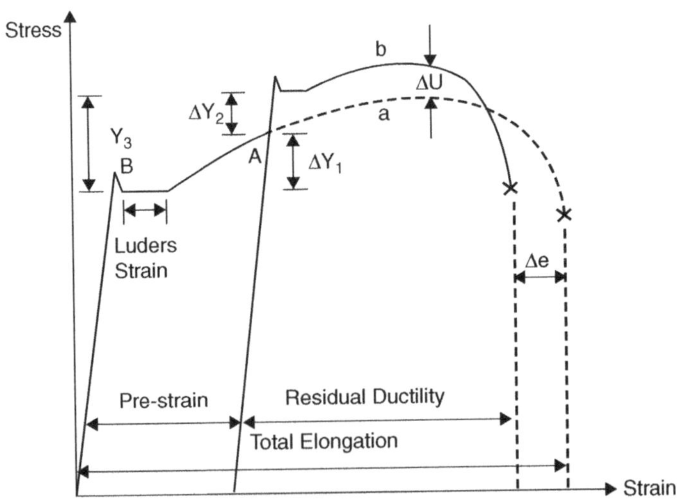

Figure 8-17 Effects of static strain aging on the mechanical properties of steels; a is the stress–strain curve of steel without prestrain, and b is the stress–strain curve of a prestrained steel specimen.

Of various affecting factors, aging time, prestrain intensity, temperature, and metallurgical composition are the primary parameters determining the strain aging behavior of steels.

Aging Time. Strain aging is a time-dependent process when temperature remains constant. Four stages are usually contained in the process [Wilson and Russell, 1960]: (1) interstitial atoms diffuse to fresh and mobile dislocations generated by prestrain and form a Cottrell atmosphere around them, locking these dislocations; (2) precipitates such as Fe_3C and Fe_4N form along the dislocations, giving rise to the resistance of steel to plastic strain due to precipitation hardening; (3) precipitates propagate around the dislocations and lead to an increase in lower yield point and ultimate tensile stress and a decrease in the elongation to fracture; and (4) the overaging phenomenon occurs due to precipitate coarsening and the hardness of coarse precipitates decreases, so the steel begins to soften.

If the aging process stops at a short aging time, such as at the first stage, a strain aging effect can be observed (i.e., a rise in the upper yield point) as well as a fall in ductility, whereas the ultimate tensile stress and elongation to fracture are rarely affected. When the aging process keeps going, more and more interstitial atoms concentrate near the dislocations and the pinning of dislocations becomes firm. When the steel cannot experience a full aging process due to a short aging time or a low aging temperature, the uncompleted process is called a *partial aging process* and the steel has not finished a complete change of its mechanical properties.

Prestrain Intensity. Prestrain is a type of plastic deformation applied on steels prior to aging (static strain aging) or during aging (dynamic strain aging). The higher the prestrain intensity, the higher the dislocation density that will be generated in steels. The work hardening effect is thus higher.

Temperature. The strain aging process depends on temperature because interstitial atom diffusion is a temperature-dependent function. The temperature provides kinetic energy to the interstitial atoms for the motion to dislocation. If the temperature is not high enough, especially when the aging time is very short, the interstitial atoms cannot obtain enough energy to overcome the obstacle from the lattices. Consequently, no strain aging occurs. The higher the temperature, the faster the diffusion rate at which the strain aging phenomenon appears. However, when the temperature is so high, work hardening decreases. This is associated with the tendency for dislocation atmospheres to be dispersed as a result of the more intense thermal vibrations at elevated temperatures [Cahn and Haasen, 1996].

Microstructure Constituents of Steels. Essentially, strain aging is a process in which interstitial atoms block dislocation movement during plastic deformation. Thus, interstitial atoms play an important role in the strain aging behavior of steels. According to the various functions of atoms, they can be classified into two types: those atoms that can form a Cottrell atmosphere and block dislocation movement, and those that can mitigate the effect of the first type on strain aging by means of the formation of chemical compounds with less mobility [Herman et al., 1987].

Generally, carbon and nitrogen atoms are regarded as the most effective interstitial atoms to generate the strain aging behavior of carbon steels. This type of atom has some characteristics, such as a high solubility limit in ferrite at most aging temperatures, a high diffusion coefficient to make atoms move rapidly, and a strong ability to pin dislocations. Particularly, nitrogen-induced strain aging is dominant at low temperatures, usually below 150°C, whereas carbon causes much less strain aging in this low-temperature regime since many fewer "free" carbon atoms exist in the ferrite. However, when the temperature is over 150°C, carbon plays a major role in inducing strain aging. If there are atoms no of the second type that would mitigate a strain aging effect in the steel, only 0.001 to 0.002 wt% carbon or nitrogen is sufficient to complete the locking of dislocations in a prestrained carbon steel. Thus, strict control of nitrogen or carbon content in steels, especially nitrogen, is the most effective way to mitigate strain aging.

In the second type, alloying elements can combine with carbon or nitrogen to form carbide or nitride, resulting in a decrease in the amount of the first type of interstitial atoms, and thus mitigate the strain aging effect. These element additions can be categorized as follows:

1. *Nitride formers:* aluminum, silicon, boron
2. *Carbide former:* molybdenum
3. *Carbide or nitride formers:* chromium, vanadium, niobium, titanium

Of these elements, aluminum, vanadium, niobium, and titanium are the most effective additions against the strain aging phenomenon. Vanadium and titanium are the strongest carbide and nitride formers. These alloying elements can reduce the quantity of "free" solute atoms and thus mitigate the strain aging effect. High-strength steels often have some of these elements and therefore have a greater resistance to strain aging than do mild steels.

However, it is usually difficult to fully eliminate the effect of strain aging on steels because a very small number of solute carbon and nitrogen atoms can cause significant strain aging. Moreover, in some cases these solute atoms act as important alloying elements of steels and cannot be removed completely. Consequently, the effect of strain aging may be mitigated, but not eliminated, by:

- Reducing the concentration of relevant solute atoms (i.e., carbon and nitrogen) during the steel smelting processes.
- Adding strong nitride and carbide formers into steels to decrease the content of solute atoms with high mobility. For example, the strain aging of steels can be "killed" with the addition of 0.03 to 0.04% aluminum, which is combined with nitrogen to form AlN. Manganese and boron can also be added in steels to increase the strain aging resistance [De Souza and Buono, 2003].

- Improving the environmental conditions that are susceptive to strain aging: for example, decreasing the intensity of prestrain and lowering the operating temperature.
- Improving the toughness of steels to offset the loss of toughness caused by strain aging.

Microstructural constituents such as ferrite, bainite, and martensite have no effect on the formation of a Cottrell atmosphere and the kinetics of strain aging. Thus, changes in mechanical properties incurred by strain aging are independent of microstructure constituents. The detrimental impact of strain aging on ductility and deformability cannot be minimized by microstructure control. Chemical composition control may be the only way to counteract this mechanism [Vodopivec, 2004; Pereloma et al., 2010].

Moreover, carbon steels that have a fine-grained microstructure show a strong response to strain aging because of the high density of dislocations. This type of steel is more susceptible to strain aging. The fine-grained microstructure is beneficial to increases in both strength and toughness, but it may damage deformability and toughness once a material experiences strain aging.

To summarize, strain aging has a strong effect on the mechanical properties (i.e., yield strength, tensile strength, and elongation behavior) of steels. Since strain aging is a time-dependent process, the mechanical properties of steels change as a function of aging time. In regard to the change in mechanical properties of steels that occurs during strain aging in response to interaction between interstitial atoms and dislocations [Wilson and Russell. 1960], it was found that after 20 min of 4% prestrain application, the lower yield stress and Lüders strain increase significantly. When the aging time increases to 35 min, the yield stress increases further, but apparently the Lüders strain does not change. However, there is a rise in the level of the stress–strain curve, and the ultimate tensile stress is increased, but elongation to the fracture of steel decreases due to the work-hardening effect. On aging for more than 150 min, the yield stress and flow stress beyond the yield point continue to increase, although this is not as apparent as the increase in ultimate tensile stress. When the aging time is longer than 10,000 min, some evidence of "overaging" is observed; that is, there is slight decrease in yield stress and ultimate tensile stress, and an increase in elongation to fracture and Lüders strain.

Strain aging would enhance the fatigue strength of steels [Thompson and Wadsworth, 1958]. Strain aging affects the fatigue strength of steels under cyclic loading, primarily by increasing the constraints imposed on deformation within active slip bands and by inhibiting the spread of plasticity ahead of the developing stress concentrators, due to dislocation locking by diffusing solute atoms [Wilson and Tromans, 1970]. However, this locking process is affected by parameters such as the cyclic loading period and temperature. When the cyclic period is very short, that is, the loading frequency is sufficiently high, the diffusion of solute atoms is limited and the dislocation movement cannot be locked in time. Thus, the enhancement of fatigue strength is not obvious. If the cyclic period is prolonged or interrupted by sufficient rest time, diffusing atoms through strain aging can lock-in the mobile dislocations associated with active slip bands, and the fatigue life can be improved considerably.

8.6.2 Strain Aging of High-Strength Pipeline Steels

Development of high-strength steel technology will enable pipeline operators to sustain high service pressure (i.e., in excess of 15 MPa) to transport energy over a long distance. However, a pipeline made of high-strength steel such as X100 steel used in Arctic and sub-Arctic areas would experience longitudinal strain aging, owing to a combination of three externally applied conditions: prestrain, temperature, and time.

A certain amount of prestrain exists in the as-received pipes as a result of manufacturing processes. However, significant strain can be generated during field service: that is, bending and ground movement. In particular, the plastic strain resulting from ground movement usually results in the highest magnitude of prestrain on pipes and thus is the primary reason that an in-service pipeline is subject to plastic strain [Wu, 2010]. To prevent pipelines from strain-induced failure, the steel is required to have an appropriate strain capacity against the strain applied. Generally, an allowable plastic strain limit for an X100 steel pipeline is designed to be a maximum of 2% [Wu, 2010].

The temperature of the pipeline steel usually ranges from 0 to 45°C [Gokhman, 1983]. More important, the coating application could elevate the pipeline temperature to 200°C or above [Sanjuan Riverol, 2008]. Thus, the temperature of X100 steel pipe could meet the aging temperature requirement. Furthermore, the minimum design life for a steel pipeline is 50 years. Apparently, there is sufficient time during pipeline service for aging to occur. Actually, it has been claimed [Wu, 2010] that the negative effect of strain aging may be one of the obstacles to the practical use of X100 steel in northern areas.

Investigations [Duan et al., 2008; Sanjuan Riverol, 2008] have demonstrated that X100 pipeline steel shows discernible strain aging behavior, due to the pipe-forming (prestrain) and thermal-coating process. The stress–strain curves measured show the yielding point elongation (YPE) phenomenon. Moreover, aged X100 steel shows increasing yielding strength and decreased elongation compared to unaged steel.

A few investigations have dealt with strain aging of X100 pipeline steel due to several minutes of thermal coating applications on the pipe [Duan et al., 2008; Ishikawa and Okatsu, 2008; Johnson et al., 2008]. Generally, the appearance of the YPE phenomenon in recorded stress–strain curves is identified as evidence that strain aging occurs on steel. Moreover, a high yield strength rise and low ultimate tensile strength rise are usually generated, resulting approximately in a 100% yield strength/tensile strength (Y/T) ratio. To solve this problem, the metallurgical design focuses on YPE removal and mechanical property improvement against strain aging.

Furthermore, the coating application processes have an obvious influence on the pipeline mechanical properties needed for strain-based design. For example, during the application of FBE coating, the pipe outer surface is, in some cases, heated up to 250°C. Strain aging can occur in this temperature range, resulting in an increase in yield strength and Y/T ratio and a decrease in ductility and strain hardening capacity [Macia et al., 2010]. Thus, the coating application process accompanying high-temperature thermal treatment is opposite to the principle of strain-based design. Consequently, a low-temperature coating technique is needed to avoid the harmful impact of strain aging on pipe steel.

The effect of strain aging on the mechanical properties of pipeline steels is dependent on the straining–stressing direction. Generally, Lüders strain, indicating the strain aging effect in a stress–strain curve at the yield point as YPE, appears at a lower aging temperature in a transverse direction than in a longitudinal direction [Nagai et al., 2010]. Moreover, the grade of the steel also affects the strain aging effect. A stress–strain curve would continue to be of round-house type (i.e., no Lüders strain phenomenon appeared at the yielding point) in X100 grade pipe steel after heat treatment at 240°C for 5 min; however, X70 grade pipe steel would show a stress–strain curve in a longitudinal direction with Lüders strain after heat treatment at the same temperature [Nagai et al., 2010].

In addition to evaluation of the strain aging behavior of high-strength pipeline steels based on stress–strain curve measurements, studies of strain aging have also focused on determining the activation energy of strain aging, and further, predicting the long-term aging behavior of steels in the field. Activation energy is an important indicator by which to assess the long-term serviceability of pipeline steels, and can be used to evaluate the susceptibility of a steel to strain aging, to predict the degradation of mechanical properties of steels during service, and to develop technologies to mitigate and prevent strain aging. The activation energy of strain aging is usually obtained by the Arrhenius equation, combined with the tensile testing of steel aged for various times and at various temperatures. For example, it was found that a group of X100 steels manufactured through specific processes had a strain aging activation energy of 22.8 kcal/mol, which is very close to that of mild steel (20 kcal/mol) and much lower than that of high-strength low-alloy steels (>30 kcal/mol) [Wu, 2010]. This low value implies that the steel is susceptible to strain aging, which may occur even at room temperature upon prestraining. The low activation energy of this group of X100 steel is due primarily to the low content of microalloying elements. Its strengthening mechanism relies more on ultrafine-grain reinforcing. Metallurgical analysis shows that ultrafine-grain microstructure leads to a high density of dislocations when steel is subject to prestrain. The dislocations would be pinned by interstitial atoms. The absence of nitride formers such as element vanadium makes nitrogen free and readily easy to diffuse in the lattice at low temperature, even at room temperature. Therefore, the reasonable restriction of the ultrafine grain size and the appropriate addition of vanadium in the steel may eliminate the susceptibility of steels to strain aging and improve the long-term serviceability.

Furthermore, the determination of activation energy of strain aging enables prediction of the long-term behavior of steels to strain aging through accelerated aging tests in the laboratory. The principle is that the aging behavior of steels at a high aging temperature for a short aging time is equivalent to those experiencing a low aging temperature and long aging time in the field. The accelerated aging time is defined as *equivalent aging time* (EAT). For example, for a specific X100 steel, an EAT of 6 days at 107°C is equivalent to 10 years of pipeline service at 40°C in the field [Wu, 2010].

8.6.3 Effect of Strain Aging on SCC of High-Strength Pipeline Steels

Strain aging being experienced by pipeline steel would affect the steel's corrosion and SCC susceptibility. Stress–strain curves of a new X100 steel and an aged X100 pipe steel taken from the field, simulating an FBE coating application up to 250°C for 5 min, are shown in Fig. 8-18. It is seen that the stress–strain behavior differs considerably for the two types of steel specimen. Whereas the new X100 steel has a round-house shape at yielding, the aged steel has an apparent yielding phenomenon, with the presence of Lüders strain during yielding. Moreover, compared to the new X100 steel, the aged steel features increased yield strength, decreased elongation, and a high Y/T ratio.

Strain aging would not alter the microstructure of the steel. As shown in Fig. 8-19, both new and aged X100 steels are featured with ferrite and bainite phases. It is

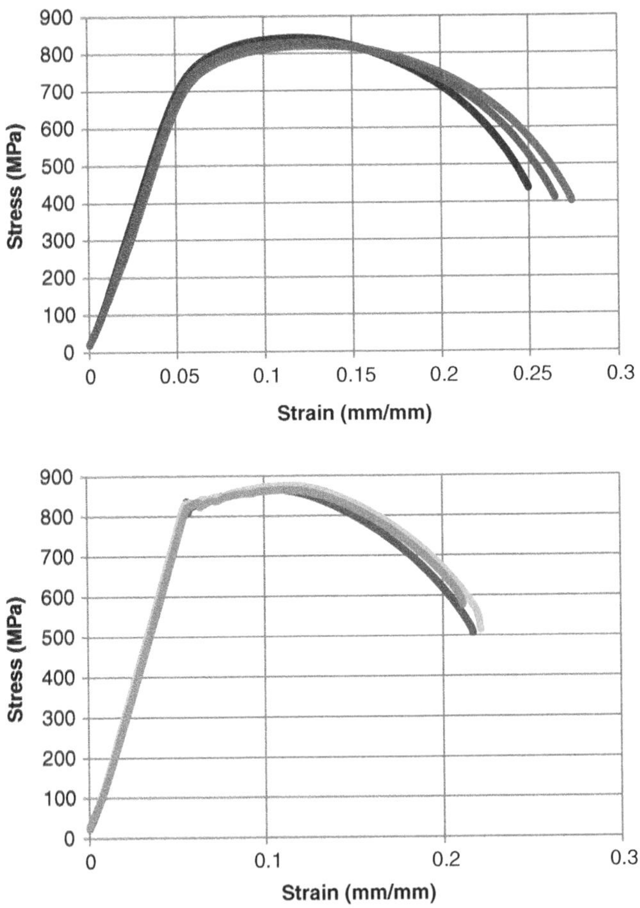

Figure 8-18 *Stress–strain curves of (a) a new X100 steel and (b) an aged X100 steel specimen. (From Sanjuan Riverol [2008].)*

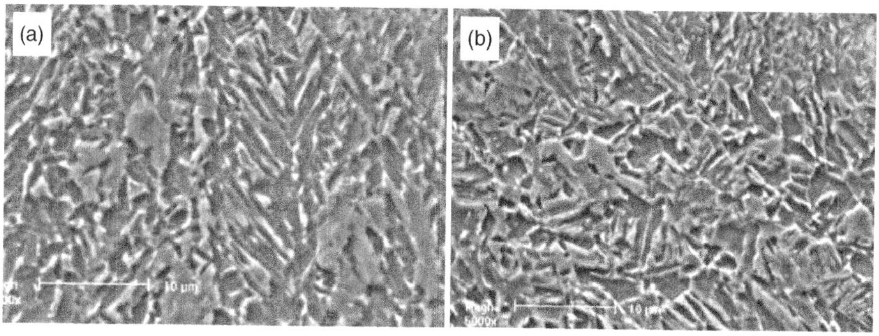

Figure 8-19 *Microstructures of (a) a new X100 steel and (b) an aged X100 steel. (From Sanjuan Riverol [2008].)*

thus assumed that the work hardening effect during strain aging is probably due to precipitate hardening rather than micro-phase change.

To investigate the effect of strain aging on the SCC susceptibility of X100 steel, slow strain rate tensile (SSRT) tests were performed on new and aged steel specimens in a nearly neutral–pH diluted bicarbonate solution that is commonly used to simulate the trapped electrolyte under a coating. The steel was under various cathodic potentials: −0.75 V($Cu/CuSO_4$, open-circuit potential), −0.85 V($Cu/CuSO_4$, normal CP potential), and −1.1 V($Cu/CuSO_4$, overprotection). The reduction in area (RRA) of two types of steel specimens after an SSRT test was calculated and is shown in Fig. 8-20. There is a smaller RRA for the aged steel at both cathodically polarized potentials than that of the new steel specimen. Since the RRA is inversely proportional to the SCC susceptibility of the steel in the environment, it is thus seen that steel

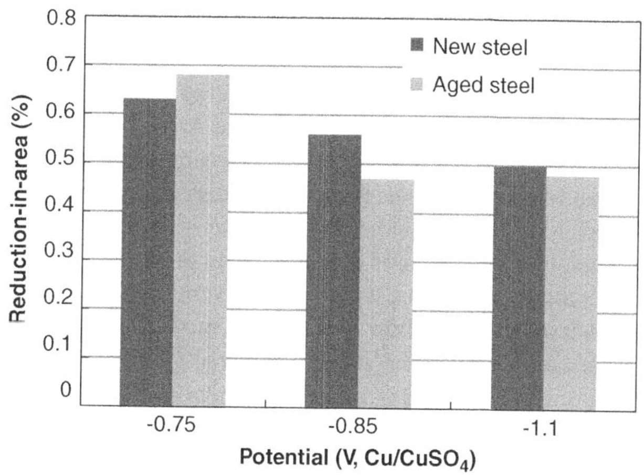

Figure 8-20 *Reduction in area of new X100 and aged X100 steel specimens under various cathodic potentials after an SSRT test in a nearly neutral–pH solution. (From Sanjuan Riverol [2008].)*

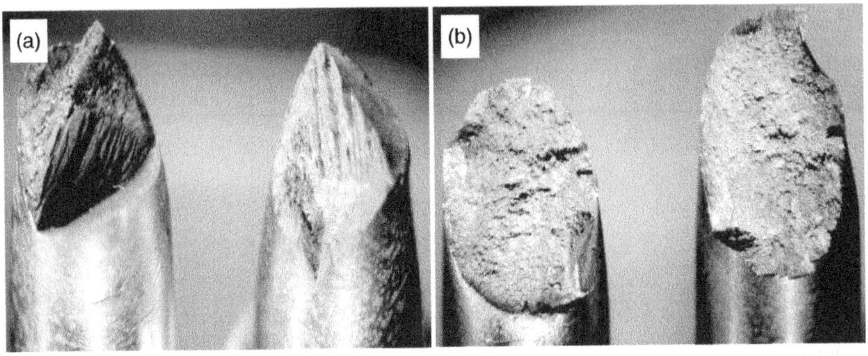

Figure 8-21 *Fractographic views of (a) new X100 and (b) aged X100 steel specimens after an SSRT test at −0.85 V(Cu/CuSO4) in a nearly neutral–pH solution. (From Sanjuan Riverol [2008].)*

under an aging process shows a high susceptibility to SCC at cathodic potentials. The fractographic views of the two specimens after SSRT at CP also confirm the effect of strain aging on the increasing SCC susceptibility of steel, as shown in Fig. 8-21.

Therefore, strain aging is expected to affect (increase) the susceptibility of high-strength steels, such as X100 steel, to SCC, especially at the CP or more negative potentials. It is suspected that hydrogen evolution is enhanced at the cathodic potential and that hydrogen is involved in the cracking process. The essential roles of strain aging in enhanced hydrogen evolution and permeation in the steel and the resulting SCC susceptibility increase are to be investigated.

8.7 STRAIN-BASED DESIGN OF HIGH-STRENGTH STEEL PIPELINES

High-strength steel pipelines have been and are being planned or constructed in geographically remote regions such as sub-Arctic and Arctic areas in North America to transport oil and gas efficiently and safely to processing plants and markets. These regions usually feature permafrost and semi-permafrost, seismic activity, and terrain susceptible to landslides or earthquake. Pipelines are subject to significant plastic deformation as a result of large ground movements. In these environments it is often impractical to meet the allowable stress limits in conventional pipeline design codes.

Strain-based design (SBD) is a limit-state design method where the pipeline longitudinal strain capacity, in addition to the transverse yield strength of pipe steel, is used as a measure of design safety for displacement-controlled axial or bending loading conditions that can result in longitudinal plastic strains in pipelines. SBD allows a more effective use of a pipeline's longitudinal strain capacity while maintaining its hoop pressure containment capacity. Hence, SBD is more suitable than a stress-based design for challenging environmental and/or operating conditions where pipelines experience significant plastic strains. A key design objective of SBD is to ensure that the safety and reliability of a pipeline are generally comparable to those of a conventional pipeline based on stress-based design. This is accomplished by selecting materials and flaw acceptance criteria that ensure that the strain capacity required is achieved [Zhou et al., 2006; Macia et al., 2010].

The SBD method shows a limit on the strain at operating conditions rather than the upper limit of stress. That is, plastic strain is allowed for pipelines in the longitudinal and transverse directions, but it must be less than the strain capacity of the pipeline steel, which is a very important property used to measure how much a steel can be deformed uniformly before necking. High uniform elongation shows high deformability, representing a high level of strain capacity, which is what the pipeline industry pursues.

8.7.1 Strain Due to Pipe–Ground Movement

As stated, pipelines can be exposed to longitudinal strain due to the movement of surrounding soil in various geologic settings, such as Arctic onshore and offshore areas. Moreover, pipelines can be strained by geotechnical phenomena, including slope instability, permafrost thaw subsidence, frost heave, seabed ice souring, subsea permafrost thaw subsidence, and seismic loading. Generally, Arctic lines can be classified into two types: one surrounded by thawing soil, which is susceptible to strain by subsidence, and a second, surrounded by increased freezing, which is susceptible to strain by frost heave [Greenslade and Nixon, 2000; Mohr, 2003].

Due to the geological varieties, the strain limit allowance varies. For example, the Norman Wells pipeline system used a limit of 0.75% longitudinal strain during construction [Mohr, 2003]. The TransAlaska pipeline used an 0.4% longitudinal strain for the main line pipe [Smith and Popelar, 1999]. The Northstar pipeline was designed with a maximum bending strain of about 1%, due to seabed ice gouging and subsea permafrost thaw subsidence [Lanan et al., 1999; Nogueira et al., 2000].

Finite element analysis (FEA) has been widely used to simulate soil movement to obtain an approximate pipe strain, although it is difficult to build a perfect soil model coincident with the true soil environment, due to the discontinuity of soil properties and complex soil–pipe interactions. Conventionally, simulation of the interactions between soil and pipe was based on a Winkler soil model [Allotey and El Naggar, 2006]. However, it has the limitations of formulation convergence and a large mesh distortion. Alternative FEA models were developed [Fredj and Dinovitzer, 2010a,b] to overcome these disadvantages using three-dimensional continuum models and smooth particle hydrodynamics (SPH) approaches to investigate the soil movement. The analysis results were compared with full-scale experimental tests and showed good agreement. However, some issues that may affect soil movement were not taken into account, such as the relative movement between soil and pipe and the variability of soil properties.

In recent years, ExxonMobil has developed advanced analysis methods for assessing the impact of soil loads on pipelines based on comprehensive experimental and numerical programs [Arslan et al., 2010]. These methods are based on nonlinear finite element models, which can be used to assess both the static and dynamic responses

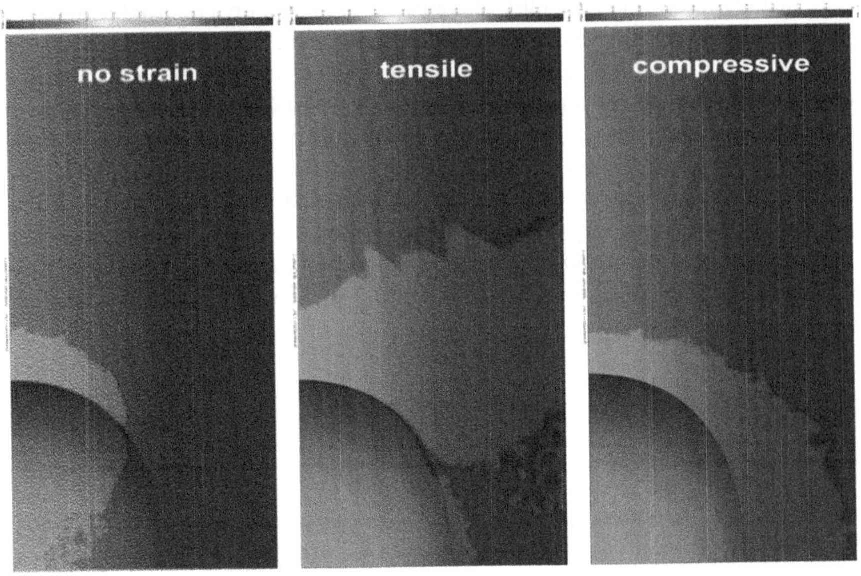

Figure 8-22 Distribution of plastic deformation of an X80 steel pipe containing a 15.28-mm-deep corrosion defect under various strain conditions (from left to right: no prestrain, 0.2% tensile strain, 0.2% compressive strain). (From Xu and Cheng [2012a].)

of a pipeline due to a variety of soil loadings. These models account for the nonlinear material behavior of a pipe and its welds, the large deformations that occur due to significant ground movements, and the nonlinear interactions between soil and pipeline as the pipe is pushed through the soil.

Recently, a new FEA model has been developed to evaluate the synergistic effect of soil strain and corrosion defect on pipelines [Xu and Cheng, 2012aa]. It was demonstrated that in the presence of corrosion defects, the plastic deformation area around the defect increases when tensile or compressive strain is applied. However, the plastic deformation zone under tensile strain is quite different from that under compressive strain. The former expands along the hoop direction and is away from the defect, while the plastic field of the latter surrounds the defect, as shown in Fig. 8-22. Therefore, longitudinal tensile strain would result in the plastic propagating along the hoop direction, leading potentially to circumferential cracking. On the contrary, longitudinal compressive strain can generate a local buckling or wrinkling effect around the corrosion defect.

8.7.2 Parametric Effects on Cracking of Pipelines Under SBD

The mechanical properties of pipeline steels can be affected dramatically by internal pressure. Generally, the internal pressure increases the resistance of pipelines to local buckling because the tensile hoop stress in a transverse direction can resist the diameter changes that contribute to local buckling. The Det Norske Veritas (DNV) recommended a standard [Det Norske Veritas, 2000] using a factor of $1 + 5\sigma_h/\sigma_{ys}$ to define the allowable critical strain, where σ_h is the hoop stress (transverse direction) due to internal pressure and σ_{ys} is the yield strength of the steel. The DNV method is more applicable for larger-diameter pipes with a hoop stress/yield strength ratio in the range 0.2 to 0.5.

Unlike internal pressure, external pressure reduces the resistance to local buckling. It can also increase the probability of crack propagation during buckling. It was found [Corona and Kyriakides, 1988] that for an inelastic pipe, the collapse pressure, maximum bending moment capacity and corresponding curvature decrease in the presence of external pressure. High internal pressure, which is associated with the increasing driving force for cracking, could reduce the critical tensile strain of pipe steels, enhancing strain-induced pipe failure.

Buckling usually occurs adjacent to a girth weld region when longitudinal compressive strain is applied to a pipe. The girth weld is also susceptible to cracking under tensile stress. The welding residual stress, metallurgical and mechanical properties, weld geometry, and misalignment across the weld need to be considered in attempts to improve the buckling and cracking resistance of the girth weld. With respect to the specifications of a girth weld for SBD, the focus should be on the material tensile properties and the tensile strain capacity of pipelines [Wang and Liu, 2010]. The girth welds tend to be the weakest links in the pipeline, due to the possible existence of weld defects and mechanical property changes from welding thermal cycles. It has been reported [Igi, 2010] that the critical global strain of a full-scale pipe specimen with a HAZ notch in the girth weld region shows only half of the value obtained in curved wide plate tests.

8.8 MECHANOELECTROCHEMICAL EFFECT OF CORROSION OF PIPELINES UNDER STRAIN

The prestrain applied on steels, simulating soil-induced strains on pipelines, causes plastic deformation, which is always associated with the strain-hardening phenomenon, as indicated by enhanced yield and ultimate tensile strengths and decreased ductility. This is due primarily to increasing dislocation density and dislocation–dislocation interaction as well as resistance to dislocation motion [Gao et al., 1999].

When the prestrain is sufficiently small to result in elastic deformation of steel, the change in electrochemical corrosion potential, $\Delta\varphi_e^0$, by the elastic load applied is calculated by

$$\Delta\varphi_e^0 = -\frac{\Delta P V_m}{nF} \tag{8-9}$$

where ΔP is the excess pressure (Pa), V_m the molar volume of the steel substrate (m^3/mol), n the charge number, and F is Faraday's constant (96,485 C/mol). Therefore, an external load ($\Delta P > 0$) would decrease the electrochemical corrosion potential of the steel and enhance its corrosion activity.

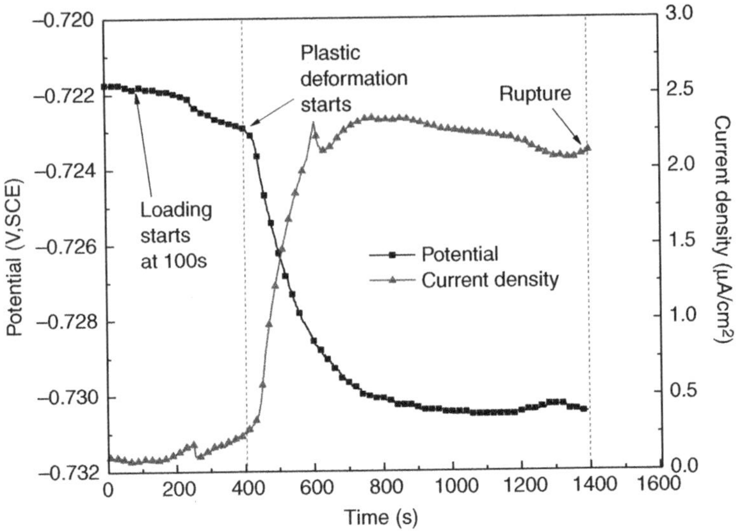

Figure 8-23 *Time dependence of the coupling potential and current density between deformed and nondeformed X100 steel specimens in a nearly neutral–pH solution. (From Xu and Cheng [2012b].)*

Shown in Fig. 3-6 is a slight negative drop of 1.1 mV(SCE) in the corrosion potential in an elastic deformation region. This small potential drop indicates that the increasing electrochemical corrosion activity of steel during elastic deformation is limited. Furthermore, the coupling potential and current density measured between deformed and nondeformed steel specimens in this region are very small, as shown in Fig. 8-23, demonstrating that the mechanoelectrochemical effect developed on the steel is small in an elastic deformation region [Xu and Cheng, 2012c].

During plastic deformation, the density of mobile dislocations is increased with increasing strain by the activation of new dislocations. The density of new dislocations, ΔN, is calculated as [Gutman, 1998]

$$\Delta N = N_0 \left[\exp\left(\frac{n_d \, \Delta\tau}{\alpha k_B N_{max} T} \right) - 1 \right] \tag{8-10}$$

where N_0 is the initial density of dislocations prior to plastic deformation, $\Delta\tau$ the hardening intensity, n_d the number of dislocations in a dislocation pile-up, α is a coefficient of 10^9 to 10^{11} cm^{-2}, k_B is Boltzmann's constant, N_{max} the maximum dislocation density, and T the temperature. The plastic strain at the hardening stage can be expressed by

$$\varepsilon_p = \frac{N_0}{\alpha\nu} \left[\exp\left(\frac{n_d \, \Delta\tau}{\alpha k_B N_{max} T} \right) - 1 \right] \tag{8-11}$$

where ν is the orientation-dependent factor, 0.4 to 0.5 for tensile deformation.

According to Gutman's mechanoelectrochemical interactions theory [Gutman, 1998], the change of electrochemical corrosion potential, $\Delta\varphi_p^0$, in plastic deformation can be expressed by

$$\Delta\varphi_p^0 = -\frac{n_d\ \Delta\tau\ R}{\alpha k_B N_{max} zF} \tag{8-12}$$

Substitute Eq. (8-11) into Eq. (8-12) to eliminate parameters n_d, $\Delta\tau$, k_B, and N_{max}, and the shift in electrochemical corrosion potential as a function of plastic strain, ε_p, is established:

$$\Delta\varphi_p^0 = -\frac{TR}{zF}\ln\left(\frac{\nu\alpha}{N_0}\varepsilon_p + 1\right) \tag{8-13}$$

To calculate the theoretical potential shift using Eq. (8-13), it is assumed that $T = 295$ K, $R = 8.314$ J/K·mol, $z = 2$, $F = 96{,}485$ C/mol, $\nu = 0.45$, and $\alpha = 1.67 \times 10^{11}$ cm^{-2}. N_0 for steel under various pre-plastic strains is assumed roughly to be 10^8 cm^{-2}, 2×10^8 cm^{-2}, 5×10^8 cm^{-2}, and 10^9 cm^{-2}. ε_p varies from 0 to 3.53% to maintain a consistent range with the experimental plastic strains (i.e., 0%, 0.597%, 2.238%, and 3.530%). A theoretical calculation of the electrochemical corrosion potential shift during plastic deformation is shown in Fig. 8-24. It is apparent that the shift in corrosion potential increases negatively with plastic strain, indicating enhanced electrochemical activity of the steel. Moreover, the enhanced corrosion activity of

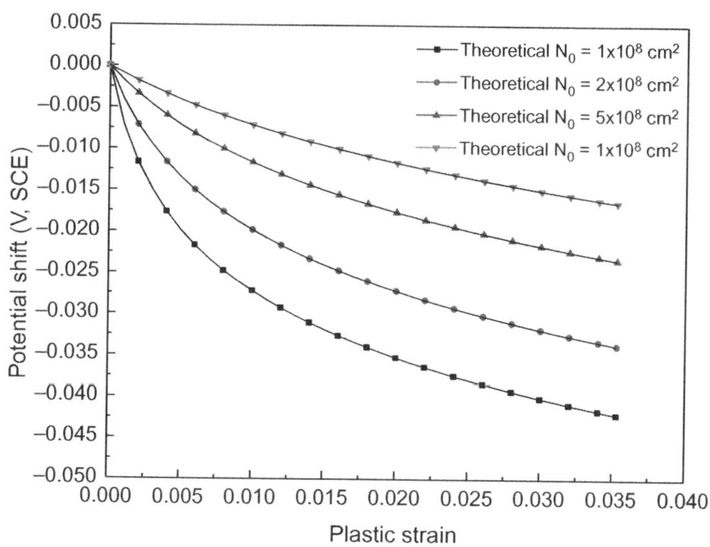

Figure 8-24 *Theoretical calculation of a corrosion potential shift as a function of the plastic strain of a steel specimen with various initial dislocation densities. (From Xu and Cheng [2012b].)*

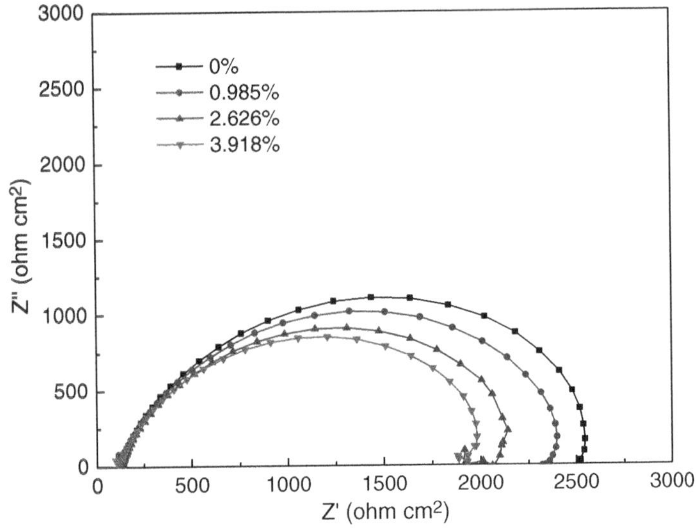

Figure 8-25 *Nyquist diagrams measured on an X100 steel specimen that is at open-circuit potential under various prestrains in a nearly neutral–pH solution. (From Xu and Cheng [2012b].)*

steel by prestrain is clearly indicated by the decreasing size of the semicircle (i.e., charge-transfer resistance for corrosion reaction) in the EIS plots in Fig. 8-25.

The mechanoelectrochemical effect exerted on a steel specimen experiencing both elastic and plastic deformation stages during tensile testing should be based on Eqs. (8-9) and (8-13) to consider simultaneously the shift in total corrosion potential, $\Delta\varphi_T^0$:

$$\Delta\varphi_T^0 = -\frac{\Delta P_m V_m}{zF} - \frac{TR}{zF}\ln\left(\frac{\nu\alpha}{N_0}\varepsilon_p + 1\right) \tag{8-14}$$

where ΔP_m is the excess pressure to elastic deformation limit. In this work, a uniaxial stress is applied on the specimen, and ΔP_m equals one-third of the yield strength of the steel [Gutman, 1998]. Furthermore, the net anodic dissolution current density, i, of a plastically stressed steel electrode is given by

$$\begin{aligned} i &= i_a \exp\left(\frac{\Delta P_m V_m}{RT} + \frac{n\,\Delta\tau}{\alpha k N_{\max} T}\right) - i_c \\ &= i_a\left(\frac{\nu\alpha}{N_0}\varepsilon_p + 1\right)\exp\left(\frac{\Delta P_m V_m}{RT}\right) - i_c \end{aligned} \tag{8-16}$$

where i_a and i_c are the anodic and cathodic current densities, respectively. Assume that the external stress on the specimen does not affect the ion activity in the solution, so that i_c remains constant. It is thus apparent that the anodic dissolution rate of the

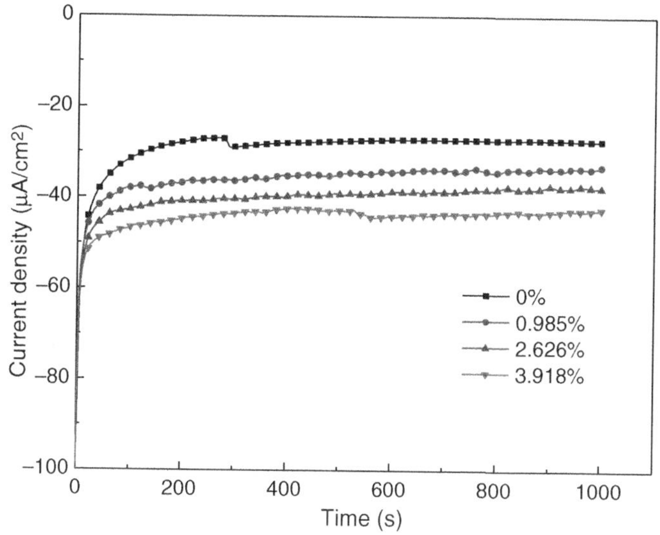

Figure 8-26 *Current densities measured on an X100 steel specimen that is polarized at −1 V(SCE) under various prestrains in a nearly neutral–pH solution. (From Xu and Cheng [2012b].)*

steel is accelerated by external stress or strain. Moreover, no mechanoelectrochemical effect is apparent in the elastic stage. However, it becomes significant in the plastic deformation stage.

Furthermore, analysis of the mechanoelectrochemical effect on corrosion of stressed steel has to consider the anodic and cathodic reactions simultaneously. As discussed, the anodic dissolution reaction would be accelerated by applied strain, which increases the electrochemical corrosion activity of the steel. The cathodic reaction (i.e., hydrogen evolution due to the reduction of water in a deoxygenated, neutral pH solution) would also be affected by external strain, which is demonstrated by the increasing cathodic current density, measured at −1 V(SCE) with prestrain applied on the steel in Fig. 8-26. It is attributed primarily to the redistribution of electrochemical heterogeneity and the increasing area of the cathodic process. An increase in slip steps, microcracks, and surface defects generated during plastic deformation would reduce the activation energy of hydrogen evolution [Gutman, 1998].

Plastic strain enhances the corrosion activity of steels. When a steel specimen contains nonuniform stress–strain distributions, it is expected that the local corrosion rate of the steel will be different. As shown in Figs. 8-27 and 8-28, when stress concentration exists at the left side of a flat tensile X100 steel specimen, the highest dissolution current density is measured by SVET at the left end and the current density decreases from left to right. In reality, a nonuniform stress–strain distribution on pipelines due to a wide variety of reasons, such as mechanical dent, corrosion, and surface defects, would cause preferential corrosion to occur locally, resulting in localized corrosion and the initiation of corrosion pits and even cracks.

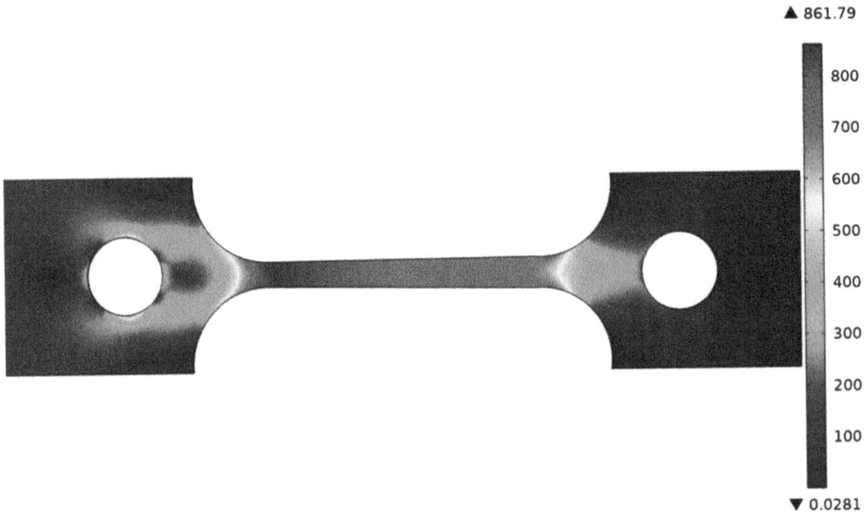

Figure 8-27 *Von Mises stress distribution of a flat plate X100 steel specimen under a tensile force of 2750 N. (From Xu and Cheng [2012b].)*

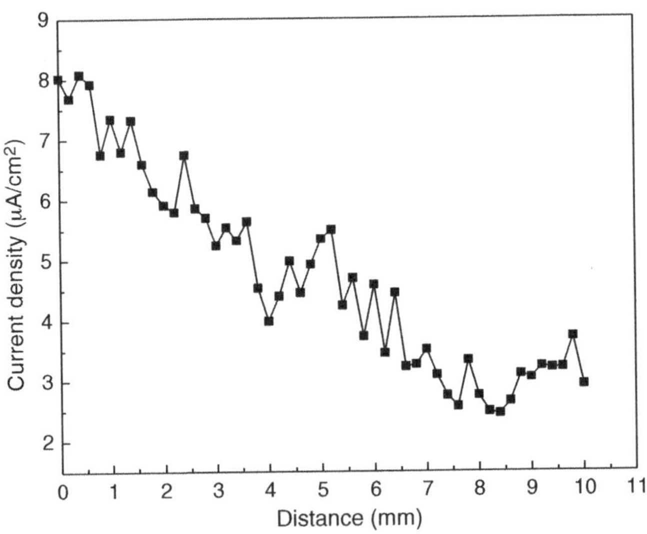

Figure 8-28 *SVET line scanning measurement of the X100 steel specimen shown in Fig. 8-27 in a nearly neutral–pH solution. (From Xu and Cheng [2012b].)*

REFERENCES

Allotey, N, El Naggar, MH (2006) Generalized dynamic Winkler model for non-linear soil-structure interaction analysis, *Can. Geotech. J.* 45, 560–573.

American Petroleum Institute (2004) *Specification for Line Pipe*, API 5L, API, Washington, DC.

Anderson, CW, Shi, G, Atkinson, HV, Sellars, CM, Yates, JR (2003) Interrelationship between statistical methods for estimating the size of the maximum inclusion in clean steels, *Acta Mater.* 51, 2331–2343.

Arslan, H, Hamilton, J, Lele, S, Minnaar, K, Albrecht, B, Cook, MF, Wong, P (2010) Strain demand estimation for pipelines in challenging arctic and seismically active regions, *Proc. 8th International Pipeline Conference*, Paper IPC2010-31505, Calgary, Alberta, Canada.

Asahi, H (2004) *Development of Ultra-High-Strength Linepipe X120*, Paper 90, Nippon Steel, Tokyo, Japan.

Bai, D, Collins, L, Hamad, F, Chen, X, Klein, R, Zhou, J (2008) X100 (Grade 690) helical-welded linepipe, *Proc. 7th International Pipeline Conference*, Paper IPC2008-64099, Calgary, Alberta, Canada.

Baker, MA, Castle, JE (1992) The initiation of pitting corrosion of stainless steels at oxide inclusions, *Corros. Sci.* 8, 1295–1312.

Banerjee, K, Chatterjee, UK (2001) Hydrogen permeation and hydrogen content under cathodic charging in HSLA 80 and HSLA 100 steels, *Scr. Mater.* 44, 213–216.

Cahn, RW, Haasen, P (1996) *Physical Metallurgy*, 4th ed., North-Holland, Amsterdam.

Capelle, J, Dmytrakh, I, Pluvinage, G (2010) Comparative assessment of electrochemical hydrogen absorption by pipeline steels with different strength, *Corros. Sci.* 52, 1554–1559.

Casanova, T, Crousier, J (1996) The influence of an oxide layer on hydrogen permeation through steel, *Corros. Sci.* 38, 1535–1544.

Cheng, YF (2007a) Analysis of electrochemical hydrogen permeation through X-65 pipeline steel and its implications on pipeline stress corrosion cracking, *Int. J. Hydrogen Energy* 32, 1269–1276.

Cheng, YF (2007b) Thermodynamically modeling the interactions of hydrogen, stress and anodic dissolution at crack-tip during near-neutral pH SCC in pipelines, *J. Mater. Sci.* 42, 2701–2705.

Corbett, KT, Bowen, RR, Petersen, CW (2004) High strength steel pipeline economics, *Int. J. Offshore Polar Eng.* 14, 75–79.

Corona, E, Kyriakides, S (1988) On the collapse of inelastic tubes under combined bending and pressure, *Int. J. Solids Struct.* 24, 505–535.

De S. Bott, I, De Souza, LFG, Teixeira, JCG, Rios, PR (2005) High-strength steel development for pipelines: a Brazilian perspective, *Metall. Mater. Trans. A* 36, 443–454.

De Souza, TO, Buono, VTL (2003) Optimization of the strain aging resistance in aluminum killed steels produced by continuous annealing, *Mater. Sci. Eng. A* 354, 212–216.

Det Norske Veritas (2000) *Submarine Pipeline System*, DNV-OS-F101, DNV, Oslo, Norway.

Devanathan, MAV, Stachurski, Z (1962) The adsorption and diffusion of electrolytic hydrogen in palladium, *Proc. R. Soc.* A 270, 90–102.

Devanathan, MAV, Stachurski, Z (1964) The mechanism of hydrogen evolution on iron in acid solutions by determination of permeation rates, *J. Electrochem. Soc.* 111, 619–623.

Devanathan, MAV, Stachurski, Z, Beck, W (1963) A technique for the evaluation of hydrogen embrittlement characteristics of electroplating baths, *J. Electrochem. Soc.* 110, 886–891.

Dong, CF, Liu, ZY, Li, XG, Cheng, YF (2009) Effects of hydrogen-charging on the susceptibility of X100 pipeline steel to hydrogen-induced cracking, *Int. J. Hydrogen Energy* 34, 9879–9884.

Duan, DM, Zhou, J, Horsley, D (2008) SA effect in high strength line pipe materials, *Proc. 7th International Pipeline Conference*, Paper IPC2008-64427, Calgary, Alberta, Canada.

Fredj, A, Dinovitzer, A (2010a) Three-dimensional response of buried pipelines subjected to large soil deformation effects: I. 3D continuum modeling using ALE and SPH formulations, *Proc. 8th International Pipeline Conference*, Paper IPC2010-31516, Calgary, Alberta, Canada.

Fredj, A, Dinovitzer, A (2010b) Three-dimensional response of buried pipelines subjected to large soil deformation effects: II. Effects of the soil restraint on the response of pipe/soil systems, *Proc. 8th International Pipeline Conference*, Paper IPC2010-31517, Calgary, Alberta, Canada.

Gao, H, Huang, Y, Nix, WD, Hutchinson, JW (1999) Mechanism-based strain gradient plasticity: I. Theory, *J. Mech. Phys. Solids*, 47, 1239–1263.

Glover, AD, Horsley, DJ, Dorling, DV (1999) High-strength steel becomes standard on Alberta gas systems, *Oil Gas J.*, Jan., 44–49.

Glover, A, Horsley, D, Dorling, DV, Takehara, J (2004) Construction and installation of XI00 pipelines, *Proc. 5th International Pipeline Conference*, Paper 0328, Calgary, Alberta, Canada.

Gokhman, MR (1983) Temperature regime of permafrost in the bed of a gas pipeline, *Soil Mech. Found. Eng.* 20, 15–17.

Graf, MK, Hillenbrand, HG, Heckmann, CJ, Niederhoff, KA (2003) High-strength large-diameter pipe for long-distance high pressure gas pipelines, *ISOPE 2003*, Paper 2003-SYMP-03, Honolulu, HI.

Greenslade, JG, Nixon, D (2000) New design concepts for pipelines buried in permafrost, *Proc. 3rd International Pipeline Conference*, Vol. 1, Calgary, Alberta, Canada, pp. 135–143.

Gutman, EM (1998) *Mechanochemistry of Materials*, Cambridge Interscience Publishing, Cambridge, UK.

Herman, WA, Erazo, MA, DePatto, LR, Sekizawa, M, Pense, AW (1987) Strain aging of micro-alloyed steels, *Weld. Res. Council Bull.* 322, 1–12.

Hillenbrand, HG, Liessem, A, Biermann, K, Heckmann, CJ, Schwinn, V (2004) Development of high strength material and pipe production technology for grade X120 linepipe, *Proc. 5th International Pipeline Conference*, Calgary, Alberta, Canada, pp. 1743–1749.

Hillenbrand, HG, Heckmann, CJ, Niederhoff, KA, online source, X80 line pipe for large-diameter high strength pipelines, http://www.europipe.com/files/x80_line_pipe_for_large_diameter_high_strength_pipeline.pdf.

Hirth, JP (1980) Effects of hydrogen on the properties of iron and steel, *Metall. Trans. A* 11, 861–890.

Horsley, D (2007) X100 field welding experience, talk given in Canberra JTM.

Hosford, WF (2005) *Mechanical Behavior of Materials*, Cambridge University Press, Cambridge, UK.

Igi, S (2010) Tensile strain capacity of X80 pipeline under tensile loading with internal pressure, *Proc. 8th International Pipeline Conference*, Paper, IPC2010-31281, Calgary, Alberta, Canada.

International Standards Organization (2004) *Method of Measurement of Hydrogen Permeation and Determination of Hydrogen Uptake and Transport in Metals by an Electrochemical Technique*, ISO 17081:2004(E), ISO, Geneva, Switzerland.

Ishikawa, N, Okatsu, M (2008) Material development and strain capacity of grade X100 high strain line-pipe produced by heat treatment online process, *Proc. 7th International Pipeline Conference*, Paper IPC 2008-64507, Calgary, Alberta, Canada.

Ishikawa, N, Okatsu, M, Endo, S, Kondo, J, Zhou, J, Taylor, D (2008) Mass production and installation of X100 linepipe for strain-based design application, *Proc. 7th International Pipeline Conference*, Paper IPC2008-64506, Calgary, Alberta, Canada.

Jack, TR, Erno, B, Krist, K (2000) Generation of near-neutral pH and high pH SC environments on buried pipelines, *Corrosion 2000*, Paper 362, NACE, Houston, TX.

Jin, TY (2011) Correlation of the Metallurgical Features of X100 High-Strength Line Pipe Steel with Its Corrosion and Cracking Behaviour, M.S. thesis, University of Calgary.

Jin, TY, Cheng, YF (2011) In-situ characterization by localized electrochemical impedance spectroscopy of the electrochemical activity of microscopic inclusions in an X100 steel, *Corros. Sci.* 53, 850–853.

Jin, TY, Liu, ZY, Cheng, YF (2010) Effect of non-metallic inclusions on hydrogen-induced cracking of API5L X100 steel, *Int. J. Hydrogen Energy* 35, 8014–8021.

Johnson, J, Hudson, M, Takahashi, N (2008) Specification and manufacturing of pipes for the X100 operational trial, *Proc. 7th International Pipeline Conference*, Paper IPC2008-64653, Calgary, Alberta, Canada.

Kalwa, C, Hillenbrand, HG, Graf, M (2002) High-strength steel pipes: new developments and applications, *Onshore Pipeline Conference*, Houston, TX; http://www.bergpipe.com/files/ep_tp_45_02en.pdf.

Kim, K, Bae, JH (2008) Metallurgical and process parameters for commercial production of high toughness API-X80 grade hot rolled strips, *Proc. 7th International Pipeline Conference*, Paper IPC2008-64249, Calgary, Alberta, Canada.

Koh, SU, Jung, HG, Kang, KB, Park, GT, Kim, KY (2008) Effect of microstructure on hydrogen-induced cracking of linepipe steels, *Corrosion* 64, 574–585.

Lanan, GA, Nogueira, AC, Evan, TM, Ennis, JO (1999) Pipeline bending limit state design for the Northstar Offshore Arctic development project, *ICAWT '99 Pipeline Welding and Technology Conference*, Paper 4.1, Galveston, TX.

Lee, WB, Hong, SG, Park, CG, Kim, KM, Park, SH (2000) Influence of Mo on precipitation hardening in hot rolled HSLA steels containing Nb, *Scr. Mater.* 43, 319–324.

Lim, YS, Kim, JS, Ahn, SJ, Kwon, HS, Katada, Y (2001) The influences of microstructure and nitrogen alloying on pitting corrosion of type 316L and 20 wt.% Mn-substituted type 316L stainless steels, *Corros. Sci.* 43, 53–68.

Liu, ZY, Li, XG, Cheng, YF (2010) In-situ characterization of the electrochemistry of grain and grain boundary of an X70 steel in a near-neutral pH solution, *Electrochem. Commun.* 12, 936–938.

Luu, WC, Wu, JK (1996) The influence of microstructure on hydrogen transport in carbon steels, *Corros. Sci.* 38, 239–245.

Macia, ML, Kibey, SA, Arslan, H (2010) Approaches to qualify strain-based designed pipelines, *Proc. 8th International Pipeline Conference*, Paper IPC2010-31662, Calgary, Alberta, Canada.

Margot-Marette, H, Bardou, G, Charbonnier, JC (1987) The application of the slow strain rate test method for the development of linepipe steels resistant to sulfide stress cracking, *Corros. Sci.* 27, 1009–1026.

Marsh, PG, Gerberich, WW (1992) *Stress Corrosion Cracking*, ASM International, Materials Park, OH.

Mohitpour, M, Glover, AG, Trefanko, W (2001) Pipeline report: Technology advances key worldwide gas pipeline developments, *Oil Gas J.* Nov., 60–65.

Mohr, W (2003) *Strain-Based Design of Pipelines*, Project. 45892GTH, U.S. Department of Interior, Minerals Management Service, and U.S. Department of Transportation, Washington, DC.

Muto, I, Kurokawa, S, Hara, N (2009) Microelectrochemical investigation of anodic polarization behavior of CrS inclusions in stainless steels, *J. Electrochem. Soc.* 156, C395–C399.

Nagai, K, Shinohara, Y, Sakamoto, S, Tsuru, E, Asahi, H, HaraT (2010) Anisotropy of the stress–strain curves for line pipe steels, *Proc. 8th International Pipeline Conference*, Paper IPC2010-31169, Calgary, Alberta, Canada.

Nanninga, N, Grochowsi, J, Heldt, L, Rundman, K (2010) Role of microstructure, composition and hardness in resisting hydrogen embrittlement of fastener grade steels, *Corros. Sci.* 52, 1237–1246.

National Energy Board (1996) *Stress Corrosion Cracking on Canadian Oil and Gas Pipelines, Report of the Inquiry*, MH-2-95, NEB, Calgary, Alberta, Canada.

Nayak, SS, Misra, RDK, Hartmann, J, Siciliano, F, Gray, JM (2008) Microstructure and properties of low manganese and niobium containing HIC pipeline steel, *Mater. Sci. Eng. A* 494, 456–463.

Nippon Steel Corporation (2006) *Integrated Mass Production of High-Strength X120-Grade Linepipe and Other High-Strength Grades with Combined Properties*, Oct., Nippon Steel, Tokyo, Japan.

Nogueira, AC, Lanan, GA, Evan, TM, Fowler, JR, Hormberg, BA (2000) Northstar Development Pipelines limit state design and experimental program, *Proc. 3rd International Pipeline Conference*, Vol. 2, Calgary, Alberta, Canada, pp. 1037–1045.

Oriani, RA, Hirth, JP, Smialowski, M (1985) *Hydrogen Degradation of Ferrous Alloys*, Noyes Publications, Park Ridge, NJ.

Palmer, T (2004) *Alaska Gas Pipeline Construction Cost Risks*, TransCanada Pipelines, Anchorage, AK.

Park, GT, Koh, SU, Jung, HG, Kim, KY (2008) Effect of microstructure on the hydrogen trapping efficiency and hydrogen induced cracking of linepipe steel, *Corros. Sci.* 50, 1865–1871.

Payandeh, Y, Soltanieh, M (2007) Oxide inclusions of different steps of steel production, *J. Iron Steel Res. Int.* 14, 39–46.

Pereloma, EV, Bata, V, Scott, RI, Smith, RM (2010) Effect of Cr and Mo on strain ageing behavior of low carbon steel, *Mater. Sci. Eng. A* 527, 2538–2546.

Pourkia, N, Abedini, M (2008) Recent developments of oil and gas transmission pipeline steels; microstructure, mechanical prosperities and sour gas resistance, *Proc. 7th International Pipeline Conference*, Paper IPC2008-64153, Calgary, Alberta, Canada.

Prolog Canada Inc. (2003) *Arctic Gas Pipeline Construction Impacts on Northern Transportation*, Prolog, Calgary, Alberta, Canada.

Qiao, LJ, Luo, JL, Mao, X (1998) Hydrogen evolution and enrichment around stress corrosion crack tips of pipeline steels in dilute bicarbonate solution, *Corrosion*, 54, 115–121.

Reed-Hill, RE (1973) *Physical Metallurgy Principles*, 2nd ed., D. van Nostrand, New York.

Reformatskaya, II, Freiman, IL (2001) Formation of sulfide inclusions in the structure of steels and their role in the local corrosion processes, *Zashch. Met.* 37, 511–516.

Sanjuan Riverol, E (2008) Studies of corrosion and stress corrosion cracking behavior of high-strength pipeline steels in carbonate-bicarbonate solutions, M.S. thesis, University of Calgary.

Shanmugam, S, Ramisetti, NK, Misra, RDK, Hartmann, J, Jansto, SG (2008) Microstructure and high strength–toughness combination of a new 700 MPa Nb-microalloyed pipeline steel, *Mater. Sci. Eng. A* 478, 26–37.

Smith, MQ, Popelar, CH (1999) A strain based rupture criterion for corroded pipelines, *ICAWT'99 Pipeline Welding and Technology Conference*, Paper 4.6, Galveston, TX.

Stalheim, DG, Muralidharan, G (2006) The role of continuous cooling transformation diagrams in material design for high-strength oil and gas transmission pipeline steels, *Proc. 6th International Pipeline Conference*, Paper IPC2006-10251, Calgary, Alberta, Canada.

Sun, W, Lu, C, Tieu, AK, Jiang, Z, Liu, X, Wang, G (2002) Influence of Nb, V and Ti on peak strain of deformed austenite in Mo-based micro-alloyed steels, *Mater. Process. Technol.* 72–76, 125–126.

Thompson, N, Wadsworth, NJ (1958) Metal fatigue, *Adv. Phys.* 7, 72–169.

Tresseder, RS (1977) *Stress Corrosion Cracking and Hydrogen Embrittlement of Iron Base Alloys*, NACE, Houston, TX.

Venegas, V, Caleyo, F, Baudin, T, Hallen, JM, Penelle, R (2009) Role of microstructure in the interaction and coalescence of hydrogen-induced cracks, *Corros. Sci.* 51, 1140–1145.

Vignal, V, Krawiec, H, Heintz, O, Oltra, R (2007) The use of local electrochemical probes and surface analysis methods to study the electrochemical behavior and pitting corrosion of stainless steels, *Electrochim. Acta* 52, 4994–5001.

Vodopivec, F (2004) Strain aging of structure steel, *Metabk* (in Croatian, Metallurgy) 43, 143–148.

Wang, YY, Liu, M (2010) Considerations of linepipe and girth weld tensile properties for strain-based design of pipelines, *Proc. 8th International Pipeline Conference*, Paper IPC2010-31376, Calgary, Alberta, Canada.

Wilson, DV, Russell, B (1960) The contribution of atmosphere locking to the strain-aging of low carbon steels, *Acta Metall.*, 8, 36–45.

Wilson, DV, Tromans, JK (1970) Effects of strain aging on fatigue damage in low carbon steel, *Acta Metall.*, 18, 1197–1208.

Wranglen, G (1974) Pitting and sulfide inclusions in steel, *Corros. Sci.* 14, 331–349.

Wu, X (2006) Weldability of high strength steels for gas transmission pipeline, assignment for ENME 619.45 Welding Metallurgy and Design, University of Calgary.

Wu, X (2010) Strain-Aging Behavior of X100 Line-Pipe Steel for a Long-Term Pipeline Service Condition, M.S. thesis, University of Calgary.

Xu, LY, Cheng, YF (2012a) Reliability and failure pressure prediction of various grades of pipeline steel in the presence of corrosion defects and pre-strain, *Int. J. Press. Vessels Piping* 89, 75–84.

Xu, LY, Cheng, YF (2012b) Corrosion of X100 pipeline steel under plastic strain in a neutral pH bicarbonate solution, *Corros. Sci.* DOI: 10.1016/j.corsci.2012.07.012.

Xu, LY, Cheng, YF (2012c) An experimental investigation of corrosion of X100 pipeline steel under uniaxial elastic stress in a nearly neutral–pH solution, *Corros. Sci.* 59, 103–109.

Xue, HB, Cheng, YF (2011) Characterization of microstructure of X80 pipeline steel and its correlation with hydrogen-induced cracking, *Corros. Sci.* 53, 1201–1208.

Xue, HB, Cheng, YF (2012) Hydrogen permeation and electrochemical corrosion behavior of the X80 pipeline steel weld, *J. Mater. Eng. Perf.* DOI: 10.1007/s11665-012-0216-1.

Yen, SK, Huang, IB (2003) Critical hydrogen concentration for hydrogen-induced blistering on AISI 430 stainless steel, *Mater. Chem. Phys.* 80, 662–666.

Yurioka, N (2001) Physical metallurgy of steel weldability, *ISIJ Int.* 41, 566–570.

Zhou, J, Horsley, D, Rothwell, B (2006) Application of strain-base design for pipelines in permafrost areas, *Proc. 6th International Pipeline Conference*, Paper IPC2006-10054, Calgary, Alberta, Canada.

9

Management of Pipeline Stress Corrosion Cracking

9.1 SCC IN PIPELINE INTEGRITY MANAGEMENT

9.1.1 Elements of Pipeline Integrity Management

Pipeline integrity management is a process to develop, implement, measure, and manage the integrity of a pipeline through assessment, mitigation, and prevention of risks to ensure a safe, environmentally responsible and reliable service [Nelson, 2002]. The objectives of effective PIM include:

- To provide systematic and structured framework to manage inspection, maintenance, and operation of pipelines.
- To provide document management policy and guidelines on pipeline design, operation, and abandonment.
- To define roles, responsibility of the practices, and compliance to the regulatory requirement.
- To develop fully computerized systems to provide an indication of the general health of pipelines.

In Canada, all oil and gas pipeline systems are designed, constructed, operated, and maintained in accordance with the latest revision of Canadian Standard Association Z662-11 [Canadian Standards Association, 2011]. Pipelines that convey liquid

Stress Corrosion Cracking of Pipelines, First Edition. Y. Frank Cheng.

hydrocarbons, oilfield water and steam, carbon dioxide used in enhanced oilfield recovery schemes, and hydrocarbon gas all fall under the requirements of the CSA, where an integrity management program (IMP) should include methods for collecting, integrating, and analyzing information related to all facets of design, construction, operations, and maintenance. The data collecting, integrating, and analyzing parts of the process have to be continuous with feedback to the user. Analysis of the data is required to monitor risk assessment, provide meaningful integrity assessment, and improve the integrity of pipelines.

The National Energy Board Section 40 Guideline [National Energy Board, 2003] suggests four major elements of an IMP, including (1) a *management system*, comprising the scope of a PIM program, description, goals, and objectives; organizational responsibility; training; qualifications; management of change methods; and measurement of the effectiveness of a PIM program; (2) a *records management system*, comprising access to pipeline records reporting data on material, construction testing, and inspection; coatings; repair history and mapping; and technical data and PIM program effectiveness; (3) *condition monitoring*, comprising characteristics of monitoring process, procedures for review, and methods involved for monitoring; (4) a *mitigation program*, comprising best practices and criteria for mitigation methods.

According to CSA Z662 [Canadian Standards Association, 2011], a PIM process includes a number of sequential steps. In particular, the IMP scope is to develop a description of the pipelines included in an integrity management program, which should have statements covering integrity-related corporate policies, objectives, and performance indicators. Companies must maintain complete, relevant records. Competency and training requirements should be developed and implemented for company personnel, contractors, and consultants to give them the appropriate knowledge and skills for performing the elements of the PIM for which they are responsible.

The management of change (MOC) process refers to changes that affect the integrity of pipelines or a company's ability to manage pipeline integrity. Such changes can include those that are initiated and controlled by the operating company, including the ownership of a pipeline; the organization and personnel of the operating company; piping and control systems; pipeline operating status; operating conditions; service fluid characteristics; methods, practices, and procedures related to pipeline integrity management; and records related to pipeline integrity management. Moreover, MOC also includes those that are not initiated and controlled by the operating company, such as changes in standards and regulations related to pipeline integrity management, and pipeline rights-of-way, adjacent land use, and development.

A procedure should be established for investigating and reporting failure and external incidences, and to identify hazards that can lead to a failure or external interference incident. Risk assessments should be performed to estimate the consequence of the incidents and significance of estimated risk, and to identify options for risk reduction. Plans and schedules for activities related to PIM will then be determined, and consequently, inspections, testing, patrols, and monitoring will be conducted. Risks should be identified and the types of corrective actions documented that will be considered for anticipated conditions or imperfections that could cause a failure

TABLE 9-1 Classification of Primary Hazards and Threats Encountered on Pipelines

Corrosion	Cracking	External Interference	Manufacturing/ Materials/ Welding	Geotechnical/ Weather	Incorrect Operations
External Internal Microbial	High-pH SCC Nearly neutral–pH SCC	First-party damage Second-party damage Third-party damage Stray current	Defective pipe seam Defective pipes Defective pipe girth weld Defective fabrication weld Wrinkle bend or buckle	Earthquake Slope movement Ground collapse Flash flood Floating pipe	Inappropriate procedures Incomplete operator training Operator error

Source: Canadian Standards Association [2011].

incident with significant consequences. PIM programs are to be reviewed and evaluated periodically to determine whether they are in accordance with the standard and be revised as necessary.

Pipeline hazards and threats are created during the design, construction, and operational stages. Some threats are time dependent, while others are independent of time. The primary threats are listed in Table 9-1 [Canadian Standards Association, 2011].

The primary integrity management approaches include [Sullivan, 2007]:

1. *A prescriptive-based approach.* In this American approach, operators must meet minimum requirements. Regulators enforce the meeting of the minimum standard, and there is no incentive to go beyond meeting a minimum requirement. There are predefined requirements (i.e., method, programs, actions, criteria, etc.) and there is little room for alternative methods. This method is effective to a certain extent.
2. *A performance-based approach.* Operators must meet prescriptive requirements and conduct case-specific assessments for developing an integrity management plan. This approach focuses on measurable benchmarks but requires customized analysis. Industry performance can be used only as a trend. This is a British approach.
3. *A goal-oriented approach.* This is a combination of a prescriptive and performance-based approach. Operators have a responsibility to improve the integrity of pipelines continually to reach the stated goals, safety and environmental health. Operators are expected to go beyond the minimum. There is a spirit of cooperation between operators and regulators. This is a Canadian approach.

An IMP process is usually valid for two or three years and is then updated to include new or modified processes, which are developed during implementation of a pipeline integrity program through multiple time-driven integrity plans. Many pipeline operators have programs to prevent pipeline failures, detect anomalies, and perform repairs to maintain and improve pipeline integrity and reliability. These programs significantly exceed all the regulatory minimums. The goal of a PIM program is to ensure that the risk is "as low as is reasonably practicable."

The PIM technologies have been envisioned as a continuous development to ensure the continued reliable performance of existing pipelines, many of which have been in service for 50 years or more, as well as of newly constructed pipelines. Moreover, there has been an increasing need to develop technologies to ensure the reliability of pipelines carrying fluids other than the traditional oil and gas for which there is limited pipelining experience, such as biofuels, carbon dioxide, and hydrogen. Finally, new technologies have been developed to be used in systems for ensuring pipeline security. Advanced monitoring technologies will be developed to meet the needs of pipeline regulators and the industry to detect third-party intrusion as well as unintended pipeline damage, which could, for example, be caused by slope instability and ground movement.

9.1.2 Initial Assessment and Investigation of SCC Susceptibility

An initial assessment of SCC susceptibility is required in every segment of a pipeline system [Canadian Energy Pipeline Association, 2007]. Prior to assessment, all records that will be able to determine the type of coating applied to pipelines must be collected. If no reliable information is available on the given segment, excavations should be conducted to determine or verify the coating type.

In addition to coatings there are several other factors to consider, since they may affect the susceptibility of a pipe segment to SCC. These include pipeline attributes (age and season of construction, manufacturer, diameter, long-seam type, grade, pipeline alignment, and stress concentrators), operating conditions (stress level, pressure cycling, temperature), environmental conditions (terrain, soil and soil drainage types, drainage pattern), and pipeline maintenance history and records (CP).

If a pipe segment is determined to be susceptible to SCC, an investigation of the presence of SCC is required. The most common method used for the investigation is the use of excavations at sites selected by SCC in-line inspection (ILI), at sites selected by SCC models, or at opportunistic sites. A SCC hydrotest is capable of detecting SCC above a minimum size.

1. *SCC ILI correlation excavations.* These provide a method for determining if SCC exists on a pipeline segment and may also be able to detect the serious consequence-resulting SCC events that have a low frequency of occurrence. This method depends on the availability of a reliable ILI tool. Dependence on the tool only, without correlation excavations and direct inspection, is not recommended.
2. *Excavations based on an SCC model.* SCC model–based excavations are designed to have the greatest chance of finding successfully SCC that may have been initiated on a pipeline surface. The amount of the pipe segment inspected should be supported by a documented SCC site-selection probability

analysis of similar pipe segments. This method can have a very high probability of detecting shallow SCC, which generally occurs with a high frequency in pipelines that have SCC. However, this method has a low probability of detecting any possible injurious SCC that may exist.

3. *Opportunistic inspections.* Opportunistic inspections for SCC within a susceptible pipe segment at locations such as corrosion or dent remediation excavations can provide a cost-effective method of accumulating data compared to an SCC model excavation. The advantage is the ability to sample a cross section of locations that may have a high, medium, or low probability of SCC occurrence. These types of excavations should not be relied on to detect injurious SCC.
4. *Hydrotesting to detect SCC.* A hydrotest is a valid way of determining if SCC, of a minimum size that will fail the hydrotest, exists on a susceptible pipe segment. However, a hydrotest that did not result in a failure does not conclude that SCC is not present on the segment, as SCC can be present with dimensions that do not fail under the hydrotest pressure.

9.1.3 Classification of SCC Severity and Postassessment

Based on the result of the SCC dimensions measured and by using modeling analysis, the SCC severity expressed as pipeline failure pressure is determined. Table 9-2 is a classification of SCC severity by the Canadian Energy Pipeline Association [Canadian Energy Pipeline Association, 2007].

Several general rules are used in determining the SCC features. For example, the failure pressure of the most severe SCC feature or feature with the lowest failure pressure determines the severity categorization above. SCC features exceeding 80% of the wall thickness must be treated as a category IV feature. SCC under 10% of wall thickness is classified under category I and is not considered a threat, but the feature should be reported and mitigated according to the management program setup by the operator. SCC data must be evaluated considering the pipe grade, wall thickness, and toughness within the pipe segment under examination regardless of where the SCC occurs. In general, category I and II features are considered not critical. However, the operators must report, investigate, and mitigate according to the regulations and policies that are set in place. Categories III and IV are generally detected by hydrotesting and/or SCC ILI tools. Generally, when there is an evidence of a category II feature, the operator must carry out exploratory examinations to further investigate the category II feature to ensure that it does not lead to discovery of a category III or IV feature.

When there is evidence of a category III feature obtained by excavations or ILI tools, care must be exercised and be conservative in assuming an SCC presence in the remainder of the pipe. A good approach would be to consider the largest remaining SCC feature that has a failure pressure above the MOP. A safe operating pressure is calculated by the maximum monitored pressure observed over a 60-day period divided by a safety factor of not less than 1.25. For hydrotesting purposes,

TABLE 9-2 SCC Severity Categories

Category	Definition		Description
I	$SCC_{failure\ pressure} \geq 110\% \times MOP^{a} \times SF^{b}$	A failure pressure greater or equal to 110% of the product of the MOP and a company safety factor (typically equaling 110% of SMYS)	SCC in this category does not reduce pipe pressure containing properties relative to the nominal pipe properties; toughness-dependent failures are not expected in the category
II	$110\% \times MOP \times SF > SCC_{failure\ pressure} \geq MOP \times SF$	A failure pressure less than 110% of the product of the MOP and a company-defined safety factor, but greater than or equal to the product of the MOP and a company-defined safety factor	No reduction in the pipe segment safety factor
III	$MOP \times SF > SCC_{failure\ pressure} > MOP$	A failure pressure less than the product of the MOP and a company-defined safety factor but greater than the MOP	A reduction in the pipe segment safety factor
IV	$MOP > SCC_{failure\ pressure}$	A failure pressure equal to or less than the MOP	An in-service failure becomes imminent as the MOP is approached

Source: Canadian Energy Pipeline Association [2007].
[a]Maximum operating pressure.
[b]Safety factor.

the safe operating pressure would be the lesser of the hydrotest pressure at failure or the maximum monitored operating pressure for 60 days. For ILI purposes, the safe operating pressure would be the lesser of the lowest calculated failure pressure of all ILI features divided by the safety factor or the maximum monitored operating pressure over 60 days.

When a category IV feature is discovered by either ILI tools or exploratory excavations, immediate pressure reduction is calculated as the maximum pressure undergone by the feature in the last 15 days, divided by an operator-decided safety factor (no less than 1.25). Further engineering assessment must be undertaken immediately to start mitigation activities.

9.1.4 SCC Site Selection

Development of reliable SCC site-selection models usually occurs after completing the prioritization of pipe segments and provides guidelines for excavation and SCC inspection as well as investigations for the presence or severity of SCC. During the site-selection process, pipeline operators experienced with SCC generally use in-line crack inspection using both low- and high-resolution tools; SCC models developed previously that incorporate terrain, CP data, historical data, and analysis of other non-SCC ILI records; results from SCC hydrotests; and complex risk analysis algorithms. Operators not experienced with SCC face more challenges in terms of the selection and prioritization required. In the beginning they need more reliance on a few well-quantified factors and many excavations. Once experienced, the excavations could be reduced. Operators with limited SCC experience will need to perform many excavations to statistically support the presence or severity of SCC on a pipeline segment.

Many methods can be used to identify sites where investigative digs should be conducted. In-line tools are used to monitor the condition of a pipeline, including the detection of cracks, using various nondestructive testing (NDT) methods to locate SCC. One of the primary objectives of ILI is to detect and ascertain the type and nature of cracks. For example, ultrasonic inspection uses sound waves of short wavelength and high frequency to detect flaws and cracks or to measure material thickness [Bickerstaff et al., online source]. Ultrasonic tools give excellent results and anomaly accuracy. They are used primarily by companies with product lines that are being inspected for SCC and other types of corrosion. Use of in-line crack inspection data increases the accuracy and efficiency of locating SCC and makes it possible to manage SCC risk in a manner similar to one used for metal loss. However, additional analysis of ILI data is needed, while taking into consideration materials science, corrosion, and fracture mechanics.

When in-line crack detection is not feasible, a model for SCC occurrence is to be developed within the subject pipeline segment. Since many factors contribute to the formation and severity of SCC, the integration of various types of data can be useful in selecting specific sites for SCC investigation. Analysis is then required to determine correlation of these data to SCC. If there is limited or no historical SCC data for use by the operator, site selection may be performed by looking for areas where common SCC-related factors occur concurrently. It is also critical for the operator to reevaluate the site-selection process after investigations are conducted and to refine the model frequently.

Moreover, hydrostatic testing, where a line pipe is pressurized with water or air, could be used to identify the SSC presence and the severity of SCC, and also as a mitigation method. The causal factors for SCC occurrence would be identified, and models for pipeline SCC can be developed to lead to the identification of additional sites to be investigated.

Opportunistic excavations can view the routine excavations as an opportunity to investigate the presence of SCC. The use of opportunistic excavation data would enhance the SCC susceptibility model and provide an opportunity to validate a lack of susceptibility at locations where SCC is not suspected.

Prioritization of SCC sites for investigation may be required if recourse is not available to excavate all locations. Generally, priority should be given to those features that exhibit the most injurious attributes, such as depth, length, and location on the pipe, based on statistical analysis models. In an SCC statistical analysis method

[Youzwishzen et al., 2004], data acquired at the investigative dig sites, such as soil conditions, drainage patterns, and local geography, is incorporated into the analysis. In addition, data acquired for the entire length of the pipeline, including geometry, metal loss features, close-interval CP readings, and operating pressures, are combined with the dig site data in the analysis process. The combined data set is analyzed using statistical regression techniques and various multivariable logistic regression models. The predictors that show the greatest correlation to SCC include the CP + ON potential, CP shift, presence of a ground depression, bend angle of the pipe, direction of the bend (vertical or horizontal), proximity to metal loss (and whether the metal loss is near a girth weld), and metal loss severity. The model is applied to the pipeline to determine probabilities of SCC at specified increments along its length. A certain number of locations with high SCC probabilities are selected for verification excavation. Moreover, one site with a lower SCC probability is also chosen for excavation. The sites under excavation are checked to determine if the prediction is consistent with the actual situation (i.e., whether the predictive SCC occurrence is confirmed by the excavation). The combined success rate of the model is calculated and checked for improvement over predictive models applied to the pipeline previously.

9.1.5 SCC Risk Assessment

Risk management is defined as a comprehensive management decision support process, implemented as a program, integrated through defined roles and responsibility into the day-to-day operation maintenance, engineering management, and regulatory decision of the operator.

The pipeline operator seeks to minimize risk by attempting to control the probability of failure occurrence, its consequence if it occurs, or a combination of both. To minimize failure probability, design code must be adhered to by applying quality assurance during the pipe fabrication and installation, and practicing sound maintenance polices. Generally, risk can be reduced by, for example:

- An effective leak detection system
- Right-of-way (ROW) monitoring
- Automated valves and check valves spacing
- Emergency response plans

There are two risk assessment approaches, qualitative and quantitative [*Wikipedia*, online source]. Whereas the former is subjective, relative, and judgmental, the latter emphasizes the objective, analytical, and absolution nature of the associated risks. The qualitative approach uses the concept of risk indices and assigns subjective scores to segments of pipes. Since it uses statistical calculation extensively, the assessment accuracy depends on historical information. The quantitative approach uses detailed information on the pipelines, and the accuracy depends on integration and analysis.

Stress corrosion cracking direct assessment (SCCDA) is a process to assess a pipe segment for the presence of SCC by examining using a magnetic particle inspection (MPI) method or equivalent [Pikas, 2002]. SCCDA requires the integration of data from historical records, indirect surveys, field examinations, and from pipe surface evaluations combined with the physical characteristics and operating history of the pipeline. Through successive applications, SCCDA should identify and address locations where SCC has occurred, is occurring, or might occur. SCCDA assists in managing SCC by selecting potential pipeline segments, choosing dig sites within those segments, inspecting the pipe, gathering and analyzing data during the dig, conducting a mitigation program, defining the reevaluation interval, and evaluating the effectiveness of the SCCDA process.

SCCDA is complementary with other inspection methods, such as ILI or hydrostatic testing, and is not necessarily an alternative or replacement for these methods. SCCDA is also a process that is complementary to internal corrosion direct assessment (ICDA) and external corrosion direct assessment (ECDA).

ICDA. Internal corrosion is the primary cause of failure of upstream pipelines. ICDA usually covers initial corrosion assessment and testing, design and planning of mitigation programs, and measurement of the efficacy of corrosion control measures. After identifying maximum risk locations using ICDA, the control of internal corrosion in pipelines is optimized by activities such as pipeline system evaluation and assessment, engineering calculations and prediction of corrosivity, laboratory analysis of gas, liquids, and solids, corrosion monitoring, and cleaning pig programs.

ECDA. External corrosion is the largest pipeline threats [American Society of Mechanical Engineers, 2001], and the relevant integrity assessment techniques include ILI, hydrostatic testing and direct assessment. In particular, ECDA is formulated to focus on maintenance activities at the most vulnerable points [National Association of Corrosion Engineers, 2008]. ECDA is a four-step process: (1) preassessment: combine a pipeline's physical characteristics, operating history, and prior inspections; (2) indirect assessment: conduct an aboveground field survey such as a direct-current voltage gradient, close-interval potential survey, pipeline current mapping, or soil resistivity measurements; (3) direct examination: expose the pipe for verification of the aboveground survey results and physical inspection of pipe coating, pipe surface, and soil electrolyte; and (4) postassessment: assess the integrity of results gathered in steps 1, 2, and 3 for overall integrity assessment, validation of the ECDA process, and determination of the interval period to redo the ECDA process [Khan, 2012]. ECDA is a proactive methodology, as it identifies whether corrosion has occurred, is occurring, or will occur. ILI detects existing corrosion, while pressure testing can cause failure of a pipeline below the pressure limit.

SCCDA. A SCC susceptibility assessment is conducted on every segment of the pipeline system to evaluate SCC risks typically when the system has never had an

SCC susceptibility assessment, especially when a segment has seen a change during condition monitoring. A wide variety of factors influence SCC susceptibility:

- Type and condition of external coating
- Pipeline attributes
  - Age and season of construction
  - Manufacturer
  - Diameter
  - Type of long seam
  - Grade
  - Pipeline alignment
  - Stress concentrators
- Operating conditions
  - Stress level and pressure cycling
  - Temperature
  - Distance downstream from a compressor and pump station
  - Environmental conditions (corrosivity)
  - Terrain
  - Soil type and soil drainage types
  - Drainage patterns
  - Pipeline/pipe segment maintenance history
  - In-line inspection data
  - CP data
  - Historical excavation records

An initial SCC susceptibility assessment is always required. If there is no record of the conditional factors listed above, an excavation is usually required. Such a susceptibility assessment of pipeline segments would provide benefits to operators, including a reduced scope of investigation within the pipeline systems, prioritization of factors investigated, and timely processing of the actual presence of SCC.

9.2 PREVENTION OF PIPELINE SCC

Prevention is a process intended to avoid risk through minimizing either threat initiation or consequence impact. Prevention must be built into pipeline systems during the initial planning, design, and construction phases by anticipating their effects on the future integrity of the system. This enables the entire life cycle of the pipeline asset to be addressed comprehensively, and optimal integrity solutions to be found.

The prevention approach must be based on proper selection of materials, sound engineering design, CP application, coatings, and operation within the maximum allowable operating pressure (MAOP). Pipeline safety and reliability begin with prevention. This means recognizing conditions that have been known to cause failures in history and then working to minimize the risk. Prevention of pipeline SCC should also start at the design stage of a pipeline project. The focus will be on the appropriate selection of pipe material, the limit of stress, and the control of environments.

9.2.1 Selection and Control of Materials

SCC can be prevented by choosing a material that is not susceptible to SCC in the service environment and by processing and fabricating it correctly. Unfortunately, it is not always quite that simple. Some environments, such as carbonate–bicarbonate solutions, will cause the SCC of most carbon steels. Mechanical requirements, such as the high yield strength of steels, could be difficult to reconcile with SCC resistance, especially where hydrogen is involved in the cracking process. Moreover, many materials are not viable economically for use on a large scale.

Carbon and low-alloy steels have been the primary materials used to fabricate long-distance transmission pipelines. However, they can suffer from SCC in a wide range of environments. In reality, the microstructure and microchemistry of steels have a small effect on the SCC behavior. With respect to high-pH pipeline SCC, neither field investigations nor laboratory studies have identified the unique metallurgical features associated with this problem. Similarly, nearly neutral–pH SCC has developed on multiple types and grades of line pipe from a variety of pipe mills. In examinations of 14 samples from susceptible pipe joints, ranging in size from NPS 8 to 42 and API X52 to X70 steels, no statistically correlation was found between the occurrence of SCC and the material factors under investigation, including chemical composition, inclusions (number, area, and composition), and local galvanic behavior [Beavers et al., 2000]. Moreover, the vast majority of line pipes for gas transmission service in examination is produced by one of the three seam-welded processes and contains varying amounts of ferrite, pearlite, and bainite with a wide variation in the crystalline grain size. There is no strong evidence that any of the items noted above either promote or inhibit SCC. However, a relationship has been found between the susceptibility to nearly neutral–pH SCC and the length of nonmetallic inclusions in the steel [Surkov, 1994].

Considering the predominant influence of the strength of steels on SCC, especially in the presence of hydrogen, limitations on the steel strength are the main selection criteria. For example, limiting the yield strength of steels used in aqueous chlorides to 689 MPa (100 ksi) is typical of this approach [Materion, 2011]. Moreover, the hardness value has been another criterion to prevent SCC of steels under certain environments. NACE's MR-0175 recommendation includes a limit on hardness levels to avoid hydrogen sulfide SCC [National Association of Corrosion Engineers, 2003]. For most steels, hardness is restricted to a maximum of 22 HRC. For pressure vessel welds, a maximum hardness of 200 HB has generally been accepted in hydrogen sulfide and sulfuric acid service.

Furthermore, there is no limit to the strength level above which problems will be experienced, as this will be a function of the amount of hydrogen in the steel, the applied stress, the severity of the stress concentration, the environment, and the composition and microstructure of the steel. As a rough guide, HE is unlikely for steels with yield strengths below 600 MPa and is likely to become a major problem above 1000 MPa [UK National Physical Laboratory, 1982].

With the development of high-strength steel pipeline technology, the metallurgical manufacturing attempts to improve the mechanical properties of steels by alloying treatment combined with controlled rolling as well as by decreasing the carbon content. There has been no convincing indicator that the new high-strength steels, such as X80 and X100 steels, may necessarily have greater or less susceptibility to SCC even though they have higher yield strengths and lower carbon contents. Actually, a more important variable for assessing initiation of SCC is the ratio of the applied stress to the actual yield strength of the steel [Parkins et al., 1981]. However, the HE or HIC of high-strength pipeline steels should not be ignored because the hydrogen-induced material degradation susceptibility usually increases with an increasing strength level in steels.

9.2.2 Control of Stress

One of the requirements for SCC occurrence is the presence of stress (tensile in nature) on the structural components. Thus, one method of prevention of pipeline SCC is to eliminate the stress from pipelines or at least to reduce it below the threshold stress for SCC. This is usually not practical for working stresses since an appropriate operating pressure should be maintained during pipeline operation. It is possible to control the residual stress introduced during welding or forming.

The internal pressure exerted on pipelines is the primary source of circumferential stress (hoop stress), which is usually the highest stress component that exists on pipe. Lowering the pipeline pressure generally reduces SCC occurrence. Moreover, the crack depth decreases with decreasing operating pressure [National Energy Board, 1996]. In particular, reducing the operating stress has the added advantages of increasing the critical flaw size as well as increasing the critical leak or rupture length [Baker, 2005]. However, there has not been a convincing threshold stress level below which SCC will not occur.

Furthermore, the pipeline pressure fluctuates continually due to loading and unloading of products, and is influenced by pump activity, which introduces cyclic stress on pipelines, resulting in pipeline cracking at stress levels even lower than that in the absence of pressure fluctuations [Parkins and Greenwell, 1977; Beavers and Jaske, 1998]. The final failure of SCC colonies can occur by pressure cycle fatigue for large deep flaws or large pressure cycles. Thus, reducing cyclic pressure fluctuations would minimize the growth rate of SCC.

There is a growing body of evidence to suggest that tensile residual stresses in the line pipe play a significant role in SCC, and that cracking can be minimized or prevented by reducing these stresses during manufacturing as well as during installation and operation. Residual stresses can be relieved by various types of stress relief for carbon steels. Shot-peening or grit-blasting tends to introduce a surface compressive stress and is beneficial for the control of SCC. The uniformity with which these processes are applied is important. If, for example, only the weld region is shot-peened, damaging tensile stresses may be created at the border of the peened area. Moreover, proper construction practices, such as minimizing fit-up stresses and avoiding stress concentration caused by dents and mechanical damage to a pipe, can reduce the likelihood of SCC initiation.

9.2.3 Control of Environments

The most direct way of controlling pipeline SCC is to control environments by changing the key component of the environment that is responsible for SCC.

Coatings. Inadequate coating performance is the key contributor to pipeline SCC. Basically, a good coating should possess the appropriate performance characteristics, including adhesion/resistance to disbonding, low water permeability, effective electrical insulator, abrasion and impact resistance, flexibility at extreme temperatures, resistance to degradation, retention of physical and mechanical properties, and non-shielding to CP if disbonded. Moreover, the constraints associated with the coating application must be considered when selecting a coating.

The most reliable method of reducing SCC occurrence on pipelines is the use of high-performance coatings combined with an effective CP. The coatings are required to be able to prevent the environment and electrolyte that may cause SCC from contacting the pipeline steel, and simultaneously, to pass CP current through the coating and thus protect the disbonded regions. Moreover, proper surface preparation prior to the coating application should alter the pipe surface condition to make it less susceptible to SCC. As recommended, FBE, liquid epoxies, and urethanes are the preferred coatings for mitigating SCC. Multilayer, yellow jacket, and extruded PE are also acceptable coatings that can mitigate the threat of SCC [Canadian Energy Pipeline Association, 2007].

Furthermore, regardless of the coating selected, the pipe steel surface should be prepared to remove mill scale and to impart sufficient residual compressive stresses to retard SCC initiation by following NACE No. 1/SSPC-SP 5 [National Association of Corrosion Engineers, 2000a] or NACE No. 2/SSPC-SP 10 [National Association of Corrosion Engineers, 2000b] standards.

Cathodic Protection. CP has been used to inhibit corrosion of underground pipelines and thus attempt to control SCC. CP is closely related to high-pH cracking since the generation of high-pH environments is due to the CP current reaching a pipeline steel surface in conjunction with dissolved carbon dioxide in the groundwater. For high-pH pipeline SCC, CP could increase the negative pipe-to-soil potential if the potential of pipe steel is more positive than the level that controls corrosion. Both the impressed current method and sacrificial anode type CP are used widely in industry to prevent high-pH SCC.

Nearly neutral–pH SCC can only occur when there is an inadequate level of CP on the pipe steel surface, where an environment resulting in pipeline SCC is formed in the absence of CP or CP is partially shielded. Thus, CP will not be effective to control SCC. The minimum level of CP required for prevention or mitigation of nearly neutral–pH SCC has not yet been established. However, it is generally accepted [National Association of Corrosion Engineers, 2002] that this form of cracking will not occur where a standard CP criterion [e.g., −850 mV($Cu/CuSO_4$)] is achieved and the coating does not promote shielding of CP current. Moreover, adequate protection should be maintained to avoid SCC and corrosion at or near coating holidays.

It is important that adequate levels of CP along the entire pipeline must be achieved to prevent SCC. However, overprotection should be avoided since it can promote coating degradation and the formation of alkaline environments conducive to high-pH SCC. Hydrogen evolution and accumulation beneath coatings are also side effects under overprotection because hydrogen gas bubbles may block the CP electric circuit. Moreover, hydrogen atoms, once permeating the steel, will cause steel brittlement and crack initiation. Seasonal fluctuations in the CP potential on pipeline should be considered to minimize the likelihood that the pipeline falls into the cracking range on a seasonal basis.

Temperature. It has been demonstrated that the effect of temperature on high-pH SCC is apparent [Fessler, 1979]. However, there is less temperature dependence for nearly neutral–pH SCC than for high-pH SCC of pipelines [National Energy Board, 1996]. A reducing pipeline temperature could reduce the likelihood of SCC occurrence by decreasing the cracking velocity, reducing the possibility of crack initiation, or improving coating performance. Since over 90% of high-pH SCC and 60% of nearly neutral–pH SCC occur close to pump stations, where the highest temperature is present, one method of temperature control that has been implemented by some operators is the installation of cooling towers [Baker, 2005].

9.3 MONITORING AND DETECTION OF PIPELINE SCC

Monitoring is defined as the regular observation and recording of activities taking place in a program. It is a process of routinely gathering information on all aspects of the program. Monitoring of pipeline operation involves the processes to gather data on changes in pipeline operational, environmental, and other conditions that assist in anticipating potential threats or consequences.

Monitoring can be conducted either at many locations along the right-of-way at a given time or continuously at discrete locations in order to observe seasonal variations. A range of environmental conditions and operating parameters can be monitored. Moreover, the extent to which these parameters change seasonally will be monitored to determine if the environment is conducive to SCC continuously or whether cracking might occur only at certain times of the year.

9.3.1 In-Line Inspections

ILI tools can be employed to detect SCC, but the use of technology and analysis that differ from ILI technology for detection and classification of wall loss is required.

Scheduling regular monitoring and ILI to check for corrosion is a useful SCC mitigation and preventive technique. Arrangement of excavation and implementation of the repair program identified by ILI will reduce SCC occurrence. In-line tools are used to monitor the condition of a pipeline using various nondestructive testing methods. The scope of ILI includes:

- Metal loss inspection
- Geometry and bend inspection
- Cracklike inspection
- Land movement and mapping inspection

Typically, inspection tools are run immediately following the chemical-cleaning stage, starting with the tool to determine the presence of any deformation or out-of-round conditions that could be considered injurious to pipelines or detrimental to the safe passage of other inspection tools as well as having onboard inertial mapping for the localization of defects. After that, a high-resolution magnetic flux leakage (MFL) tool is run in each section of pipes to determine the presence of any metal loss, corrosion, or other metallurgical type of anomaly and/or irregularity. Ultrasonic crack detection (UTCD) is usually the final ILI tool to be run. It is an extremely sophisticated device, capable of detecting and measuring fine longitudinal cracking such as SCC. UTCD has been used extensively to identify critical, and, even to a large extent, subcritical SCC.

Several devices and tools have been used at various resolutions for pipeline ILI, including:

1. *Magnetic flux leakage (axial and circumferential).* The basic principle of detection of corrosion and SCC by magnetic flux is that a powerful magnet is used to magnetize the steel. At areas where there is corrosion or metal loss, the magnetic field "leaks" from the steel. In an MFL tool, a magnetic detector is placed between the poles of the magnet to detect the leakage field. Analysts interpret the chart recording of the leakage field to identify damaged areas and to estimate the depth of metal loss. MFL tools have been used for many years in the detection of three-dimensional defects in pipelines, such as corrosion, defects, and mechanical damage. Recently, it has been used for the detection of longitudinally oriented defects, such as cracks, longitudinal weld defects, and narrow axial corrosion.

   MFL transverse field inspection (TFI) has been used in gas pipelines to attempt detection of SCC. However, it has not been acknowledged extensively by pipeline operators. One of the most significant variables affecting a TFI tool's sensitivity to SCC is the size of the crack opening. Typically, longer and deeper cracks have a greater crack opening, which normally increases with increasing hoop stress level. Therefore, maintaining the hoop stress at the highest practical level during TFI runs should improve the sensitivity of the TFI tool.

2. *Ultrasonic tools (UT).* Ultrasonic inspection uses sound waves of short wavelength and high frequency to detect flaws or to measure material thickness. The UT method has been applied for the detection of SCC in liquid pipelines, but its use is more challenging when applied to gas pipelines, due to the need for a liquid to provide a pathway for transmission of the ultrasonic wave to the pipe wall. UTs basically give excellent results and anomaly accuracy. Examples of such tools include the ultrasonic crack detection tool, the elastic wave tool, and electromagnetic acoustic emission.

   Detection of anomalies oriented in the longitudinal direction is best accomplished with the shear wave UT tool, which introduces shear waves in the circumferential direction. Liquid coupled tools are the most accurate and most commonly used tools for crack detection. UT ILI is most commonly used in liquid pipelines, where the fluid being transported acts as the couplant between the ultrasonic transducers and the pipe wall. The working principles and applications of two types of ILI shear wave UT tools, the wheel-coupled tool and the liquid-coupled tool, have been introduced in detail [Baker, 2005] and are not repeated here. The primary shortcoming of UT-based ILI is that these tools have size limitations and are not available for small-diameter pipelines. Moreover, they can detect cracks accurately only above a certain threshold dimension.

   Additional details regarding MFL and UT are explored in Section 9.3.2.

3. *Electromagnetic acoustic transducer.* The newest technology being applied in ILI equipment is EMAT, which depends on the generation of an ultrasound pulse using a magnetic field at the internal surface of the pipe wall [National Association of Corrosion Engineers, 2000c]. This technique has begun to be utilized in pipeline inspection.

4. *Eddy current testing.* Eddy current testing is an electromagnetic NDT technique that can only be used on conductive materials. Its applications range from crack detection to the rapid sorting of small components for flaws, size variations, or material variation.

5. *Geometry tools.* Geometry tools use ultrasonic waves, mechanical arms, or other electromechanical means to measure the bore of pipes. They identify dents, deformation, changes in the ovality, and may occasionally detect bends in the pipe. Ultrasonic geometry tools record reflections off the inside pipe wall, similar to sonar and radar applications. Resolution of the tools is related to the number of sensors and their firing frequencies. This type of tool can be used in liquid or gas pipelines. Mechanical geometry tools use arms that ride on the interior surface of pipeline. There are often fewer mechanical gauges on geometry tools than the number of sensors on electromagnetic or ultrasonic ILI tools. As a result, geometry tools are not as sensitive as electromagnetic and ultrasonic tools to internal corrosion or metal loss on the inside pipe surface and cannot be used as a low-cost substitute for detecting small flaws.

6. *NOVAProbe.* Environmental conditions at the pipe depth can be monitored using the NOVAProbe system, which is available in both a portable version and a permanent installation. Both versions measure parameters such as pipe-to-soil

potential, soil resistivity, and soil pH. The permanent probe is often installed as part of a pipeline environment and CP monitoring system (PECPMS). In addition to the environmental parameters, the PECPMS unit also monitors the air temperature, open-circuit potential, and the on- and instant-off potentials.

9.3.2 Intelligent Pigs

Pipelines are normally pigged for the purpose of cleaning, batching, displacing, or inspection. The intelligent pig has become an important tool for inspecting and assessing the condition of a pipeline, and is set to become an integral part of pipeline maintenance.

Pigging has been around for a long time, and the earliest application in retrieving information from a pig originated in 1959 [Woodley, 2011], where a calliper tool was used to detect dents in pipelines using the MFL technique. The first commercial job was for Shell in 1965, and the tool inspected only the bottom 90° of the pipeline. There were only four sensors, and location was identified by digging down and placing permanent magnets on the pipe before the pig passed. Early inspection pigs were delivering a service to pipeline operators that enabled them to target maintenance more effectively.

By 1980, British Gas built its own inspection pigs based on the MFL principle, and the tools produced reliable data that needed to be interpreted as metal loss values. MFL is an inferred measurement system, where any feature detected causes a disturbance in the magnetic field around the sensors. The disturbance is related to the percentage of metal loss, based on sophisticated analysis.

Ultrasonic technique pigs are advanced inspection tools, where the remaining wall thickness is determined by time-of-flight ultrasound, and an array of UT probes is packed into a pig-borne system. Whereas MFL is a system that predictes metal loss, UT is a direct measurement and thus a more accurate tool. However, UT needs a liquid medium to transmit sound waves into the pipe steel. It is thus restricted to use in oil lines. It can also be deployed into gas lines, but this requires that it be within a liquid slug, which results in the project being more difficult and expensive.

Standard MFL pigs are sensitive to wide areas of metal loss and to small isolated pits, but they are relatively blind to areas of axially long narrow corrosion. However, TFI pigs can be used in pipelines where narrow axial external corrosion is known or suspected.

Axial cracking caused by a variety of factors, such as SCC, is another problem that pipeline operators have to tackle. However, detection of this type of SCC is not possible with standard tools (neither MFL nor UT). The TFI tool has had some limited success in detecting cracks, but usually only in areas where the cracking is relatively severe. Ultrasonic tools have been developed that employ a technique whereby sound waves are fired into the steel at an angle to propagate around the pipe. These tools have proved to be very successful in detecting cracks, but are still limited by the fact that they need a liquid medium in which to operate.

In 1974, a system known as the *elastic wave* was developed in the UK for crack detection in gas pipelines. A sound wave is fired into a pipe wall from an angled UT probe which is carried inside a liquid-filled wheel that runs along the pipe. However, the pigs are difficult to maintain and expensive to run. EMAT tools are the latest development in the accurate detection of crack size, although the method is still in its infancy.

9.3.3 Hydrostatic Inspection

Hydrostatic testing is an alternative to ILI. It has been used to locate stress corrosion cracks of critical sizes at the test pressure and, when properly implemented, assures that such critical defects are removed at the time of the test. Hydrotesting has been the most commonly utilized technique to ensure the integrity of the pipeline. The ideal test process includes filling with fluid (water or air), pressurizing, a spike test (139% of MAOP for 0.5 h), a pressure test (125% of MAOP for 4 h), a leak test (110% of MAOP for 44 h), dewatering, and drying.

At present, the hydrostatic retest is the only certain means of locating SCC-affected pipe, but it only identifies sections with SCC that would be nearly critical under operating conditions. Moreover, pressure testing does not provide information about either the presence or severity of cracks that survive a test. Hydrostatic testing failures could occur when cracks reduce the load-carrying capability of a pipeline, depending on the fracture toughness of the pipe material. Hydrostatic testing ruptures do not propagate a significant distance, and the stress level drops rapidly after a rupture occurs.

Periodic hydrostatic testing is also a common method used to ensure the integrity of operating pipelines that contain growing defects, such as general or pitting corrosion, corrosion fatigue, or SCC. Typically, a desired pressure range is established, with the minimum pressure selected to ensure integrity and the maximum test pressure designed to minimize the failure of noninjurious features in the pipeline. Factors considered in the selection of a pressure range include the estimated population of defects in the pipeline, the estimated growth rate of these defects, and the MAOP/MOP of the pipeline. If there are a large number of slow-growing defects and the MAOP/MOP of the pipeline is relatively low compared to the SMYS, it may be desirable to establish a relatively low maximum test pressure to avoid a large number of hydrostatic test failures. Furthermore, a relatively high minimum test pressure is needed to avoid frequent retesting for fast-growing defects and high operating pressures.

Hydrostatic testing has the advantage of remove axial defects, regardless of geometry, that have critical dimensions at the test pressure. It can also open up incipient leaks so that they can be detected. Hydrostatic testing will blunt and impart a compressive residual stress at the crack tip of subcritical defects that remain in the pipeline to inhibit subsequent fatigue or SCC crack growth [Hohl and Knauf, 1999].

Following a hydrostatic test, subcritical cracks will remain in the pipeline and, potentially, may be just smaller than the size that would have failed in a hydrostatic test. Moreover, hydrostatic testing can cause tearing of these subcritical flaws leading to a pressure reversal, where the pipeline fails in service or at a lower pressure in a subsequent hydrostatic test. Hydrostatic testing is not effective against circumferential flaws because the maximum axial stress produced by internal pressurization is less than one-half the circumferential stress. Although hydrostatic testing is capable of locating leaks, it is not effective in removing short flaws that ultimately will produce leaks. Compared to ILI, hydrostatic testing may be expensive, since the pipeline must be taken out of service.

Although an effective hydrostatic test is able to eliminate critical cracklike flaws and blunt the subcritical crack tip to suppress further SCC propagation, plastic deformation could result from the high stress intensity at the crack tip due to hydrostatic testing pressure. With increased hydrotesting pressure, nearly critical defects in a typical pipeline material could grow by a mechanism of plasticity-induced crack growth or ductile tearing [Zheng et al., 1998]. Some of these nearly critical cracks that survived hydrostatic testing could pose a direct threat to pipeline integrity by a phenomenon known as *pressure reversal* [Kiefner and Maxey, 2000]. Even using a modest hydrostatic test pressure below the SMYS, the stress concentration developed or stress–strain intensity at some of the tips of stress corrosion cracks is inevitable. For example, a high-resolution imaging by a focused ion beam (FIB) microscope shows that plastic deformation can be induced near crack tips by hydrostatic tests [Li et al., 2008]. Thus, caution must be taken when making life predictions on pipelines that have been hydrotested.

9.3.4 Pipeline Patrolling

Pipeline SCC prevention also depends on a constant round of inspections to the potential trouble spots along the pipeline right-of-way. Using aircraft, land vehicles, or foot patrols, regular monitoring of the pipeline route is possible. This will provide evidence of potentially damaging activities such as unauthorized digging and construction. As a new trend in place of traditional pipeline patrol inspection, Google Earth can be used to support the process. Many studies are on going to develop a pipeline patrol inspection system based on a global positioning system (GPS) and Google Earth.

9.4 MITIGATION OF PIPELINE SCC

Mitigation is a process intended to reduce risk through decreasing either probability of failure from identified threats, or consequences, or both. When SCC mitigation methods are planned and implemented, an engineering assessment will be performed to understand the commonalities of SCC defects, the potential mechanism of the SCC, and the SCC growth rate. The time line for mitigation depends on the severity of the SCC and is determined on the basis of engineering assessment. Generally, the time required for SCC mitigation is between one and four years.

Some SCC mitigation methods are described below.

1. *Excavation and repair.* Once the condition of the pipeline is known by excavation, suitable repair methodologies can be selected and the work programs implemented on pipelines by field crews. A planned maintenance program will

then be adopted to maintain the rehabilitated condition. Modeling software tools predict corrosion and crack growth rate on pipelines, which can be in turn be used to develop SCC mitigation tactics.

Pipe containing SCC defects can be repaired using one or more of the acceptable methods given in CSA Z662 [Canadian Standards Association, 2011], including grinding and buffing repair, a pressure containment sleeve, pipe replacement, a steel reinforcement repair sleeve, a steel compression reinforcement repair sleeve, a fiberglass reinforcement repair sleeve, a hot tap, and direct deposition welding.

2. *Recoating.* Mitigation of pipeline SCC depends on recoating the existing pipelines. Several factors must be considered in selection of the repair coating, including the ambient weather and environmental conditions required for application, the surface conditions and treatment of pipe steel, compatibility with existing coatings, equipment requirements, and access to the field site and pipeline as well as the permeability of coatings to CP [Canadian Energy Pipeline Association, 2007]. Coatings selected for recoating of pipelines must resist cathodic disbondment, adhere well to pipe, resist mechanical damage, and resist moisture degradation. Moreover, these coatings should not shield CP if they do disbond.
3. *Repair sleeves.* A repair sleeve may be installed if grinding of an excavated section of pipeline results in a wall thickness below the minimum required for the MAOP/MOP. The ground area should be filled with incompressible filler when a repair sleeve is used to mitigate SCC. Only full-encirclement sleeves, such as reinforcing, pressure containing, bolt-on, and composite reinforced, should be used for repair of SCC [Baker, 2005].
4. *Chemical cleaning and pigging.* SCC can be controlled with in-line devices such as "pigs" to inspect and clean pipelines from the inside. The program is designed to ensure that a pipeline's inner surface would be sufficiently clean for a successful ILI while minimizing disruption to normal operations. Proprietary chemical compounds such as Superclean and a specially designed online cleaning procedure are used for this method. Superclean can run in batches mixed with diesel and contained between cleaning pigs. The pigs, pipeline debris, and chemical batches are collected at each receiving station and analyzed to determine the level of cleanliness obtained. This will ensure that the inside wall of pipe will be free of crude oil residues that could prevent ultrasonic sensors in the crack detection procedure from detecting potentially dangerous anomalies. Moreover, a clean pipe wall will avoid the deposit of rust and dirt that could cause localized corrosion.

   The movement of equipment used for the chemical cleaning stage from one section to the next is challenging, especially considering that the inspection tools have to be run immediately after each cleaning. Strict safety measures ensure successful execution of this procedure.

   For "unpiggable" pipelines, defined as pipelines that cannot be inspected by standard intelligent pigs, some companies have initiated an entirely new direction in research and development. Unpiggable pipelines may operate at pressures not compatible with standard inspection pigs, or have only one point of entry. Solutions to inspect unpiggable pipelines range from pigs that can be launched through hot taps into the pipeline flow to robotic tractor units that can tow inspection vehicles. The inspection modules employ a variety of technologies: magnetic, MFL or remote eddy current, ultrasound, time of flight or transverse, and even visual, using cameras and laser light sources [Woodley, 2011].

5. *Low-plasticity burnishing* (LPB) *method*. The LPB method introduces compressive residual stresses into carbon steels to mitigate SCC and improve the mechanical strength of the steel dramatically [Kiefner et al., 1994]. This is a method of metal improvement that provides deep, stable surface compressive stress with little cold work for improved damage tolerance and stress corrosion performance.

   The basic LPB tool is a ball that is supported in a spherical hydrostatic bearing. A machine tool coolant is used to pressurize the bearing with a continuous flow of fluid to support the ball. The ball is loaded at a normal state to the surface of pipeline steel with a hydraulic cylinder that is in the body of the tool. The ball rolls across the surface of the steel, and the tool path and normal pressure applied are designed to create a distribution of compressive residual stress. The form of the distribution is designed to counter applied stresses and optimize fatigue and stress corrosion performance. Since no shear is being applied to the ball, it is free to roll in any direction. As the ball rolls over the steel surface, the pressure from the ball causes plastic deformation in the steel. Since the bulk of the steel constrains the deformed area, the deformed zone is left in compression after the ball passes.

   The LPB is a reliable and reproducible method of producing deep compressive residual stresses in complex geometric components. LPB produces a very smooth surface finish, which aids in nondestructive inspection and examination. The LPB process is thus employed to treat components such as pipeline steels to provide a substantial increase in service life.

REFERENCES

American Society of Mechanical Engineers (2001) *Managing System Integrity of a Gas Pipeline*, ASME B31.8S-2001, ASME, New York.

Baker, M, Jr. (2005) *Final Report on Stress Corrosion Cracking Study*, Integrity Management Program Delivery Order DTRS56-02-D-70036, Office of Pipeline Safety, U.S. Department of Transportation, Washington, DC.

Beavers, JA, Jaske, CE (1998) Near-neutral-pH SCC in pipelines: effects of pressure fluctuations on crack propagation, *Corrosion 1998*, Paper 98257, NACE, Houston, TX.

Beavers, JA, Johnson, JT, Sutherby, RL (2000) Materials factors influencing the initiation of near-neutral-pH SCC on underground pipelines, *Proc. 4th International Pipeline Conference*, Paper. 047, Calgary, Alberta, Canada.

Bickerstaff, R, Vaughn, M, Stoker, G, Hassard, M, Garrett, M, Online source, Review of sensor technologies for in-line inspection of natural gas pipelines, http://netl.doe.gov/technologies/oil-gas/publications/Status_Assessments/71702.pdf.

Canadian Energy Pipeline Association (2007) *Stress Corrosion Cracking: Recommended Practices*, 2nd ed., CEPA, Calgary, Alberta, Canada.

Canadian Standards Association (2011) *Oil and Gas Pipeline Systems*, Z662-11, CSA, Reydale, Ontario, Canada.

Fessler, RR (1979) Stress corrosion cracking temperature effects, *Proc. 6th Symposium on Line Pipe Research*, Paper L30175, PRCI, Falls Church, VA.

Hohl, G, Knauf, G (1999) Field hydrotesting and its significance to modern pipelines, *Proc. EPRG/PRCI 12th Biennial Joint Technical Meeting on Pipeline Research*, Paper 32.

Khan, AJ (2012) Successful implementation of ECDA methodology, *Mater. Perf.* 51, 32–34.

Kiefner, JF, Maxey, WA (2000) Periodic hydrostatic testing or in-line inspection to prevent failures from pressure-cycle-induced fatigue, *API's Annual Pipeline Conference Cybernetics Symposium*, New Orleans, LA.

Kiefner, JF, Bruce, WA, Stephens, DR (1994) *Pipeline Repair Manual*, L51716, PRCI, Falls Church, VA.

Li, J, Elboujdaini, M, Gao, M, Revie, RW (2008) Investigation of plastic zones near SCC tips in a pipeline after hydrostatic testing, *Mater. Sci. Eng. A* 486, 496–502.

Materion (2011) Chloride stress corrosion cracking of high performance copper alloys, www.materion.com.

National Association of Corrosion Engineers (2000a) *White Metal Blast Cleaning*, Joint Surface Preparation Standard NACE 1/SSPC-SP 5, NACE, Houston, TX.

National Association of Corrosion Engineers (2000b) *Near-White Metal Blast Cleaning*, Joint Surface Preparation Standard NACE 2/SSPC-SP 10, NACE, Houston, TX.

National Association of Corrosion Engineers (2000c) *In-Line Nondestructive Inspection of Pipelines*, NACE, Houston, TX.

National Association of Corrosion Engineers (2002) *Control of External Corrosion on Underground or Submerged Metallic Piping Systems*, NACE Standard RPO 169-2002, NACE, Houston, TX.

National Association of Corrosion Engineers (2003) *Methods for Sulfide Stress Cracking and Stress Corrosion Cracking Resistance in Sour Oilfield Environments*, NACE MR 0175, NACE, Houston, TX.

National Association of Corrosion Engineers (2008) *Pipeline External Corrosion Direct Assessment Methodology*, NACE SP0502-2008, NACE, Houston, TX.

National Energy Board (1996) *Stress Corrosion Cracking on Canadian Oil and Gas Pipelines*, MH-2-95, NEB, Calgary, Alberta. Canada.

National Energy Board (2003) *Guidance Notes for the National Energy Board Processing Plant Regulations*, Section 40, Plant Hazard and Safety Management, NEB, Calgary, Alberta, Canada.

Nelson, BR (2002) Pipeline integrity: program development, risk assessment and data management, the *11th Annual GIS for Oil and Gas Conference*, Houston, TX.

Parkins, RN, Greenwell, BS (1977) The interface between corrosion fatigue and stress corrosion cracking, *Met. Sci.* 11, 405–416.

Parkins, RN, Slattery, PW, Poulson, BS (1981) The effects of alloying additions to ferritic steels upon stress-corrosion cracking resistance, *Corrosion* 37, 650–659.

Pikas, J (2002) Direct assessment, data integration important in establishing pipeline integrity, *82nd Annual Gas Processors Association Convention*, Dallas, TX.

Sullivan, J (2007) Pipeline integrity: enhancing integrity management, Enspiria Solutions Inc., http://www.esintial.com/Article%20pdfs/Pipeline%20Integrity.pdf.

Surkov, JP (1994) Corrosion crack initiation in gas pipelines, *Phys. Met. Metallogr.* 78, 102–104.

UK National Physical Laboratory (1982) *Guide to Good Practice in Corrosion Control: Stress Corrosion Cracking*, NPL, London.

Wikipedia, online source, Risk assessment, http://en.wikipedia.org/wiki/Risk_assessment.

Woodley, D (2011) The origin of intelligent pigs, *Pipeline Int.* 10, 26–28.

Youzwishzen, OO, Van Aelst, A, Ehlers, PF, Nettel, A (2004) A statistical model for the prediction of SCC formation along a pipeline, *Proc. 5th International Pipeline Conference*, Paper IPC04-0267, Calgary, Alberta, Canada.

Zheng, W, Tyson, WR, Revie, RW, Shen, G, Brade, JEM (1998) Effects of hydrostatic testing on growth of stress-corrosion cracks, *International Pipeline Conference*, Vol. 1, Calgary, Alberta, Canada, pp. 459–472.

Index

Stress Corrosion Cracking of Pipelines, First Edition. Y. Frank Cheng.

Printed and bound by CPI Group (UK) Ltd, Croydon, CR0 4YY
16/09/2026
14963417-0001